Molecular Water Oxidation Catalysis

Molecular Water Oxidation Catalysis

A Key Topic for New Sustainable Energy Conversion Schemes

Editor

ANTONI LLOBET

Institute of Chemical Research of Catalonia, Tarragona, Spain

WILEY

This edition first published 2014
© 2014 John Wiley & Sons, Ltd

Registered office

John Wiley & Sons Ltd, The Atrium, Southern Gate, Chichester, West Sussex, PO19 8SQ, United Kingdom

For details of our global editorial offices, for customer services and for information about how to apply for permission to reuse the copyright material in this book please see our website at www.wiley.com.

Wiley also publishes its books in a variety of electronic formats. Some content that appears in print may not be available in electronic books.

Designations used by companies to distinguish their products are often claimed as trademarks. All brand names and product names used in this book are trade names, service marks, trademarks or registered trademarks of their respective owners. The publisher is not associated with any product or vendor mentioned in this book.

Limit of Liability/Disclaimer of Warranty: While the publisher and author have used their best efforts in preparing this book, they make no representations or warranties with respect to the accuracy or completeness of the contents of this book and specifically disclaim any implied warranties of merchantability or fitness for a particular purpose. It is sold on the understanding that the publisher is not engaged in rendering professional services and neither the publisher nor the author shall be liable for damages arising herefrom. If professional advice or other expert assistance is required, the services of a competent professional should be sought

The advice and strategies contained herein may not be suitable for every situation. In view of ongoing research, equipment modifications, changes in governmental regulations, and the constant flow of information relating to the use of experimental reagents, equipment, and devices, the reader is urged to review and evaluate the information provided in the package insert or instructions for each chemical, piece of equipment, reagent, or device for, among other things, any changes in the instructions or indication of usage and for added warnings and precautions. The fact that an organization or Website is referred to in this work as a citation and/or a potential source of further information does not mean that the author or the publisher endorses the information the organization or Website may provide or recommendations it may make. Further, readers should be aware that Internet Websites listed in this work may have changed or disappeared between when this work was written and when it is read. No warranty may be created or extended by any promotional statements for this work. Neither the publisher nor the author shall be liable for any damages arising herefrom.

Library of Congress Cataloging-in-Publication Data

Molecular water oxidation catalysis : a key topic for new sustainable energy conversion schemes / editor, Antoni Llobet.
    pages cm
  Includes bibliographical references and index.
  ISBN 978-1-118-41337-1 (cloth)
  1. Energy harvesting. 2. Water–Purification–Oxidation–By-products. 3. Renewable energy sources. 4. Electric power production from chemical action. I. Llobet, Antoni, 1960- editor of compilation.
  TJ808.M65 2014
  621.31′242 – dc23

                                                                                                    2014004158

A catalogue record for this book is available from the British Library.

ISBN: 9781118413371

Set in 10/12pt TimesLTStd by Laserwords Private Limited, Chennai, India
Printed and bound in Singapore by Markono Print Media Pte Ltd

1   2014

Contents

4. Towards the Visible Light-Driven Water Splitting Device: Ruthenium Water Oxidation Catalysts with Carboxylate-Containing Ligands 51

Lele Duan, Lianpeng Tong, and Licheng Sun

5. Water Oxidation by Ruthenium Catalysts with Non-Innocent Ligands 77

Tohru Wada, Koji Tanaka, James T. Muckerman, and Etsuko Fujita

6. Recent Advances in the Field of Iridium-Catalyzed Molecular Water Oxidation 113

James A. Woods, Stefan Bernhard, and Martin Albrecht

7. Complexes of First Row d-Block Metals: Manganese   135
Philipp Kurz

8. Molecular Water Oxidation Catalysts from Iron   153
W. Chadwick Ellis, Neal D. McDaniel, and Stefan Bernhard

List of Contributors

Martin Albrecht, School of Chemistry & Chemical Biology, University College Dublin, Dublin, Ireland

Shoshanna M. Barnett, Department of Chemistry, University of Washington, Seattle, WA, USA

Victor S. Batista, Department of Chemistry, Yale University, New Haven, CT, USA

Stefan Bernhard, Department of Chemistry, Carnegie Mellon University, Pittsburgh, PA, USA

Roger Bofill, Department of Chemistry, Autonomous University of Barcelona, Barcelona, Spain

Gary W. Brudvig, Department of Chemistry, Yale University, New Haven, CT, USA

Christopher J. Cramer, Department of Chemistry, Chemical Theory Center and Supercomputing Institute, University of Minnesota, Minneapolis, MN, USA

Holger Dau, Department of Physics, Free University of Berlin, Berlin, Germany

Lele Duan, Department of Chemistry, School of Chemical Science and Engineering, KTH Royal Institute of Technology, Stockholm, Sweden

Jennifer L. DuBois, Department of Chemistry and Biochemistry, Montana State University, Bozeman, MT, USA

W. Chadwick Ellis, Department of Chemistry and Chemical Biology, Baker Laboratory, Cornell University, Ithaca, NY, USA

Mehmed Z. Ertem, Department of Chemistry, Chemical Theory Center and Supercomputing Institute, University of Minnesota, Minneapolis, MN, USA

Lluis Escriche, Department of Chemistry, Autonomous University of Barcelona, Barcelona, Spain

Laia Francàs, Institute of Chemical Research of Catalonia (ICIQ), Tarragona, Spain

Etsuko Fujita, Chemistry Department, Brookhaven National Laboratory, Upton, NY, USA

Laura Gagliardi, Department of Chemistry, Chemical Theory Center and Supercomputing Institute, University of Minnesota, Minneapolis, MN, USA

Jordi García-Antón, Department of Chemistry, Autonomous University of Barcelona, Barcelona, Spain

Yurii V. Geletii, Department of Chemistry, Emory University, Atlanta, GA, USA

Karen Goldberg, Department of Chemistry, University of Washington, Seattle, WA, USA

Craig L. Hill, Department of Chemistry, Emory University, Atlanta, GA, USA

Katharina Klingan, Department of Physics, Free University of Berlin, Berlin, Germany

Philipp Kurz, Institute for Inorganic and Analytical Chemistry, Albert-Ludwigs-University of Freiburg, Freiburg, Germany

Antoni Llobet, Institute of Chemical Research of Catalonia (ICIQ), Tarragona, Spain; Department of Chemistry, Autonomous University of Barcelona, Barcelona, Spain

Hongjin Lv, Department of Chemistry, Emory University, Atlanta, GA, USA

James M. Mayer, Department of Chemistry, University of Washington, Seattle, WA, USA

Neal D. McDaniel, Phillips 66, Bartlesville, OK, USA

Pere Miró, Department of Chemistry, Chemical Theory Center and Supercomputing Institute, University of Minnesota, Minneapolis, MN, USA

James T. Muckerman, Chemistry Department, Brookhaven National Laboratory, Upton, NY, USA

Jared C. Nesvet, Department of Chemistry, University of Washington, Seattle, WA, USA

Marcel Risch, Department of Physics, Free University of Berlin, Berlin, Germany; Electrochemical Energy Laboratory, Massachusetts Institute of Technology, Cambridge, MA, USA

Ivan Rivalta, Department of Chemistry, Yale University, New Haven, CT, USA

Xavier Sala, Department of Chemistry, Autonomous University of Barcelona, Barcelona, Spain

Margaret L. Scheuermann, Department of Chemistry, University of Washington, Seattle, WA, USA

Jordan M. Sumliner, Department of Chemistry, Emory University, Atlanta, GA, USA

Licheng Sun, Department of Chemistry, School of Chemical Science and Engineering, KTH Royal Institute of Technology, Stockholm, Sweden; State Key Lab of Fine Chemicals, DUT-KTH Joint Education and Research Center on Molecular Devices, Dalian University of Technology, Dalian, China

Koji Tanaka, Institute for Integrated Cell-Material Sciences, Kyoto University, Kyoto, Japan

Lianpeng Tong, Department of Chemistry, School of Chemical Science and Engineering, KTH Royal Institute of Technology, Stockholm, Sweden

James W. Vickers, Department of Chemistry, Emory University, Atlanta, GA, USA

Tohru Wada, Department of Chemistry, College of Science and Research Center for Smart Molecules, Rikkyo University, Toshima, Tokyo, Japan

Christopher R. Waidmann, Department of Chemistry, University of Washington, Seattle, WA, USA

James A. Woods, Department of Chemistry, Carnegie Mellon University, Pittsburgh, PA, USA

Ivelina Zaharieva, Department of Physics, Free University of Berlin, Berlin, Germany

Preface

The evolution of humanity is directly linked to our access to energy resources. In the last few decades, the importance of oil access and processing to modern society has become clear, basically dictating world economic politics and countries' welfare. The extraordinarily high rate of fossil fuel consumption, the realization that these fossil fuel resources are not limitless, and the consequences of substantially increasing the CO_2 concentration in the atmosphere clearly advocate for a change of energy source. Sunlight is a highly desirable one as it is the only one that is truly sustainable in the long term.

Nature has been harvesting sunlight as a source of energy for a few billion years through photosynthetic processes like the ones carried out by green plants and algae. Nature's strategy consists in the oxidation of H_2O to dioxygen and the reduction of CO_2 and generation of carbohydrates. Water splitting is a promising strategy for capturing the sun's energy. It involves coupling light harvesting to the oxidation of water to dioxygen, in the same manner as in the natural system, and the reduction of protons to hydrogen. The design of devices aimed at energy harvesting, based on sunlight and water splitting, requires efficient water oxidation catalysts (WOCs). Thus special attention needs to be paid to understanding this reaction in order to come up with technologically useful WOCs. In addition, the knowledge gained from molecular WOCs is also very useful in shedding light on how complex systems such as the oxygen evolving complex (OEC) of photosystem II (PSII) work and how heterogeneous systems used as intermediates are much more difficult to characterize.

This book is exclusively dedicated to the field of molecular water oxidation catalysis in the homogeneous phase, which has experienced spectacular development in the last 5 years. The first chapter, by Rivalta, Brudvig, and Batista, is dedicated to the OEC-PSII of the natural system, to some Mn complexes that act as mimics of the OEC-PSII, and to the theoretical description of these systems based on density functional theory (DFT). Chapter 2, by DuBois, looks at chlorite dismutase, a heme protein that is capable of making O−O bonds, one of the crucial reactions in producing molecular oxygen. Chapters 3 and 4 concern low-molecular-weight second-row transition-metal complexes, which have been reported to be able carry out water oxidation; these chapters are written by two teams: Chapter 3 by Francàs, Bofill, García-Antón, Escriche, Sala, and Llobet and Chapter 4 by Duan, Tong, and Sun. Chapter 5 reports a WOC based on Ru but containing non-redox-innocent ligands; it is authored by Wada, Tanaka, Muckerman, and Fujita. Chapter 6, by Woods, Bernhard, and Albrecht, focuses on Ir complexes that act as WOCs. The next four chapters are dedicated to first-row WOCs. Kurz reports on Mn, Ellis, McDaniel, and Bernhard on Fe, Risch, Klingan, Zaharieva, and Dau on Co, and Barnett, Waidmann, Scheuermann, Nesvet, Goldberg, and Mayer on Cu. Chapter 11, by Sumliner, Vickers, Lv, Geletii, and Hill, is dedicated to polyoxometalates, which can be considered low−molecular-weight models of surfaces. Finally, the book closes with a theoretical DFT description of a variety of Ru, Co, and Fe complexes, presented by Miró, Ertem, Gagliardi, and Cramer.

1

Structural Studies of Oxomanganese Complexes for Water Oxidation Catalysis

Ivan Rivalta, Gary W. Brudvig, and Victor S. Batista
Department of Chemistry, Yale University, New Haven, CT, USA

1.1 Introduction

Photosystem II (PSII) is a 650 kDa protein complex embedded in the thylakoid membrane of green plant chloroplasts and the internal membranes of cyanobacteria. It is responsible for catalyzing oxygen evolution by water splitting into oxygen, protons and electrons. The catalytic site is the oxygen-evolving complex (OEC) embedded in the protein subunit D1, an oxomaganese cuboidal core comprising earth-abundant metals (Ca^{2+} and $Mn^{3+/4+}$) linked by μ-oxo bridges. The reaction is initiated upon light absorption by an antenna complex, in a process that oxidizes the chlorophyll a species P680 and forms the radical cation $P680^{+\bullet}$, a strong oxidizing species that in turns oxidizes tyrosine Y_Z, a redox-active amino acid residue located in close proximity to the oxomanganese cluster. The oxidized Y_Z is able to oxidize Mn, storing oxidizing equivalents in the inorganic core of the OEC. This photocatalytic process is repeated multiple times while evolving the OEC through five oxidation storage states (S_0–S_4) along the catalytic cycle (the so-called Kok cycle) [1, 2]. In the fully oxidized S_4 state, the Mn cluster catalyzes oxygen evolution, completing the four-electron water oxidation reaction that splits water into molecular oxygen, protons and electrons, as follows:

$$2H_2O \rightarrow O_2 + 4e^- + 4H^+ \tag{1.1}$$

The characterization of the OEC structure and overall structural rearrangement during the multistep photocatalytic cycle is crucial for understanding the reaction mechanism and

Molecular Water Oxidation Catalysis: A Key Topic for New Sustainable Energy Conversion Schemes,
First Edition. Edited by Antoni Llobet.
© 2014 John Wiley & Sons, Ltd. Published 2014 by John Wiley & Sons, Ltd.

for the design of biomimetic catalytic systems. X-ray spectroscopy has been largely used to reveal the atomistic details of the OEC structure, with several X-ray crystal models proposed in the past decade [3–6]. The most recent breakthroughs in the field have resolved the OEC structure at 1.9 Å resolution [7], including the complete coordination of metal centers by water ligands and proteinaceous side chains. However, the high doses of X-ray radiation necessary for data collection are thought to have reduced the Mn centers, changing the geometry of the OEC and leaving uncertain the actual geometry of the oxomanganese complex in its dark-adapted (S_1) state [8–11]. A model of the OEC in the S_1 state consistent with both high-resolution spectroscopy and X-ray diffraction (XRD) data has been obtained using quantum mechanics/molecular mechanics (QM/MM) hybrid methods, implemented at the density functional theory (DFT) level [12]. The model has been validated through simulations of extended X-ray absorption fine structure (EXAFS) spectroscopy and direct comparisons with experimental measurements [12].

The oxomanganese complex of the OEC of PSII has inspired the development of biomimetic catalysts for water oxidation with high-valent Mn centers linked by μ-oxo bridges, as in the Mn_4CaO_5 cluster ligated by terminal waters and surrounding amino acid residues. Several complexes with common structural features have been synthesized [13–23] to investigate the structure/function relations responsible for catalytic water oxidation and to provide fundamental insight into the use of catalysts for artificial photosynthetic devices based on earth-abundant metals [24–26]. One of these biomimetic complexes is the Mn–terpy dimer $[H_2O(terpy)Mn(\mu-O)_2Mn(terpy)OH_2]^{3+}$ **(1)** (terpy = $2, 2'$: $6', 2''$-terpyridine), which has terminal water molecules bound to the oxomanganese core, in close analogy to the OEC. Complex **1** also catalyzes water oxidation upon activation by a primary oxidant in homogeneous solutions [27, 28] and when deposited on TiO_2 thin films [24, 25, 29] or immobilized in clays [30, 31]. Mechanistic aspects of water oxidation catalyzed by complex **1** are thought to be common to the OEC of PSII, where deprotonation of terminal waters and oxidation of the Mn core is thought to give rise to the formation of a hot oxyl radical intermediate that is susceptible to nucleophilic attack by substrate water. DFT studies of the water splitting catalyzed by **1** have also shown the non-innocent role of acetate buffer as proton acceptor centers during the O–O bond formation step [32]. This chapter reviews these recent advances in computational structural studies of the OEC and biomimetic oxomanganese complexes, including the structural characterization of the OEC in the S_1 state and mechanistic studies of water oxidation catalyzed by the OEC and by complex **1** in solution or covalently attached to nanoparticulate semiconductor surfaces.

1.2 Structural Studies of the OEC

An XRD model of the OEC of PSII at 3.5 Å resolution [5] suggested a cuboidal cluster $CaMn_4$, with the metal centers at the vertices of a cuboidal frame and a "dangling" Mn linked by μ-oxo bridges (Figure 1.1). While the X-ray model was partially consistent with EXAFS and electron paramagnetic resonance (EPR) studies [33], structural disorder and radiation damage prevented a complete characterization of the complex, including coordination of water and proteinaceous ligands to the metal centers. Consequently, several computational studies were performed to build realistic models with a complete coordination of the metal centers [23, 34–48], including QM/MM structural models with an explicit

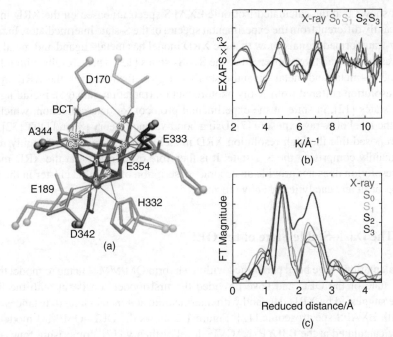

Figure 1.1 *Structural models of the OEC of PSII. (a) Superposition of the OEC in the XRD models of PSII at 3.5 Å (red) and 1.9 Å (blue) resolution. (b–c) Comparison between experimental isotropic EXAFS spectra of S_0 (green), S_1 (light blue), S_2 (dark gray), S_3 (brown), and calculated EXAFS spectra of the high-resolution XRD model (blue), including the k^3-weighted EXAFS spectra (b) and the corresponding Fourier transform (FT) magnitudes (c). See color plate*

treatment of the protein environment [23, 35–39]. Several possible ligation schemes were proposed, including models with terminal water ligands bound to Ca and the dangling Mn, as proposed by DFT–QM/MM computational models [6, 7]. Remarkably, the most recent XRD model of PSII at 1.9 Å resolution [7] has confirmed the coordination of terminal water molecules bound to Ca^{2+} and the dangling Mn, the presence of an additional μ-oxo bridge linking the dangling Mn to the cuboidal $CaMn_3$ cluster, the coordination of carboxylate groups bridging the metal centers (Figure 1.1), and the proximity of chloride to the OEC. In addition, the latest XRD data introduce new features that have not been previously proposed by either empirical or computational models, including the bidentate coordination of the D170 side chain bridging the dangling Mn and Ca (Figure 1.1a compares the two XRD models). These advances have stimulated new studies of structural changes of the OEC along the Kok catalytic cycle. In particular, the first challenge was to establish whether the OEC model proposed by the XRD structure at 1.9 Å resolution could be assigned to the S_1 resting state or if it was perhaps more representative of a mixture of S-state intermediates along the photocatalytic cycle. This question was first addressed by simulations of EXAFS spectroscopy [12] and direct comparisons to experimental data characterizing the structural modifications in the OEC cluster along the S_0–S_3 transitions [10, 49, 50].

Figure 1.1 provides a comparison of EXAFS spectra for the S_0–S_3 states and the spectrum calculated for the XRD model using the *ab initio* real-space Green function approach

[51]. It shows that the calculated isotropic EXAFS spectrum based on the XRD model is significantly different from the experimental spectra of the S-state intermediates. In fact, the Fourier-transformed signals show that the XRD model has metal–ligand and metal–metal distances larger than those observed in the S_0–S_3 states [12]. These results with relatively large metal–metal and metal–ligand distances suggest that the even the XRD model at 1.9 Å resolution suffered from X-ray photoreduction (radiation damage), including the S_1 and S_2 states [12], in spite of the experimental protocol for data collection, which minimized the level of X-ray exposure [7]. Using bond-valence theory [8] and DFT [52], it was later proposed that the high-resolution XRD model must be a mixture of highly reduced states mainly comprising the S_{-3} state. It is therefore established that the XRD model at 1.9 Å resolution does not provide an accurate description of the OEC cluster in the resting S_1 state, or in any catalytically active form.

1.3   The Dark-Stable State of the OEC

DFT calculations have been performed within a hybrid QM/MM scheme to model the resting S_1 state of the OEC and have provided the first model consistent with the ligation scheme suggested by XRD data and with intermetallic and metal–ligand distances consistent with EXAFS spectroscopic [12]. Figure 1.2 shows the DFT–QM/MM model of the S_1 state calculated at the B3LYP/LACVP* level of theory [12], comprising four terminal water ligands (including the two substrate water molecules W_1^* and W_2^* bound to Ca and Mn(4), respectively), six carboxylate ligands (Asp170, Glu189, Glu333, Asp342, Ala344, and Glu354), and one imidazole ligand (His332). The QM/MM model included a proper description of the hydrogen-bonding network surrounding the OEC (Figure 1.2), including important amino acid residues next to the oxomanganese cluster, which were previously proposed to be critical acid/base-redox cofactors (e.g. Tyr161 (YZ), Asp61, Lys317, and chloride (Cl^-)).

DFT–QM/MM geometry optimization of the Mn_4 cluster in the S_1 state reduced the intermetallic and metal–ligand distances in the oxomanganese core with respect to the XRD model (Figure 1.2b), providing good agreement with the simulated EXAFS spectra when compared to the experimental data for the cluster with formal oxidation numbers IV, III, IV, and III, assigned respectively to the Mn centers 1, 2, 3, and 4, as labeled in Figure 1.1. Table 1.1 shows that the structural rearrangements were nevertheless quite modest, with the Mn–Mn distances reduced by less than 0.1 Å within the Mn(1)–Mn(2)–Mn(3) cuboidal base and by ca. 0.2 Å between Mn(3) and the dangling Mn(4). Remarkably, the isotropic EXAFS spectrum based on the S_1 DFT–QM/MM model is in much better agreement with experimental measurements than the simulated spectrum based on the XRD model, with small differences probably caused by the limitation of the DFT methods. A structural refined protocol [38] has been successfully employed to obtain a quantitative agreement between the experimental and calculated isotropic and polarized EXAFS spectra [12]. Table 1.1 reports small differences between the Mn–Mn distances in the DFT–QM/MM and the R-QM/MM, indicating that the DFT–QM/MM model requires a minimal structural refinement to achieve quantitative agreement with X-ray absorption spectroscopic data. Figure 1.2 shows a comparison of the calculated isotropic EXAFS spectra obtained for the R-QM/MM model, as compared to the calculated spectrum based on the

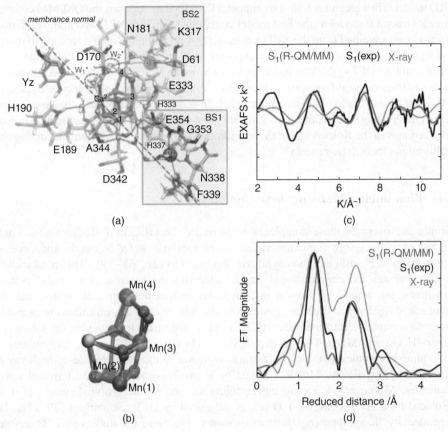

Figure 1.2 *DFT–QM/MM model of the dark-adapted S_1 state of the OEC of PSII. (a) Ligation scheme of the OEC oxomanganese cluster, including substrate water molecules (W_1^* and W_2^*), six carboxylate ligands (D170, E189, E333, D342, A344, and E354), one imidazole ligand (H332), surrounding amino acid residues (Y161 (YZ), D61, and K317) and the chloride binding sites (BS1 and BS2). (b) Superposition of the oxomanganese core of the high-resolution XRD (magenta) and the DFT–QM/MM (colored) models. (c–d) Comparison between experimental (black) isotropic EXAFS spectra of S1 and calculated spectra of the refined R-QM/MM model (red) and the high-resolution XRD model (magenta). See color plate*

Table 1.1 *Comparison of intermetallic Mn–Mn distances (Å) in the X-ray model (chain A/a), the DFT–QM/MM model obtained at the B3LYP/LACVP* level, and the refined R-QM/MM model*

	XRD (A/a)	QM/MM	R-QM/MM
Mn(1)–Mn(2)	2.84/2.76	2.80	2.76
Mn(1)–Mn(3)	2.89/2.91	2.80	2.71
Mn(2)–Mn(3)	3.29/3.30	3.38	3.25
Mn(3)–Mn(4)	2.97/2.91	2.75	2.76

XRD model. The agreement with experimental EXAFS data supports the QM/MM computational structural model as the first model consistent with both XRD and high-resolution spectroscopy, obtained from the XRD model at 1.9 Å resolution, which is thought to represent a photoreduced state generated by exposure to X-ray radiation during data collection.

The resulting DFT–QM/MM model of the OEC in the S_1 resting state has been used for studies of the structural/functional role of chloride in PSII by means of molecular dynamics (MD) and Monte Carlo (MC) techniques [53]. In addition, the structural characterization of the OEC resting state represents a valuable point of departure for the characterization of intermediates in the Kok cycle and the assignment of structural rearrangements during the multistep photocatalytic process.

1.4   Biomimetic Oxomanganese Complexes

Biomimetic oxomanganese complexes inspired by the OEC of PSII have been studied in order to explore fundamental aspects of catalytic water oxidation and develop photoanodes for artificial photosynthetic devices [13–26, 35–39]. The combination of experimental and computational studies addresses the nature of the oxidation-state transitions, the acidity of protons in μ-hydroxo bridges and terminal waters, and the influence of ligands and their environment on the acid–base/redox transitions responsible for proton-coupled electron transfer (PCET) and O_2 evolution. In particular, the Mn–terpy dimer $[H_2O(terpy)Mn(\mu-O)_2Mn(terpy)OH_2]^{3+}$ (**1**) (terpy = 2, 2′ : 6′, 2″-terpyridine) is a biomimetic complex with several features common to the OEC, including high-valent Mn centers (Mn(III) and Mn(IV)), linked by μ-oxo bridges, and bound terminal water molecules. Moreover, in acetate buffer solutions, one of the terminal waters of **1** is displaced by acetate (complex **1-OAc**), as suggested by DFT calculations [20, 21], pH-dependent cyclic voltammetry (CV) experiments [18], and EPR studies [54]. Therefore, structure/function relations in the Mn–terpy dimer are expected to provide valuable insights into catalytic water oxidation chemistry common to the OEC (Figure 1.3).

DFT studies have determined the regulatory effect of oxidation-state transitions on the pKas of oxo ligands [20] and the influence of (negatively charged) carboxylate ligands on the redox potential of Mn centers in complex **1** [21]. Similar redox effects might be expected in the PCET mechanism of the OEC, in which the oxidation-state transitions along the Kok cycle are known to be coupled with the release of protons in the lumen and the redox leveling of the four oxidation-state transitions at ca. 1 V can be mainly the result of the presence of several carboxylate ligands, such as Asp170, Glu189, Glu333, (C-terminal)Ala344, Asp342, and Glu354. DFT studies of complex **1** have also focused on the elucidation of the O–O bond formation mechanism during catalytic water oxidation in solution [55, 56]. According to numerous computational studies of the water oxidation mechanism in PSII [5, 23, 37, 57–60] and biomimetic oxomanganese complexes [55, 56, 61], a key intermediate along the reaction pathway of O_2 evolution is the Mn(IV)-O· oxyl radical, a very active species that initiates the fast O–O bond formation process. Figure 1.3 shows how the oxyl radical might be formed upon partial oxidation and deprotonation of a terminal water ligand in both the OEC in PSII and complex **1-OAc** in a buffered solution. The main difference between the formation of the oxyl radical in the OEC and that in complex **1-OAc** is that in the former the oxidation states of the Mn centers are advanced by multiple photoinduced one-electron transfer processes, while in the latter

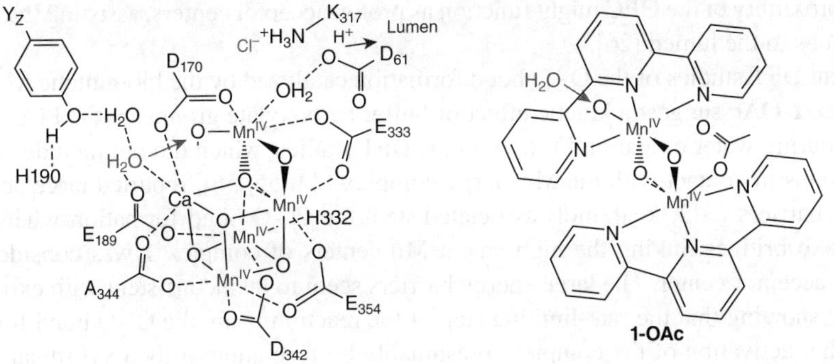

$$[Mn^{IV}-Mn^{IV}-O^{\bullet}]^{+Y}+H_2O+B \rightarrow [Mn^{III}-Mn^{IV}-O-OH]^{+(Y-1)}+[B-H]^{+}$$

(c)

Figure 1.3 *Schematic representation of (a) the Mn(IV)-O· oxyl radical in the oxidized forms of the OEC of PSII and (b) complex **1-OAc**, showing an analogous nucleophilic attack mechanism to that of a substrate water molecule (red). (c) O–O bond formation mechanism, involving production of the hydroperoxo Mn(IV)–Mn(III) species and proton release to a generic basic center (B). See color plate*

the oxidation of Mn centers is obtained using a sacrificial chemical oxidant (e.g. oxone). The Mn(IV)–Mn(IV)–O· oxyl radical formed upon oxidation is susceptible to nucleophilic attack by a substrate water molecule, giving rise to the formation of the O–O bond, a reaction step in which a PCET involves electron injection into the Mn core and deprotonation of the substrate water. Thus, the O–O bond formation produces a hydroperoxo Mn(IV)–Mn(III) species, releasing a proton to an acceptor center. Different basic centers can work as proton acceptor centers during the O–O bond formation process, including μ-oxo bridges, solvent water molecules and nearby Lewis bases. In the next section we analyze the role of basic centers in the kinetics of the O–O bond formation reaction of complex **1-OAc**, which has implications for the water oxidation mechanism in PSII.

1.5 Base-Assisted O–O Bond Formation

The PSII protein environment surrounding the OEC is thought to facilitate electron and proton transfer, establishing a redox-leveling PCET mechanism that allows for the accumulation of multiple oxidizing equivalents in the OEC, using the oxidizing power of the radical cation P680$^{+\bullet}$ with about 1.25 V. Understanding the role of the acid–base and redox cofactors that mediate the transfer of electrons and protons in PSII would establish the nature of the PCET mechanism that allows for multiple oxidation-state transitions of the OEC at low overpotentials. Such insight could provide valuable guidelines for the rational design of artificial photosynthetic architectures.

MD and MC simulations have explored the structure/function relations that might be essential for the PCET mechanism of the OEC catalytic cycle. These computational studies have analyzed the interactions of chloride in close contact with amino acid residues along a postulated proton translocation channel in PSII and have shown how carboxylate moieties in the proximity of the OEC might function as proton acceptor centers, assisting the transfer of protons to the lumen [26].

Recent DFT studies of the O–O bond formation catalyzed by the biomimetic Mn–terpy complex **1-OAc** suggest a kinetic effect of buffer carboxylate groups on the PCET mechanism during water oxidation [32]. Previous DFT studies, which did not include carboxylate groups in contact with the Mn–terpy complex [43, 55, 56], reported large activation energy barriers (> 20 kcal/mol) associated with the O–O bond formation when one of the μ-oxo bridges linking the high-valent Mn centers of complex **1** was considered the proton-acceptor center. The large energy barriers seem to be inconsistent with experimental data showing that the rate-limiting step of the reaction is not the O–O bond formation but rather activation of the complex (presumably by formation of the oxyl rdical), which forms up to 1.5 O_2 molecules/second. The more recent studies [32] suggest however that other molecular species in solution could be available as proton-acceptor centers, including the conjugate of the primary oxidant (oxone) and the basic acetate buffer.

The pKa of the Mn(IV)–Mn(III) μ-hydroxo bridges has been determined to be around 2.0 pH units [18], while the pKa of the hydronium cation is much lower (<0) and that of acetic acid is much higher (4.75 pH units). This indicates that the carboxylates in the buffer solution should be much better proton acceptors than the Mn(IV)–Mn(III) μ-oxo bridges.

Computational modeling has shown that the basicity of acetate is significantly stronger than that of μ-oxo bridges in the oxomanganese core, leading to a significant reduction in the activation energy barrier associated with deprotonation of a substrate water molecule during the O–O bond formation mechanism. Figure 1.4 compares the activation energy barriers calculated at the DFT level of theory for the O–O bond formation step in the absence and presence of a buffer acetate proton acceptor. The presence of acetate has a dramatic effect on the kinetics of O–O bond formation, with a decrease of the activation barrier of ca. 20 kcal/mol with the transfer of a proton to a μ-oxo bridge. A similar kinetic effect has been previously observed in Ru single-site water oxidation catalysis [62] and in a base-assisted proton–electron transfer reaction in tyrosine oxidation [63]. Therefore, these results are expected to have profound implications for the water oxidation mechanism in PSII, in which carboxylate groups of titratable residues surrounding or bound to the OEC (such as Asp170, Glu189, Glu333, Glu354, (C-terminal)Ala344, Asp342, and in particular Asp61) could play a crucial role in the PCET mechanism.

1.6 Biomimetic Mn Catalysts for Artificial Photosynthesis

Photocatalytic solar cells can mimic photosynthetic organisms and use solar light to extract reducing equivalents from water (e.g. protons and electrons), which could be used to generate fuels with carbon-neutral atmospheric footprint (e.g. H_2 by proton reduction, liquid fuels such as methanol generated by CO_2 reduction). Synthetic oxomanganese catalysts, such as the Mn–terpy dimer discussed in previous sections, can mimic the OEC of PSII and split water into O_2, protons and electrons when immobilized on a photoanode surface (Figure 1.5), or in clay suspensions when coupled by a redox mediator (Figure 1.6). Therefore, deposition of oxomanganese catalysts on to inexpensive semiconductor electrodes, such as TiO_2 or silica, has been considered an attractive option for large-scale applications (Figure 1.5) [24, 25, 64]. The design, optimization, and assembly of molecular components

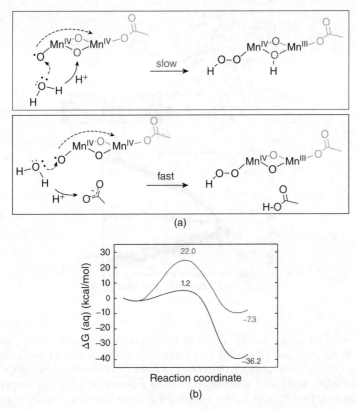

Figure 1.4 *(a) O–O bond formation step during O$_2$ evolution catalyzed by **1-OAc** in the absence (red box) and presence (blue box) of an acetate proton acceptor. (b) Comparison of the free energy profiles for the O–O bond formation mechanism depicted in (a). See color plate*

that can actually demonstrate the feasibility of these devices is the subject of much current research. Such research requires a fundamental understanding of the kinetic and thermodynamic factors that regulate the reaction mechanism for efficient light absorption and the use of the harvested solar energy for direct water oxidation.

Complex **1**, either attached to TiO$_2$ nanoparticles (NPs) [24, 25, 29] or immobilized in clays [30, 31], has been shown to preserve its catalytic activity for water oxidation when activated with Ce^{4+} as a primary chemical oxidant. In addition, it has been shown that the oxidation state of the Mn(III–IV) dimer can be advanced to Mn(IV,IV) by visible light photoexcitation, with concomitant interfacial electron injection into the conduction band of the semiconductor. It has also been shown that complex **1** can be directly deposited or *in situ* synthesized on TiO$_2$ NPs [25]. Figure 1.5 shows a DFT model of the **1**–TiO$_2$ assembly, in which the Mn–terpy dimer complex is bound to the semiconductor surface by substitution of one water ligand with the NPs, leaving a terminal water molecule as substrate for O$_2$ evolution. EPR measurements and DFT–QM/MM modeling have also suggested that the dimeric mixed-valence (III–IV) state covalently binds to the near-amorphous TiO$_2$ surface.

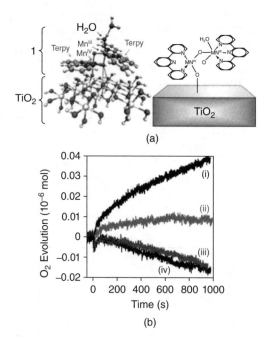

(a)

(b)

Figure 1.5 (a) DFT cluster model of a **1**–TiO$_2$(NP) covalent assembly, with a water ligand exchanged by the nanoparticles (NPs) and a terminal water exposed to the solvent. (b) O$_2$ evolution during water oxidation catalyzed by the **1**–TiO$_2$(NP) assembly, using Ce^{4+} as a single-electron oxidant. Complex **1** was loaded on three different TiO$_2$ (50 mg) samples: (i) P25, (ii) D450, and (iii) D70, as well as (iv) a control test using bare P25 NPs as the catalyst. See color plate

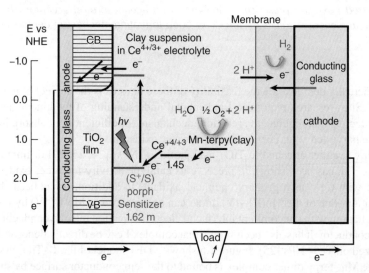

Figure 1.6 Schematic diagram of a photocatalytic cell used for direct solar water oxidation, based on the Mn–terpy catalyst in a clay suspension coupled to the sensitized TiO$_2$ electrode surface by a redox couple Ce$^{4+/3+}$ mediator. See color plate

Oxygen evolution was observed for **1**–TiO$_2$ hybrid assemblies upon chemical oxidation using Ce^{4+} as a primary oxidant (Figure 1.5b), with high reactivity observed for the semicondutor material composed of 85% anatase (P25), mild activity when deposited on anatase NPs sintered at 450 °C (D450), and no activity for TiO$_2$ with low crystallinity (D70) or bare P25 NPs. Structural characterization of the **1**–TiO$_2$ assemblies suggests that the (III, IV) Mn dimer is not the predominant form on the surface of well-crystallized TiO$_2$ nanoparticles, probably due to dimerization of complex **1** and formation of Mn(IV) tetramers. The degree of crystallinity plays a crucial role in promoting photooxidation of the biomimetic catalyst anchored to the NPs [24]. These studies have therefore reported evidence of the successful assembly of biomimetic oxomanganese catalysts on inexpensive semiconductor materials. The outstanding challenge is to exploit such heterogeneous assemblies based on the immobilization of effective homogeneous catalysts, in order to achieve photocatalytic water oxidation for fuel production from water.

1.7 Conclusion

Structural and mechanistic studies of the OEC of PSII and biomimetic oxomanganese complexes are essential to gaining fundamental insights into catalytic water oxidation mechanism in natural and artificial photosynthesis. Recent advances in X-ray crystallography, high-resolution spectroscopy, and computational modeling have provided valuable contributions towards understanding the critical structure/function relations for catalytic activity based on oxomanganese complexes.

Computational studies based on DFT–QM/MM hybrid methods have established a structural model of the OEC in the S$_1$ resting state that is fully consistent with XRD and high-resolution spectroscopy. Comparative studies of water oxidation catalyzed by biomimetic oxomanganese complexes have also provided insights into the mechanism that might be common to oxomanganese catalysts in general, including the essential role of acid/base and redox cofactors that influence both the thermodynamics and the kinetics of water-splitting. In particular, DFT calculations have shown that the surrounding buffer or deprotonated side chains of amino acid residues in close contact with the OEC may play essential regulatory roles by tuning the redox potential of Mn centers during activation of the catalytic complex or expediting O–O bond formation when participating as proton acceptors in the nucleophilic attack of the oxyl radical by substrate water molecules.

The resulting insights into catalytic water oxidation, obtained by combining experimental and computational studies, should be particularly valuable for the design and optimization of molecular assemblies for solar water splitting. Biomimetic oxomanganese complexes could be used as homogeneous catalysts adsorbed on semiconductor electrodes, or in suspensions of clays mediated by one-electron redox couples. In fact, O$_2$ evolution has already been observed for oxomanganese complexes deposited on TiO$_2$ NPs or in clay suspensions when activated by a sacrificial primary oxidant. However, visible-light photoactivation of oxomanganese catalysts for water oxidation has yet to be demonstrated. Such advances could establish photocatalytic solar cells for artificial photosynthesis based on inexpensive and earth-abundant materials.

Acknowledgments

We acknowledge support by US Department of Energy Grant DE-FG02-07ER15909 (GWB), DE-SC0001423 (VSB), and computer resources from NERSC and Yale University. The authors acknowledge Dr Sandra Luber, Dr Leslie Vogt, Dr Mehmed Z. Ertem, and Dr Rhitankar Pal for valuable discussions.

References

[1] Joliot, P., Barbieri, G., Chabaud, R. (1969) Photochem. Photobiol., 10: 309 ff.

[2] Kok, B., Forbush, B., McGloin, M. (1970) Photochem. Photobiol., 11: 457 ff.

[3] Rhee, K. H. (2001) Annu. Rev. Biophys. Biomol. Struct., 30: 307–328.

[4] Zouni, A., Witt, H. T., Kern, J., Fromme, P., Krauss, N., Saenger, W., Orth, P. (2001) Nature, 409: 739–743.

[5] Ferreira, K. N., Iverson, T. M., Maghlaoui, K., Barber, J., Iwata, S. (2004) Science, 303: 1831–1838.

[6] Loll, B., Kern, J., Saenger, W., Zouni, A., Biesiadka, J. (2005) Nature, 438: 1040–1044.

[7] Umena, Y., Kawakami, K., Shen, J.-R., Kamiya, N. (2011) Nature, 473: 55–60.

[8] Grundmeier, A., Dau, H. (2012) Biochim Biophys Acta-Bioenergetics, 1817: 88–105.

[9] Yano, J., Kern, J., Irrgang, K., Latimer, M., Bergmann, U., Glatzel, P., Pushkar, Y., Biesiadka, J., Loll, B., Sauer, K., Messinger, J., Zouni, A., Yachandra, V. (2005) Proc. Natl. Acad. Sci. U. S. A., 102: 12 047–12 052.

[10] Dau, H., Liebisch, P., Haumann, M. (2004) Phys. Chem. Chem. Phys., 6: 4781–4792.

[11] Grabolle, M., Haumann, M., Muller, C., Liebisch, P., Dau, H. (2006) J. Biol. Chem., 281: 4580–4588.

[12] Luber, S., Rivalta, I., Umena, Y., Kawakami, K., Shen, J.-R., Kamiya, N., Brudvig, G. W., Batista, V. S. (2011) Biochemistry, 50: 6308–6311.

[13] Ruettinger, W., Yagi, M., Wolf, K., Bernasek, S., Dismukes, G. C. (2000) J. Am. Chem. Soc., 122: 10 353–10 357.

[14] Chen, H. Y., Tagore, R., Das, S., Incarvito, C., Faller, J. W., Crabtree, R. H., Brudvig, G. W. (2005) Inorg. Chem., 44: 7661–7670.

[15] Poulsen, A. K., Rompel, A., McKenzie, C. J. (2005) Angew. Chem. Int. Ed., 44: 6916–6920.

[16] Wiechen, M., Berends, H. M., Kurz, P. (2012) Dalton Trans., 41: 21–31.

[17] Cady, C. W., Crabtree, R. H., Brudvig, G. W. (2008) Coord. Chem. Rev., 252: 444–455.

[18] Cady, C. W., Shinopoulos, K. E., Crabtree, R. H., Brudvig, G. W. (2010) Dalton Trans., 39: 3985–3989.

[19] Tagore, R., Chen, H. Y., Crabtree, R. H., Brudvig, G. W. (2006) J. Am. Chem. Soc., 128: 9457–9465.

[20] Wang, T., Brudvig, G., Batista, V. S. (2010) J. Chem. Theory Comput., 6: 755–760.

[21] Wang, T., Brudvig, G. W., Batista, V. S. (2010) J. Chem. Theory Comput., 6: 2395–2401.

[22] Luo, S., Rivalta, I., Batista, V., Truhlar, D. G. (2011) J. Phys. Chem. Lett., 2: 2629–2633.

[23] Sproviero, E. M., Gascon, J. A., McEvoy, J. P., Brudvig, G. W., Batista, V. S. (2008) Coord. Chem. Rev., 252: 395–415.

[24] Li, G., Sproviero, E. M., McNamara, W. R., Snoeberger, R. C. III, Crabtree, R. H., Brudvig, G. W., Batista, V. S. (2010) J. Phys. Chem. B, 114: 14 214–14 222.

[25] Li, G., Sproviero, E. M., Snoeberger, R. C., III, Iguchi, N., Blakemore, J. D., Crabtree, R. H., Brudvig, G. W., Batista, V. S. (2009) Energy Env. Sci., 2: 230–238.

[26] Rivalta, I., Brudvig, G. W., Batista, V. S. (2012) Curr. Opin. Chem. Biol., 16: 11–18.

[27] Limburg, J., Vrettos, J. S., Liable-Sands, L. M., Rheingold, A. L., Crabtree, R. H., Brudvig, G. W. (1999) Science, 283: 1524–1527.

[28] Limburg, J., Vrettos, J. S., Chen, H. Y., de Paula, J. C., Crabtree, R. H., Brudvig, G. W. (2001) J. Am. Chem. Soc., 123: 423–430.

[29] McNamara, W. R., Snoeberger, R. C., Li, G., Schleicher, J. M., Cady, C. W., Poyatos, M., Schmuttenmaer, C. A., Crabtree, R. H., Brudvig, G. W., Batista, V. S. (2008) J. Am. Chem. Soc., 130: 14 329–14 338.

[30] Yagi, M., Narita, K. (2004) J. Am. Chem. Soc., 126: 8084–8085.

[31] Narita, K., Kuwabara, T., Sone, K., Shimizu, K.-I., Yagi, M. (2006) J. Phys. Chem. B, 110: 23 107–23 114.

[32] Rivalta, I., Brudvig, G. W., Batista, V. S. (2013) J. Phys. Chem. Lett., submitted.

[33] Britt, R., Campbell, K., Peloquin, J., Gilchrist, M., Aznar, C., Dicus, M., Robblee, J., Messinger, J. (2004) Biochim Biophys Acta-Bioenergetics, 1655: 158–171.

[34] Cox, N., Rapatskiy, L., Su, J. H., Pantazis, D. A., Sugiura, M., Kulik, L., Dorlet, P., Rutherford, A. W., Neese, F., Boussac, A., Lubitz, W., Messinger, J. (2011) J. Am. Chem. Soc., 133: 14 149–14 149.

[35] Sproviero, E. M., Gascon, J. A., McEvoy, J. P., Brudvig, G. W., Batista, V. S. (2006) J. Chem. Theory Comput., 2: 1119–1134.

[36] Sproviero, E. M., Gascon, J. A., McEvoy, J. P., Brudvig, G. W., Batista, V. S. (2007) Curr. Opin. Struct. Biol., 17: 173–180.

[37] Sproviero, E. M., Gascon, J. A., McEvoy, J. P., Brudvig, G. W., Batista, V. S. (2008) J. Am. Chem. Soc., 2: 3428–3442.

[38] Sproviero, E. M., Gascon, J. A., McEvoy, J. P., Brudvig, G. W., Batista, V. S. (2008) J. Am. Chem. Soc., 130: 6728–6730.

[39] Sproviero, E. M., Shinopoulos, K., Gascon, J. A., McEvoy, J. P., Brudvig, G. W., Batista, V. S. (2008) Phil. Trans. Royal Soc. B – Biol. Sci., 363: 1149–1156.

[40] Lundberg, M., Siegbahn, P. E. M. (2004) Phys. Chem. Chem. Phys., 6: 4772–4780.

[41] Siegbahn, P. E. M., Lundberg, M. (2005) Photochem. Photobiol. Sci., 4: 1035–1043.

[42] Siegbahn, P. E. M. (2006) Chemistry – A European Journal, 12: 9217–9227.

[43] Siegbahn, E. M. (2008) Chemistry – A European Journal, 14: 8290–8302.

[44] Siegbahn, P. E. M. (2008) Phil. Trans. Royal Soc. B – Biol. Sci., 363: 1221–1228.

[45] Siegbahn, P. E. M. (2008) Inorg. Chem., 47: 1779–1786.

[46] Siegbahn, P. E. M. (2009) J. Am. Chem. Soc., 131: 18 238 ff.

[47] Siegbahn, P. E. M. (2009) Acc. Chem. Res., 42: 1871–1880.

[48] Schinzel, S., Schraut, J., Arbuznikov, A. V., Siegbahn, P. E. M., Kaupp, M. (2010) Chemistry – A European Journal, 16: 10 424–10 438.

[49] Sauer, K., Yano, J., Yachandra, V. K. (2008) Coord. Chem. Rev., 252: 318–335.

[50] Haumann, M., Muller, C., Liebisch, P., Iuzzolino, L., Dittmer, J., Grabolle, M., Neisius, T., Meyer-Klaucke, W., Dau, H. (2005) Biochemistry, 44: 1894–1908.

[51] Ankudinov, A. L., Ravel, B., Rehr, J. J., Conradson, S. D. (1998) Physical Review B, 58: 7565–7576.

[52] Galstyan, A., Robertazzi, A., Knapp, E. W. (2012) J. Am. Chem. Soc., 134: 7442–7449.

[53] Rivalta, I., Amin, M., Luber, S., Vassiliev, S., Pokhrel, R., Umena, Y., Kawakami, K., Shen, J.-R., Kamiya, N., Bruce, D., Brudvig, G. W., Gunner, M. R., Batista, V. S. (2011) Biochemistry, 50: 6312–6315.

[54] Milikisiyants, S., Chatterjee, R., Lakshmi, K. V. (2011) J. Phys. Chem. B, 115: 12 220–12 229.

[55] Lundberg, M., Siegbahn, P. E. M. (2005) Chem. Phys. Lett., 401: 347–351.

[56] Lundberg, M., Blomberg, M. R. A., Siegbahn, P. E. M. (2004) Inorg. Chem., 43: 264–274.

[57] Pecoraro, V. L., Baldwin, M. J., Caudle, M. T., Hsieh, W. Y., Law, N. A. (1998) Pure Appl. Chem., 70: 925–929.

[58] Limburg, J., Szalai, V. A., Brudvig, G. W. (1999) J. Chem. Soc. – Dalton Trans., 1353–1361.

[59] Vrettos, J. S., Limburg, J., Brudvig, G. W. (2001) Biochim Biophys Acta – Bioenergetics, 1503: 229–245.

[60] Hillier, W., Messinger, J., Wydrzynski, T. (1998) Biochemistry, 37: 16 908–16 914.

[61] Sameera, W. M. C., McKenzie, C. J., McGrady, J. E. (2011) Dalton Trans., 40: 3859–3870.

[62] Chen, Z., Concepcion, J. J., Hu, X., Yang, W., Hoertz, P. G., Meyer, T. J. (2010) Proc. Natl. Acad. Sci. U. S. A., 107: 7225–7229.

[63] Fecenko, C. J., Thorp, H. H., Meyer, T. J. (2007) J. Am. Chem. Soc., 129: 15 098 ff.

[64] Weare, W. W., Pushkar, Y., Yachandra, V. K., Frei, H. (2008) J. Am. Chem. Soc., 130: 11 355–11 363.

2

O–O Bond Formation by a Heme Protein: The Unexpected Efficiency of Chlorite Dismutase

Jennifer L. DuBois
Department of Chemistry and Biochemistry, Montana State University,
Bozeman, MT, USA

2.1 Introduction

Natural processes for O–O bond generation are exceedingly rare. Outside of photosynthesis, nature offers little inspiration to chemists aspiring to make dioxygen. Recent work, however, has identified an unusual heme protein-catalyzed process whereby some bacteria detoxify chlorite (ClO_2^-), converting it to Cl^- and O_2. This reaction necessitates forming an O–O bond from a high-valent metal–oxo intermediate in an aqueous environment, without the need to navigate a series of accompanying electron and proton transfers. As such, it offers insights to the chemist aiming to understand specifically the metal-mediated O–O joining step in the water oxidation process. This chapter reviews structural, kinetic, and mechanistic information about the heme-dependent chlorite dismutase (Cld) enzymes that catalyze this unusual reaction, with the objective of summarizing key features of the catalyst that are important in promoting the formation of the O–O bond.

2.2 Origins of O_2-Evolving Chlorite Dismutases (Clds)

Bacteria must respond, effectively and sometimes rapidly, to changes in their environment. Environmental selection pressures are often supplied by human activity, illustrated dramatically by the prevalence and rapid spread of antibiotic resistance. By analogy, the deposition

Molecular Water Oxidation Catalysis: A Key Topic for New Sustainable Energy Conversion Schemes,
First Edition. Edited by Antoni Llobet.
© 2014 John Wiley & Sons, Ltd. Published 2014 by John Wiley & Sons, Ltd.

of a host of chemical agents in the environment requires environmental microbes – the living component in so many of the biogeochemical element cycles of the earth – to respond or perish. Efficient efflux provides one means of coping with toxic agents; chemical catalysis provides another.

A small handful of bacteria have adopted the latter, more interesting strategy for contending with a specific class of compounds: the oxochlorates. These are low-molecular-weight anionic species used on the bulk/industrial scale as oxidants, biocides, and bleaches. Several bacterial species, mostly from phylum Proteobacteria, have been identified that have evolved the ability to use the higher-oxidation-state oxochlorates – perchlorate (ClO_4^-) and chlorate (ClO_3^-) – as electron acceptors [1–7]. These anions are reduced at the terminus of the respiratory electron transport chain by a molybdopterin-dependent perchlorate reductase with clear evolutionary and structural relationships to widespread respiratory nitrate reductases [6, 8–13]. The end product of this unusual metabolic pathway is chlorite (ClO_2^-), which is detoxified by Cld via conversion to harmless Cl^- and O_2.

Unlike perchlorate reductase, the Cld enzyme lacks a clear evolutionary origin. Moreover, it is part of a large and taxonomically diverse family of proteins comprising a cluster of orthologs (COG) and a component of an even larger structural superfamily (Figure 2.1) [14, 15]. The biological roles of the majority of the members of the family and superfamily, which almost certainly do not relate to perchlorate or chlorite, are just beginning to be clarified. In most cases, genes encoding a bona fide Cld co-localize in the bacterial genome with genes encoding a perchlorate reductase, and the two metabolic activities are obligately linked. In general, although with possible interesting exceptions [15, 16], nonperchlorate respirers that have Cld-encoding genes are more likely from a biological standpoint to be using the protein for some other function. These Clds from nonrespirers are currently under investigation.

2.3   Major Structural Features of the Proteins and their Active Sites

Four O_2-evolving Clds have been crystallographically characterized to date. Three of them, from the bacteria *Dechloromonas aromatica* [17], *Azospira oryzae* [18], and *Nitrospira defluvii* [15], share a highly similar homopentameric structure. Each monomer is > 200 amino acids in length, with a pseudotwofold axis at the center of two structurally similar domains. Each domain consists of an α-helical bundle and a series of β-sheets with a cavity in the middle. Only the C-terminal domain of each monomer binds heme.

Although the majority of structurally characterized Clds are of the bidomain-monomer/pentameric-protein type, a second distinct category exists, represented by a fourth Cld structure (from *Nitrobacter winogradskyi*) that has an N-terminally truncated monomer containing the tail end of one domain attached to a complete, heme-binding one at the C-terminus [16]. These "short" Cld monomers do not assemble as pentamers but rather form dimers. Their monomer–monomer interfaces are moreover completely different from those found in the homopentameric Clds. Interestingly, they are predominantly found among Proteobacteria, the same phylum from which most of the characterized perchlorate respiring bacteria derive. However, the small-monomer/dimeric Clds appear to come primarily or exclusively from nonperchlorate respirers.

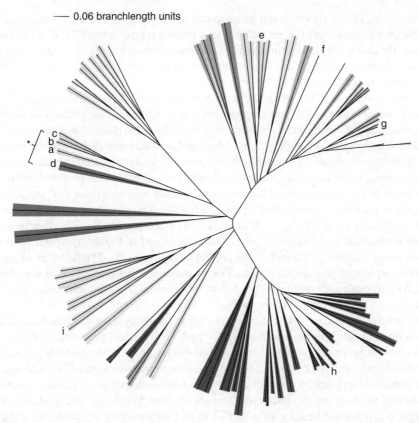

— 0.06 branchlength units

Figure 2.1 *Phylogenetic tree of Cld sequences from diverse hosts. The phylum/kingdom affil-iation of each species is indicated by color: Proteobacteria, yellow; Firmicutes, orange; Nitro-spirae, red; Actinobacteria, blue; Archaea, light blue; Deinococcus-Thermus, gray; Chloroflexi, green; Planctomycetes, dark purple; Verrucomicrobia, light purple; Acidobacteria, pink. The Halobacteriaceae, pictured near the bottom of the tree, form their own group distinct from the other archaea. Species known to carry out chlorite detoxification are indicated with a bracket/asterisk. Representative crystallographically characterized Clds are labeled as follows: (a) D. aromatica, (b) D. agitata, (c) Ideonella dechloratans, (d) N. defluvii, (e) Halobac-terium sp. NRC-1, (f) T. thermophilus HB8, (g) G. stearothermophilus, (h) M. tuberculosis, and (i) T. acidophilum. A three-iteration PSI-BLAST search was performed using DaCld as the bait sequence. The top 500 result sequences were aligned by ClustalX and a phylogenetic tree was constructed. Representative sequences from each phylum were chosen for the display. Set-tings used for the tree building were random number = 111 and bootstrap maximum = 1000. The iTOL (Interactive Tree of Life) program was used for branch coloring and figure generation (http://itol.embl.de/, last accessed 13 December 2013). Reprinted with permission from [14]. Copyright © Elsevier, 2011. See color plate*

The relative stability of the pentameric and dimeric Clds was assessed by comparing representative proteins from *Nitrospira defluvii* and *Nitrobacter winogradskyi*, respectively [19]. The thermal and chemical unfolding of the proteins was examined using a number of methods, including differential scanning calorimetry (DSC), electronic circular dichroism, and ultraviolet/visible (UV/vis) and fluorescence spectroscopies. It was demonstrated that the pentamer is significantly more stable, with a melting temperature of 92 °C at pH 7.0; by contrast, the dimer begins to unfold at 53 °C. Because of their greater thermal stability and turnover numbers with chlorite (see 2.4), the pentameric Clds have gained more attention as oxygen-generating catalysts.

Among O_2-evolving Clds, the amino acid composition of the active site (Figure 2.2) is strongly conserved [20]. The heme's proximal side is defined by the helical bundle of the C-terminal domain, from which a histidine residue extends and binds to the heme iron. This histidine is hydrogen bonded through its other imidazole nitrogen to a conserved glutamic acid. In heme peroxidases, a similar hydrogen bonding interaction lends anionic character to the proximal histidine ligand. Imidazolate character strengthens the Fe–His bond in turn [21]. Resonance Raman spectra of the ferrous *Dechloromonas aromatica* Cld (*Da*Cld) indicate that, in spite of the presence of a His–Glu hydrogen bonding interaction, the Fe–His bond is relatively weak ($v_{Fe-His} = 221\,cm^{-1}$) [22]. A comparable Fe–His vibrational frequency is observed in myoglobins, where the heme ligand is weakly hydrogen bonded to the oxygen atom from a carbonyl on the protein's backbone [23]. The structure of the Cld from the sequenced perchlorate respirer *Da*Cld shows that the glutamic acid is hydrogen bonded via its other carboxylate oxygen to a water molecule. This may attenuate its strength as a hydrogen bond acceptor relative to the Fe–His.

Functionally, the weak Fe–His interaction leads to relatively less ligand-derived electron density on the iron. According to the classic "push–pull" model proposed by Poulos and Kraut for heme peroxidases, this in turn affords the iron less "push" toward heterolytic bond cleavage (e.g. in a ferric–heme–OOH complex) on the heme's distal face [24]. Additionally, three conserved tryptophan residues are found in the proximal pocket. The substitution of either of the two strictly conserved Trps (W155 and W156) for phenylalanine leads to strongly diminished heme affinity in *Da*Cld and instability of the pentameric state of

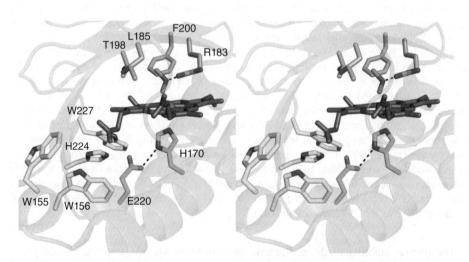

Figure 2.2 *Active site of nitrite-bound DaCld. The proximal His–Glu pair is shown in green, the hydrophobic Thr/Leu/Phe triad in salmon, the Trp/His network in yellow, and the distal Arg in orange. Adapted with permission from [17] and generated using PyMOL. Reprinted from [17] with kind permission from Springer Science + Business Media. See color plate*

the protein. A similar though smaller-magnitude effect is seen for the W227F mutant. These residues are apparently important to the structural integrity of the heme pocket and protein as a whole.

The heme's distal side, the side with an open coordination position where chemistry takes place, is defined by β-sheets which create a sterically confined, low-lying pocket above the heme plane. Steric confinement is likely to be important in enclosing reactive intermediates and enforcing the recombination of those intermediates with leaving groups. A lone arginine residue is an exception to an otherwise highly hydrophobic distal pocket and a residue of clear catalytic importance [15, 25]. Consistent with this hydrophobicity, the heme's Soret peak is red shifted relative to what would be observed in a more hydrophilic, histidine- and arginine-containing pocket like that found in canonical peroxidases [26–29]. The heme moreover converts from a five-coordinate high-spin ferric form to a six-coordinate low-spin ferric-hydroxide near pH 9, 2–3 units lower than for the formation of the analogous complex in peroxidases [22]. In spite of the weak Fe–His bond noted from the resonance Raman, the Fe–OH vibrational frequency is exceedingly low [22]. This suggests that the Fe–OH bond is weakened via strong hydrogen bond donation from the distal arginine. Substitution of the distal arginine for the (roughly) isosteric but neutral amide, glutamine, leads to dramatically reduced affinity for anions and complete loss of the so-called alkaline transition to form the stable ferric-hydroxide. In keeping with these observations, in the available crystal structures, the arginine side chain frequently hydrogen bonds to a heme ligand: nitrite [17], thiocyanate [18], imidazole [15], or cyanide [15]. The presence of the bound ligand is critical for successful crystallization of the *Da*Cld, likely due to motion of the arginine in its absence. It is also critical for efficient catalysis, although O_2 production is still observed in the glutamine-substituted mutant [15, 25]. Notably, the major effect of this substitution on catalysis is observed for K_M, and k_{cat} is only minimally perturbed. This further suggests a critical role for the distal residue in forming and/or stabilizing the Fe^{III}–chlorite complex.

Resonance Raman spectra of the ferrous carbonyl complex of *Da*Cld suggests that the distal arginine's side chain can exist in two conformers with respect to the carbonyl oxygen atom: one strongly interacting and directed towards the heme plane (in) and one weakly interacting and directed away from the heme (out) (Figure 2.2) [22]. The "in" conformer predominates at low pH and the "out" at high, with a pK_a near 6.6. Replacement of the distal arginine with structurally similar but neutral glutamine led to a four-orders-of-magnitude reduction in k_{cat}/K_M(chlorite) as well as dramatically reduced affinity of the iron for anions [15, 29]. These results collectively indicate a role for the distal arginine in recruiting and stabilizing the anionic substrate in the distal pocket.

As they possess a distal arginine without an accompanying distal histidine, Clds have a more hydrophobic pocket with less positive charge than canonical heme peroxidases. This structural distinction is consistent with *Da*Cld's relatively low Fe(III)/Fe(II) reduction potential (−23 mV versus normal hydrogen electrode (NHE)), which is at the less negative end of the range measured for peroxidases (−0.180 to −0.280 V). In heme peroxidases, the arginine likewise serves to stabilize ligands bound to Fe(III) via hydrogen bonding and to polarize the O–O bond of the coordinated hydroperoxide. That is, it helps supply part of the "pull" of the push–pull model [30, 31]. The presence of an arginine in the Cld pocket suggests it could serve an analogous role in promoting Cl–O bond cleavage for a coordinated chlorite. The peroxidase/Cld analogy and its mechanistic implications will be explored more thoroughly later in the chapter.

2.4   Efficiency, Specificity, and Stability

Despite still poorly understood selection pressures driving their evolution Clds catalyze chlorite decomposition with the efficiency of true detoxification enzymes. Values for k_{cat}/K_M measured by monitoring initial rates of O_2 evolution or $ClO_2{}^-$ disappearance in the steady state span $10^6 – 10^8\,M^{-1}/s$, suggestive of a reaction that is operating near the diffusion limit. The $ClO_2{}^-$ bond cleavage process is fast and the frothing due to O_2 production is vigorous enough to preclude monitoring of *Da*Cld's heme via standard transient UV/vis methods.

However, the emitted O_2 can be detected in real time by trapping it at a reduced metal center. Measurements of this type were carried out using the ribonucleotide reductase from *Chlamydia trachomatis*, with the enzyme's $Mn^{IV}Fe^{IV}–O_2$ intermediate forming from the reaction of the Mn^{II}/Fe^{II} complex and O_2 [33]. These measurements gave a second-order rate constant for the metallocenter/O_2 reaction of $3 \times 10^3\,M^{-1}/s$. They moreover showed that transiently supersaturating O_2 solutions could be generated at concentrations up to ~ 9 mM, with solutions ≤ 6 mM in O_2 forming in less than 15 ms. *Da*Cld's ability to generate concentrated aqueous solutions of O_2 in such short pulses has made it a highly desirable reagent for the generation and trapping of reactive O_2 intermediates (Figure 2.3) [34].

*Da*Cld is also remarkably specific. In the presence of chlorite and large numbers of equivalents of potential cosubstrates used by other heme/oxidant-reactive proteins – including reductants, hydrogen atom transfer reagents, oxygen atom acceptors, and chlorine atom acceptors – the only observed reaction is O_2 generation (Figure 2.4) [26]. This may reflect the speed of the O_2-producing reaction relative to competing processes, or it might signal

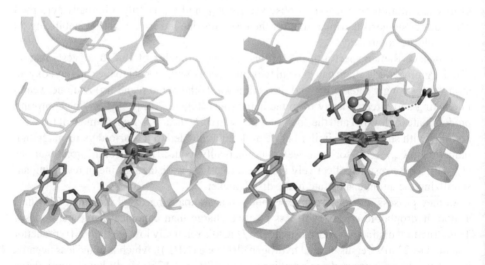

Figure 2.3   *Active sites of the nitrite-bound D. aromatica Cld (left) [17] and water-bound N. winogradskyi Cld (right) [16], illustrating the conformational flexibility of the distal arginine. In the image on the left, this residue is hydrogen bonded to the axial nitrito ligand, and in that on the right, to water and an asparagine. These images represent the proposed "in" and "out" configurations of the arginine, respectively. Reproduced with permission from [36]. Copyright © 2012, American Chemical Society. See color plate*

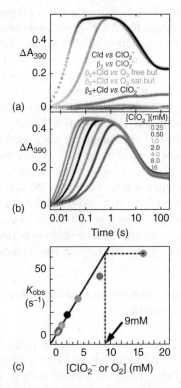

Figure 2.4 *A transient enzyme-O_2 intermediate (with absorbance at 390 nm) was prepared using DaCld and ClO_2^- to generate the O_2. (a) Absorbance versus time traces were measured following rapid mixing at 5 °C of two solutions in a stopped-flow spectrometer. Syringe 1 contained an anaerobic/reduced solution of 0.2 mM of the Mn- and Fe-dependent ribonucleotide reductase β_2 subunit from Chlamydia tracomatis, 0.6 mM Mn^{II}, 0.2 mM Fe^{II}, and 0.01 mM DaCld (molarity given per heme-containing subunit). This was mixed with an equal volume of either 20 mM ClO_2^- (black trace), O_2 – saturated 100 mM HEPES buffer (pH 7.6) (green trace), or O_2 – free buffer (orange trace). The steep upward curve represents formation of the Mn^{IV}/Fe^{IV} intermediate and the subsequent downward curve shows its decomposition; the plateau region in between therefore represents the intermediate's lifetime, which is extended when DaCld/ClO_2^- rather than O_2 – saturated buffer is used as the O_2 generator, because higher O_2 concentrations are attainable. Traces from control reactions, from which DaCld and β_2 were omitted, are colored red and blue, respectively. (b) Delineation of the O_2 concentration dependence of the β_2 activation reaction by variation of ClO_2^- concentration. Reactions were conducted as described for the black trace in panel (a), but with the concentration of the ClO_2^- reactant solution varied to give the final ClO_2^- concentrations noted in the figure. Traces were analyzed by nonlinear regression using the equation for two exponential phases to extract observed first-order rate constants for formation of the Mn^{IV}/Fe^{IV} intermediate (k_{obs}). (c) Plot of these observed first-order rate constants versus ClO_2^- or O_2 concentration. The points at ≤ 4 mM ClO_2^- were fit by the equation for a line. Extrapolation of the k_{obs} for the reaction with 16 mM ClO_2^- to the linear-fit line (dashes) in this case gave an effective O_2 concentration of 9 mM (arrow). Reproduced with permission from [34], Copyright © 2008, American Chemical Society. See color plate*

the absence of reactive intermediates such as a ferryl porphyrin cation radical ($Fe^{IV} = O$ por$^{\bullet+}$, compound I) or hypochlorite (OCl^-) that could otherwise be trapped by these reagents. Consistent with absent or short-lived intermediate species, mass spectrometry has shown that the oxygen from solvating ^{18}O-labeled water does not emerge in the evolved O_2 [35]. High-valent intermediates and full catalytic cycles have, however, been observed in the presence of other oxidants that do not evolve O_2. These are described below (2.5).

Like most oxidation catalysts, Clds undergo a limited number of turnovers before inactivating. Early experiments with *Da*Cld showed that the progress of reaction curves often terminated before all the chlorite in the reaction mixture was consumed. This was the first indication of possible oxidative damage to the enzyme. Careful measurements of residual activity following exposure to chlorite showed that the enzyme could undergo ~ 20 000 turnovers per heme prior to complete loss of activity [26]. A similar number was measured by titrating the UV/vis spectrum of the heme with chlorite. The spectrum gradually diminished, consistent with the chlorite-dependent destruction of the heme. The end product has not been characterized but does not appear to be verdoheme (the chromophoric breakdown product of the canonical heme oxygenase (i.e. non-IsdG) reaction).

2.5   Mechanistic Insights from Surrogate Reactions with Peracids and Peroxide

The heme chromophore supplies a powerful diagnostic of the ligation state and electronic structure of the iron; ideally, the mechanism of O–O bond formation by *Da*Cld should therefore be probed via a direct, real-time method such as UV/vis stopped-flow spectroscopy. The efficiency of the O_2 formation reaction here is problematic, however, as the reaction with chlorite is too fast to monitor by conventional techniques. However, related O–O bond-breaking reactions catalyzed by *Da*Cld, using either peroxides or peracids as the oxidants, can be readily examined. Their study provides some insights into O_2 generation, as there are many analogies between O–O bond-breaking and -forming processes and the proposed mechanisms favored for biological systems are the microscopic reverse of one another, as illustrated in Scheme 2.1.

*Da*Cld reacts sluggishly with H_2O_2 and high-pK_a organic peroxides such as cumylhydroperoxide or *tert*-butyl-hydroperoxide, either alone in the transient state or in the steady state when a one-electron oxidizable cosubstrate is present (e.g. guiacol, ascorbate, 2, 2′-azinobis-3-ethylbenzthiazoline-6-sulfonic acid (ABTS)). Accordingly, the first intermediate observed in the *Da*Cld/H_2O_2 reaction ($\geq$ 200 equiv.) via stopped flow is neither compound I nor another isoelectronic, two-electron-oxidized ferryl–protein–radical species, as one would expect for a heme peroxidase. Rather, it is an apparent hydroperoxy- or superoxy-ferric species (compound 0 or compound III), based on the characteristic UV/vis spectra for either adduct. At acidic pH (pH 6), the second-order reaction is exceedingly slow (~ 120 M^{-1}/s^2); it accelerates by an order of magnitude at pH 8 [36]. Heme peroxidases depend on the presence of an active site base (typically histidine, sometimes aspartic acid) to activate H_2O_2 [37]. Substitution of the active site base with alanine leads to a dramatic reduction in the second-order rate constant for compound I formation of up to five orders of magnitude. *Da*Cld lacks a basic residue in the distal pocket, which may limit the heme's reactivity with H_2O_2.

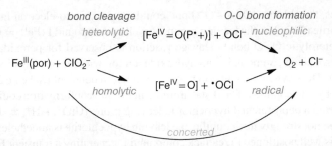

Scheme 2.1 *Three possible mechanisms for O_2 generation from ClO_2^-. The pathway along the top begins with the two-electron heterolytic cleavage of the (O) Cl–O^- bond, yielding the two-electron oxidized ferryl porphyrin cation radical (compound I) intermediate and hypochlorite. Nucleophilic attack of the anionic leaving group upon the electron deficient oxygen of the ferryl will lead to a short-lived [Fe^{III}–$OOCl^-$] species, breaking down to O_2, Cl^-, and the ferric heme. The middle pathway begins with homolytic, one-electron cleavage of the (O) Cl–O^- bond, yielding a ferryl species (compound II) and hypochloryl radical. These react through a radical coupling mechanism to give the [Fe^{III}–$OOCl^-$] species and subsequently O_2, Cl^-, and the ferric heme. Finally, a concerted reaction with no observable intermediates is possible.*

By contrast, rapid peroxidase activity with ascorbate is observed at acidic pHs in the steady and transient states when peracetic acid (PAA) is the oxidant. Intermediate formation from the Fe(III)/PAA reaction in the transient state is likewise rapid ($\geq 10^5 \, M^{-1}/s$), potentially due to increased deprotonation of PAA ($pK_a = 8.2$) to yield the reactive anion. At weakly acidic pH, compound I ([$Fe^{IV} = O$] with an exchange-coupled porphyrin cation radical) is the initially observed intermediate, based on a strongly diminished Soret band intensity and the appearance of characteristic visible absorbance bands. Compound I forms from the heterolytic, two-electron cleavage of the Fe^{III}–O–OAc bond to yield acetate. Hence, at acidic pHs, DaCld reacts with PAA very much like a peroxidase would react with H_2O_2.

Under alkaline conditions (pH ≥ 8), however, peroxidase reactivity with PAA/ascorbate slows significantly in the steady state and the first observable intermediate in the transient state is not compound I. Rather, the UV/vis spectrum of a ferryl ([$Fe^{IV} = O$]) species is observed. This might be due to a change in mechanism at alkaline pHs to *homolytic* cleavage of the Fe^{III}–O–OAc bond, yielding [$Fe^{IV} = O$] and acetyl radical. Alternatively, the intermediate could be isoelectronic with compound I, but with the organic radical on the protein and uncoupled from the ferryl. Such an intermediate would be expected to have a UV/vis spectrum more or less indistinguishable from the simple [$Fe^{IV} = O$]. Current data point toward the latter model. If proven, such a model would suggest that heterolytic cleavage is the major reaction pathway at all pH values, with compound I forming and the radical migrating away from the porphyrin more rapidly at higher pH.

2.6 Possible Mechanisms

There are three simple ways to envision generating O_2 and Cl^- from a single molecule of ClO_2^- (Scheme 2.1). First, after coordinating to the ferric iron through one of the oxygen atoms, the chlorite (O) Cl–O$^-$ bond could break in a two-electron fashion, yielding hypochlorite (ClO$^-$) as the leaving group, as well as compound I (Fe^{IV} = O por$^{\bullet+}$). An analogous heterolytic O–O bond cleavage reaction is observed for peroxidases reacting with H$_2$O$_2$, producing compound I and hydroxide ion (or, with the participation of a proton donor, water). The oxygen atom in compound I is electron deficient (Scheme 2.2) and subject to attack by nucleophiles. The distal arginine in its "in" conformation could help stabilize and position a deprotonated hypochlorite leaving group (ClO$^-$ + H$^+$ ⇌ HClO pK_a = 7.5). Although not strongly nucleophilic, the leaving hypochlorite is nonetheless negatively charged and is well positioned to reattack compound I, generating a transient Fe^{III}–OOCl$^-$ species. The coordinated anion would then spontaneously break down to yield Cl$^-$ and O$_2$. One of the proposed mechanisms for the O–O joining step in photosystem II (PSII) has an analogous high-valent manganese–oxo intermediate undergoing nucleophilic attack from an adjacent calcium-hydroxide (Scheme 2.2).

As a second possible mechanism, the (O) Cl–O$^-$ bond could break homolytically, yielding hypochloryl radical and compound II (Fe^{IV} = O). Homolytic cleavage of the (O) Cl–O$^-$ bond might be expected to be less exergonic than heterolytic cleavage, lowering the overall driving force for the reaction. However, a homolytic pathway may still be kinetically preferred. The distal arginine in its "out" configuration (Scheme 2.2) would create a very hydrophobic distal pocket, which would help to stabilize the neutral radical leaving group and possibly the transition state along this pathway. The hypochloryl radical and compound II would subsequently recombine to produce the same transient Fe^{III}–OOCl$^-$ species, breaking down to O$_2$, Cl$^-$, and Fe^{III}. This mechanism resembles the second proposed O–O bond joining process for PSII (Scheme 2.2). A radical reaction of this type was also proposed for water-soluble synthetic heme catalysts that mediate O$_2$ formation from chlorite, albeit with low efficiency [38].

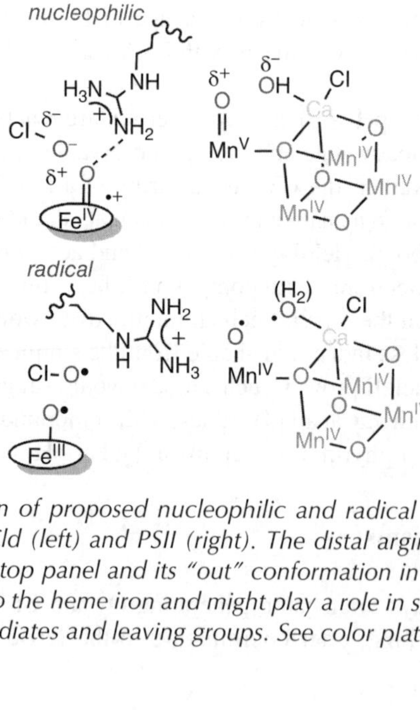

Scheme 2.2 *Comparison of proposed nucleophilic and radical mechanisms for the O–O bond joining step in DaCld (left) and PSII (right). The distal arginine residue is drawn in its "in" conformation in the top panel and its "out" conformation in the bottom. The positively charged residue is close to the heme iron and might play a role in stabilizing or positioning the various proposed intermediates and leaving groups. See color plate*

Finally, O_2 formation could occur in a concerted process without any observable high-valent iron intermediates. Indeed, no intermediates are observed in the reaction of *Da*Cld and chlorite at 4 °C under stopped-flow conditions [36]. Mass spectrometric measurements of the *Da*Cld reaction carried out in ^{18}O-labeled water likewise failed to reveal any ^{18}O label in the evolved O_2. This suggests either that no Fe/O intermediates form or, if they do, that the oxygen atom in the intermediates is not exchangeable with the water-derived oxygen.

2.7   Conclusion

Studies of the *Da*Cld reaction have suggested that several important features of its structure promote rapid O_2 production. These include a near-neutral histidine ligand to the five-coordinate, high-spin heme iron, which would be expected to provide a relatively weak push towards cleavage of a coordinated peroxide O–O or (O) Cl–O$^-$ bond [39]. On the heme's distal face, a relatively low, hydrophobic ceiling lies above the porphyrin plane. The distal arginine is of clear importance in stabilizing bound anionic ligands to the Fe^{III}, a role for which its spectroscopically observed conformational flexibility is likely critical. The rapidity of O_2 evolution by *Da*Cld prohibits the direct observation of the single turnover reaction by transient kinetic methods. However, the surrogate reaction with peracids suggests a preference for heterolytic bond cleavage that may likewise carry over to the natural chlorite substrate.

*Da*Cld's "champion" performance makes it a model for efficient O_2 generation catalyzed by a simple metallocenter. That same champion status has made it a useful reagent for generating aqueous O_2 solutions, rapidly and even at supersaturating concentrations. Less efficient O_2-generating Clds may likewise prove interesting, as they may permit direct observation of the single turnover reaction by stopped-flow methods. They may also shed light on the evolutionary origins of this fascinating, unusual chemistry.

Acknowledgements

This work was funded by National Institutes of Health grant R01GM090260. We wish to thank our collaborators and all the lab members who have contributed to this project.

References

[1] Bruce, R. A., Achenbach, L. A., Coates, J. D. (1999) Env. Microbiol., 1(4): 319–329.

[2] Coates, J. D., Michaelidou, U., Bruce, R. A., O'Connor, S. M., Crespi, J. N., Achenbach, L. A. (1999) App. Env. Microbiol., 65(12): 5234–5241.

[3] Bender, K. S., O'Connor, S. A., Chakraborty, R., Coates, J. D., Achenbach, L. A. (2002) App. Env. Microbiol., 68(10): 4820–4826.

[4] Chaudhuri, S. K., O'Connor, S. M., Gustavson, R. L., Achenbach, L. A., Coates, J. D. (2002) App. Env. Microbiol., 68(9): 4425–4430.

[5] Coates, J. D., Achenbach, L. A. (2004) Nat. Rev. Microbiol., 2(7): 569–580.

[6] Bender, K. S., Shang, C., Chakraborty, R., Belchik, S. M., Coates, J. D., Achenbach, L. A. (2005) J. Bacteriol., 187(15): 5090–5096.

[7] Wallace, W., Ward, T., Breen, A., Attaway, H. (1996) J. Ind. Microbiol., 16(1): 68–72.

[8] Kengen, S. W. M., Rikken, G. B., Hagen, W. R., van Ginkel, C. G., Stams, A. J. M. (1999) J. Bacteriol., 181(21): 6706–6711.

[9] Giblin, T., Frankenberger, W. T. (2001) Microbiol. Res., 156(4): 311–315.

[10] Okeke, B. C., Frankenberger, W. T. (2003) Microbiol. Res., 158(4): 337–344.

[11] Xu, J. L., Trimble, J. J., Steinberg, L., Logan, B. E. (2004) Water Res., 38(3): 673–680.

[12] Nozawa-Inoue, M., Jien, M., Hamilton, N. S., Stewart, V., Scow, K. M., Hristova, K. R. (2008) App. Env. Microbiol., 74(6): 1941–1944.

[13] Bansal, R., Deobald, L. A., Crawford, R. L., Paszczynski, A. J. (2009) Biodegradation, 20(5): 603–620.

[14] Goblirsch, B., Kurker, R. C., Streit, B. R., Wilmot, C. M., DuBois, J. L. (2011) J. Mol. Biol., 408(3): 379–398.

[15] Kostan, J., Sjöblom, B., Maixner, F., Mlynek, G., Furtmüller, P. G., Obinger, C., Wagner, M., Daims, H., Djinović-Carugo, K. (2010) J. Struct. Biol., 172(3): 331–342.

[16] Mlynek, G., Sjöblom, B., Kostan, J., Füreder, S., Maixner, F., Gysel, K., Furtmüller, P. G., Obinger, C., Wagner, M., Daims, H., Djinović-Carugo, K. (2011) J. Bacteriol., 93(10): 2408–2417.

[17] Goblirsch, B. R., Streit, B. R., DuBois, J. L., Wilmot, C. M. (2010) J. Biol. Inorg. Chem., 15(6): 879–888.

[18] de Geus, D. C., Thomassen, E. A. J., Hagedoorn, P. L., Pannu, N. S., van Duijn, E., Abrahams, J. P. (2009) J. Mol. Biol., 387(1): 192–206.

[19] Hofbauer, S., Gysel, K., Mlynek, G., Kostan, J., Hagmueller, A., Daims, H., Furtmüller, P. G., Djinović-Carugo, K., Obinger, C. (2012) Biochim. Biophys. Acta – Proteins Proteomics, 1824(9): 1031–1038.

[20] Mayfield, J. A. (2012) PhD thesis, University of Notre Dame, Notre Dame, IN.

[21] Goodin, D. B., McRee, D. E. (1993) Biochem., 32(13): 3313–3324.

[22] Streit, B. R., Blanc, B., Lukat-Rodgers, G. S., Rodgers, K. R., DuBois, J. L. (2010) J. Am. Chem. Soc., 132(16): 5711–5724.

[23] Adachi, S., Nagano, S., Ishimori, K., Watanabe, Y., Morishima, I., Egawa, T., Kitagawa, T., Makino, R. (1993) Biochem., 32(1): 241–252.

[24] Finzel, B. C., Poulos, T. L., Kraut, J. (1980) J. Biol. Chem., 259(21): 3027–3036.

[25] Blanc, B., Mayfield, J. A., McDonald, C. A., Lukat-Rodgers, G. S., Rodgers, K. R., DuBois, J. L. (2012) Biochem., 51(9): 1895–1910.

[26] Streit, B. R., DuBois, J. L. (2008) Biochem., 47: 5271–5280.

[27] Thorell, H. D., Karlsson, J., Portelius, E., Nilsson, T. (2002) Biochim. Biophys. Acta – Gene Struct. Express., 1577(3): 445–451.

[28] Hagedoorn, P. L., de Geus, D. C., Hagen, W. R. (2001) Eur. J. Biochem., 269(19): 4905–4911.

[29] Battistuzzi, G., Bellei, M., Bortolotti, C., Sola, M. (2010) Arch. Biochem. Biophys., 500: 21–36.

[30] Howes, B. D., Rodriguez-Lopez, J. N., Smith, A. T., Smulevich, G. (1993) Biochem., 36(6): 1532–1543.

[31] Neri, F., Indiani, C., Welinder, K. G., Smulevich, G. (1998) Eur. J. Biochem., 251(3): 830–838.

[32] DuBois, J. L., Mayfield, J. A. (2012) In Handbook of Porphyrin Science, Vol. 19. World Scientific, Singapore, p. 232.

[33] Tagore, R., Crabtree, R. H., Brudvig, G. W. (2008) Inorg. Chem., 47(6): 1815–1823.

[34] Dassama, L. M. K., Yosca, T. H., Conner, D. A., Lee, M. H., Blanc, B., Streit, B. R., Green, M. T., DuBois, J. L., Krebs, C., Bollinger, J. M. Jr, (2012) Biochem., 51(8): 1607–1616.

[35] Lee, A. Q., Streit, B. R., Zdilla, M. J., Abu-Omar, M. M., DuBois, J. L. (2008) Proc. Nat. Acad. Sci. U. S. A., 105(41): 15 654–15 659.

[36] Mayfield, J. A., Blanc, B., Rodgers, K. R., Lukat-Rodgers, G. S., DuBois, J. L. (2013) Submitted.

[37] Erman, J. E., Vitello, L. B., Miller, M. A., Shaw, A., Brown, K. A., Kraut, J. (1993) Biochem., 32(37): 9798–9806.

[38] Keith, J. M., Abu-Omar, M. M., Hall, M. B. (2011) Inorg. Chem., 50(17): 7928–7930.

[39] Poulos, T. L., Fenna, R. E. (1994) In Metal Ions In Biological Systems, Vol 30. Dekker, New York, NY. p. 25.

3

Ru-Based Water Oxidation Catalysts

Laia Francàs[1], Roger Bofill[2], Jordi García-Antón[2], Lluis Escriche[2],
Xavier Sala[2] and Antoni Llobet[1,2]

[1]Institute of Chemical Research of Catalonia (ICIQ), Tarragona, Spain
[2]Department of Chemistry, Autonomous University of Barcelona, Barcelona, Spain

3.1 Introduction

The enormous and increasing anthropogenic consumption of fossil fuels, together with the progressive decrease of their global reserves [1], has created an urgent need for a cheap and sustainable energy source in the near future. It is essential that this new source be clean and carbon-free in order to stop global warming, due to the increasing atmospheric CO_2 concentration originating in fossil-fuel combustion.

For the last 2400–3000 million years of evolution, nature has harvested sunlight as an energy source through the photosynthetic processes carried out by green plants, algae and cyanobacteria. An enormous advance in our knowledge of the molecular machinery involved in photosynthesis has taken place in the last decade. Two families of electronically coupled protein complexes, photosystem I (PSI) and photosystem II (PSII), are involved in photosynthesis. In PSII, four protons and four electrons are removed from two water molecules during photosynthesis in a thermodynamically unfavorable reaction, thanks to sunlight energy absorbed by chlorophyll P680 (Equation 3.1). This process generates dioxygen and a gradient of electrons and protons that ends up with two equivalents of NADPH and three of ATP, which constitute the required reducing equivalents and the energy needed for PSI to generate carbohydrates from CO_2 (Equation 3.2) [2].

$$2\ H_2O + 2\ NADP^+ + 3\ ADP + 3\ Pi + 8\ h\upsilon \rightarrow O_2 + 2\ NADPH + 2\ H^+ + 3\ ATP \qquad (3.1)$$

$$CO_2 + 2\ NADPH + 3\ ATP + 2\ H^+ \rightarrow 1/n\ (CH_2O)_n + 2\ NADP^+ + 3\ ADP + 3\ Pi + H_2O \qquad (3.2)$$

Molecular Water Oxidation Catalysis: A Key Topic for New Sustainable Energy Conversion Schemes,
First Edition. Edited by Antoni Llobet.
© 2014 John Wiley & Sons, Ltd. Published 2014 by John Wiley & Sons, Ltd.

From a chemical standpoint, one of the most interesting processes in photosynthesis occurs at the oxygen-evolving complex (OEC) of PSII, where the thermodynamically uphill oxidation of water takes place in an Mn_4CaO_5 cluster in the dark. The electrons released are transferred to a $Tyr_Z O\cdot$ radical (Tyr161), formed after oxidative quenching of the excited P680*, which is stabilized in turn by the presence of a proximal His residue (His190) [3–5]. Later on, these electrons flow from Tyr_Z through a channel of electronic transport that consecutively involves P_{680}, pheophytin (Phe), and quinones A and B (Q_A, Q_B), until they finally reach PSI.

Recently, the structure of the Mn_4CaO_5 cluster has been solved at 1.9 Å resolution (Figure 3.1). It shows the coordination of five carboxylates from surrounding acidic amino acids binding in a bidentate manner to the Mn and Ca atoms. Additionally, four water molecules coordinate to the cluster, two to the Ca ion [6]. The presence of mono- and di-μ-oxo ligands completes the first coordination sphere. The anionic and σ-donating nature of these ligands is essential for the accessibility of the higher oxidation states of the metal centers that will promote the O–O bond formation, which in turn will finally lead to the generation of dioxygen. The role of the Ca metal is a subject of current discussion. It has been suggested that not only is the Ca ion a structural cofactor but that it might also play an important role as a Lewis acid during the oxidation of water, by affecting the nucleophilic reactivity of its bounded water molecules [7–9]. In addition, it has recently been proposed to strongly affect the electronic properties of the rest of the Mn transition metals in the cluster, based on related model complexes [10]. Detailed knowledge of how the Mn_4CaO_5 works at a molecular level is thus of paramount importance from a biological perspective, as well as for the design of low-molecular-weight functional analogs.

One of the potential solutions to the current energy crisis is water splitting by sunlight:

$$2\,H_2O \rightarrow O_2 + 2\,H_2 \quad \Delta G^\circ = 113.5\,kcal/mol \tag{3.3}$$

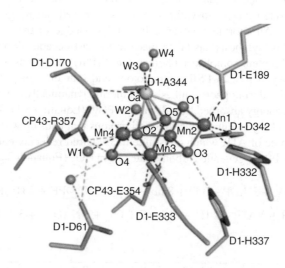

Figure 3.1   *X-ray structure of the OEC-PSII at 1.9 Å resolution. Reprinted by permission from Macmillan Publishers Ltd: Nature, [6], copyright © 2011. See color plate*

From a redox point of view, this reaction can be divided into water oxidation and proton reduction, as indicated in the following equations (all redox potentials reported in this work are referenced to SSCE unless explicitly indicated otherwise):

$$2\,H_2O \rightarrow O_2 + 4H^+ + 4e^-\quad E^\circ = 0.94\,\text{V at pH}\,1.0 \tag{3.4}$$

$$4\,H^+ + 4\,e^- \rightarrow 2\,H_2 \tag{3.5}$$

Incidentally, the water oxidation reaction needed for water splitting is identical to that occurring at the OEC-PSII.

3.2   Proton-Coupled Electron Transfer (PCET) and Water Oxidation Thermodynamics

Water oxidation is a challenging task for a catalyst. There are two main reasons for this difficulty: one is the large endothermicity of the reaction and the other is the large molecular complexity from a mechanistic point of view. Recently, important developments have been made in the design of new transition-metal complexes capable of carrying out the water oxidation reaction in a catalytic manner. These molecular complexes are characterized by different sets of ligand architectures and by their nuclearity. They show the variety of strategies that can be used to achieve water oxidation, as well as the numerous problems that can present themselves [11]. A fundamental piece of knowledge for the development of this field is an understanding of the reaction mechanism through which the reaction can proceed, since this will help in the design of efficient and rugged catalysts. Therefore, there is an urgent need to characterize reactive intermediates, as well as decomposition pathways that should be avoided [12]. A fundamental problem is the unavoidable use of water as solvent, due to the limited temperature range in which reactions can be studied and the high absorptivity. Another problem is the limited solubility of the catalysts and catalyst precursors in water. The high thermodynamic redox potential required for water oxidation permits the catalyst to oxidize a broad range of organic and inorganic substrates, which means that the presence of organic solvents can lead to undesired deactivation pathways [13, 14].

The oxidation of water leads to a range of species, according to the number of electrons removed. The thermodynamics of these are summarized in Figure 3.2. As can be seen, the more electrons transferred, the lower the thermodynamic potential. Thus, a $1e^-$ oxidation process has a prohibitive thermodynamic barrier of 2.5 V at pH 1.0. The four

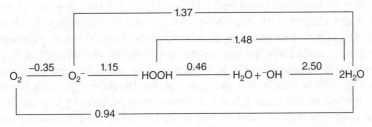

Figure 3.2   *Latimer diagram for water. The redox potentials are indicated at pH 1.0 versus SSCE*

electron transfer (ET) that occurs at the OEC-PSII [15] has the lowest thermodynamic barrier (Equation 3.4). This lowering of thermodynamics contrasts with the increase of molecular complexity. In the 4e$^-$ process, four O–H bonds from two water molecules must be broken and an O–O bond must be formed. Thus, the transition-metal complexes that can be considered as candidates for carrying out this reaction catalytically must deal with multiple ET processes, accompanied by proton-transfer management.

These requirements are met by Ru–OH$_2$ polypyridyl complexes, discovered by T. J. Meyer's group about 3 decades ago [16]. An example of this type of complex is *cis*-[Ru^{II}(bpy)$_2$(py)(H$_2$O)]$^{2+}$ (**1**) (bpy = 2, 2'-bypyridine; py = pyridine), which has been shown to be able to lose two protons and two electrons and thus reach higher oxidation states within a relatively narrow potential range at pH 7:

$$[Ru^{III}(bpy)_2(py)(HO)]^{2+} + 1\ e^- + 1\ H^+ \rightarrow [Ru^{II}(bpy)_2(py)(H_2O)]^{2+}\ E^\circ = 0.42\ V \quad (3.6)$$

$$[Ru^{IV}(bpy)_2(py)(O)]^{2+} + 1\ e^- + 1\ H^+ \rightarrow [Ru^{III}(bpy)_2(py)(HO)]^{2+}\ E^\circ = 0.53\ V \quad (3.7)$$

The higher oxidation states are easily accessible, mainly because of the σ- and π-donating character of the oxo group. From a mechanistic perspective, the simultaneous loss of protons and electrons precludes an otherwise highly destabilized scenario with highly charged species. Thus the proton-coupled electron transfer (PCET) type of process provides energetically reasonable reaction pathways that avoid high-energy intermediates. For instance, for the comproportion reaction of [LRu^{II}–OH$_2$]$^{2+}$ (L = (bpy)$_2$(py)) and [LRu^{IV} = O]$^{2+}$ to give two molecules of [LRu^{III}–OH]$^{2+}$, the energy penalty for a stepwise process with regard to the PCET is higher than 12.6 kcal/mol for the ET–PT process and higher than 13.6 kcal/mol for the PT–ET process, whereas the concerted pathway is downhill by −2.5 kcal/mol. Furthermore, the energy of activation for the concerted electron–proton transfer (EPT) process is 10.1 kcal/mol, which is lower than the thermodynamic value of any of the stepwise pathways [17].

Recently, one more oxidation process has been found in mononuclear complexes. It formally involves the oxidation of Ru(IV) to Ru(V) [18], as indicated in the following equation, in which the polypyridylic ligands are not shown:

$$Ru^V = O + 1e^- \rightarrow Ru^{IV} = O \quad (3.8)$$

One complex that displays this behavior is [Ru(tpm)(bpy)(H$_2$O)]$^{2+}$ (**2**) (tpm = tridentate facial ligand tris-(1-pyrazolyl)methane, L14; see Figure 3.5 for a drawing of the ligands discussed in this chapter) [19, 20], whose structure and Pourbaix diagram are shown in Figure 3.3. The latter is a graphical representation of the thermodynamic parameters of the species derived from the loss of protons and electrons from the initial Ru(II)–OH$_2$ complex.

For purposes of simplicity and to help keep track of electron counting, formal metal oxidation states will be used in the redox-active atoms. However, it has to be borne in mind that the oxidation can take place at both the metal center and the oxygen atom. The relative electron distributions at these locations will depend on the rest of the auxiliary ligands. As an example, for the Ru–OH$_2$ complexes, the complete bond description will be represented as a combination of two extreme resonance forms, such as [Ru(IV) = O ↔ Ru(III)–O·].

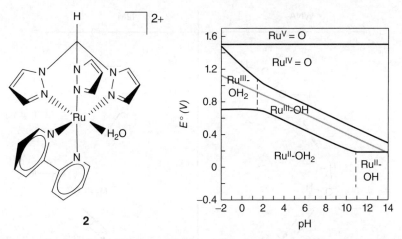

Figure 3.3 *Left: structure of [Ru(tpm)(bpy)(H₂O)]²⁺ (**2**). Right: Pourbaix diagram for **2**. E°
are referred to the SSCE electrode. The gray line represents the redox potential for the couple
H₂O/O₂*

3.3   O–O Bond Formation Mechanisms

The formation of an O–O bond promoted by transition-metal complexes can be classified
according to whether or not an unbound free water molecule participates in it. According
to this perspective, two possibilities exist: the solvent water nucleophilic attack (WNA)
mechanism and the interaction of 2 M–O entities (I2M). Both are shown in Scheme 3.1.

- *WNA*: This has a four-ET demand, which is quite stringent for a mononuclear complex.
  One solution is to share the burden with more metal centers, provided there is a bridging
  ligand that can couple them electronically and thus allow the generation of a coopera-
  tive effect. Another option for a metal complex is to cycle up and down through similar
  oxidation states of different species (see Section 3.4). It is interesting to point out here
  that both WNA and I2M mechanisms have been proposed for the OEC-PSII [7, 21, 22];
  the WNA is the inverse reaction and is proposed to occur in the reduction of dioxygen in the heme-iron protein Cyt-P450 [23]. The existence of this mechanism has been
  recently shown to occur in an Mn–porfirine model complex [24]. As will be discussed
  in Section 3.4, the nature of the complex can be mononuclear or polynuclear and it can
  have one or several Ru–O groups. In the case of the latter, they will have radically different jobs; while one will be responsible for the O–O formation bond, the other(s) will
  be responsible for facilitating electron trafficking so that the 4e⁻ acceptance process can
  be shared among the different metal centers.
- *I2M*: This is described as a reductive elimination but, depending on the oxidation states
  of both the metal center and the oxygen atoms, the O–O bond forming step can also be
  a radical–radical coupling reaction. As in the previous mechanism, the nuclearity of the
  complex can be variable.

Additionally, the organic auxiliary ligands can potentially participate in the electron trafficking process related to the water oxidation reaction and in the O–O bond formation step.

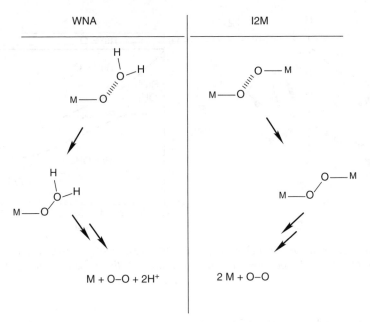

Scheme 3.1 *Potential O–O bond formation pathways promoted by transition-metal complexes*

When this happens, the ligands are generally termed "redox non-innocent." Examples of these cases will be discussed in the following section.

3.4   Polynuclear Ru Water Oxidation Catalysts

In 1982, Meyer's group reported the first molecularly well-characterized dinuclear Ru complex *cis, cis*-[(bpy)$_2$(H$_2$O)Ru(μ-O)Ru(H$_2$O)(bpy)$_2$]$^{4+}$ (**3**) (Figure 3.4) capable of oxidizing water to dioxygen, which is also called the "blue dimer" because of its absorption properties at $\lambda = 637$ nm ($\epsilon = 21\,100$) at pH 1.0 [25]. This complex showed a turnover frequency (TOF) as low as 0.24 min^{-1} and a turnover number (TON) of 13.2 when using Ce(IV) as a sacrificial oxidizing agent at pH 1.0 [26, 27]. The reductive cleavage of the oxo bridge is one of the deactivation pathways suffered by this catalyst [28]. From a mechanistic perspective, this dimer has been studied experimentally by the groups of Meyer [29–31] and Hurst [32–37] and has displayed a high degree of complexity. Meyer and coworkers reported a detailed analysis of the potential mechanism [38], which has been further extended in related and more recent papers [39, 40]. It is proposed that the catalytically active species [(bpy)$_2$(O)Ru^{V}ORu^V(O)(bpy)$_2$]$^{4+}$, {(O)Ru^{V}ORu^V(O)}$^{4+}$, is formed upon PCET oxidation of the initial {(H$_2$O)Ru^{III}ORu^{III}(OH$_2$)}$^{4+}$ compound with Ce(IV). The rapid reaction of the Ru^{V}ORu^V species with water via WNA creates a peroxidic intermediate, {(HO)Ru^{IV}ORu^{IV}(OOH)}$^{4+}$, that ends up forming the O–O bond, as can be seen in Scheme 3.2. In addition, under excess of Ce(IV) the peroxidic intermediate can be further oxidized to {(HO)Ru^{IV}ORu^V(OOH)}$^{5+}$, which finally releases dioxygen very quickly.

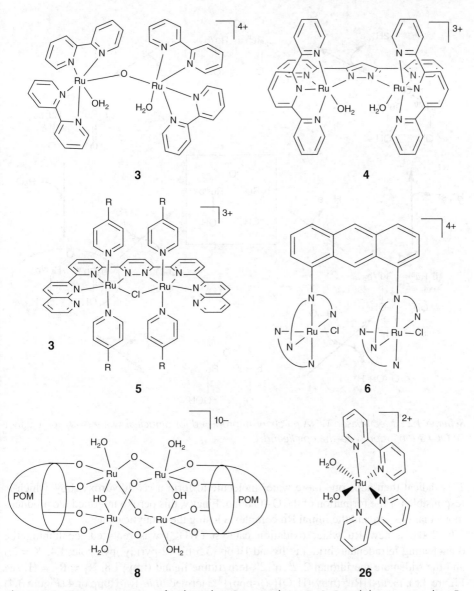

Figure 3.4 *Drawn structures of polynuclear Ru complexes 3–6, 8 and the mononuclear Ru complex 6. The three N atoms connected with arcs in 6 represent the tpry ligand and the two N atoms connected with an arc represent the bpy. The POM label represents the tetradentate O4 ligand γ-SiW$_{10}$O$_{36}$$^{8-}$.*

For the blue dimer, Hurst's group [41] suggests that besides the WNA mechanism described earlier, there is a part of the oxidation process that involves the bpy ligands in the formation of the O–O bond, as depicted Scheme 3.3. They propose that this mechanism occurs when the dimer reaches its Ru^V = O form and that at this point a water-solvent molecule adds to one of the pyridyl rings of a bpy, forming a coordinated bpy radical.

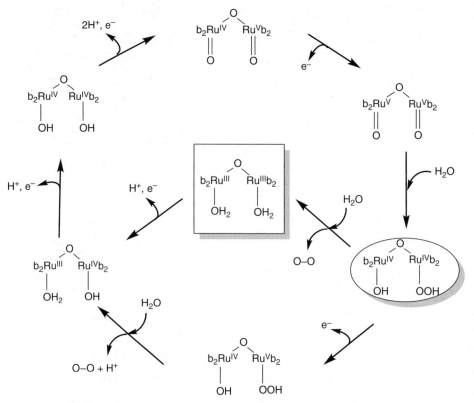

Scheme 3.2   *Metal-based WNA mechanism proposed for dinuclear water oxidation catalyst (WOC)* **3** *("b" represents the bpy ligand)*

This radical then adds one more water molecule to form a *cis*-dihydroxyl-bpy, which is responsible for the formation of the O—O bond. Finally, this peroxo intermediate produces molecular oxygen and the initial Ru complex, closing the catalytic cycle.

In 2004, a new Ru water oxidation catalyst (WOC) was reported, containing the dinucleating tetradentate bridging ligand Hbpp (3,5-bis-(2-pyridyl)pyrazole; L4, X = N) and the tridentate meridional 2, 2′:6′, 2″-terpyridine ligand (trpy; L8, $R_1 = R_2 = H$; see Figure 3.5), *in, in*-{[Ru^{II}(trpy)(H$_2$O)]$_2$(μ-bpp)}$^{3+}$, termed *in,in*-Ru-Hbpp or **4** (Figure 3.4) [42]. The flexible μ-oxo bridge of the blue dimer was replaced by the anionic and more rigid bpp$^-$, while two bpy were replaced by trpy ligands. This coordination environment forced a close disposition of the O atoms of the two aqua groups, producing a throughspace supramolecular interaction [43]. Addition of excess Ce(IV) to this complex generated dioxygen efficiently, showing TONs and TOFs of 512.0 and 0.78 min^{-1}, respectively, under optimized conditions [44–46].

Kinetic analysis and ^{18}O labeling studies were then employed to study the water oxidation mechanism taking place for this complex [46, 47]. The {(H$_2$O)Ru^{II}-Ru^{II}(OH$_2$)}$^{3+}$ species is sequentially oxidized by four 1e$^-$ processes, with Ce(IV) losing 4H$^+$ in addition, up to the IV,IV oxidation state {(O)Ru^{IV}-Ru^{IV}(O)}$^{3+}$. At this stage, intramolecular O—O bond

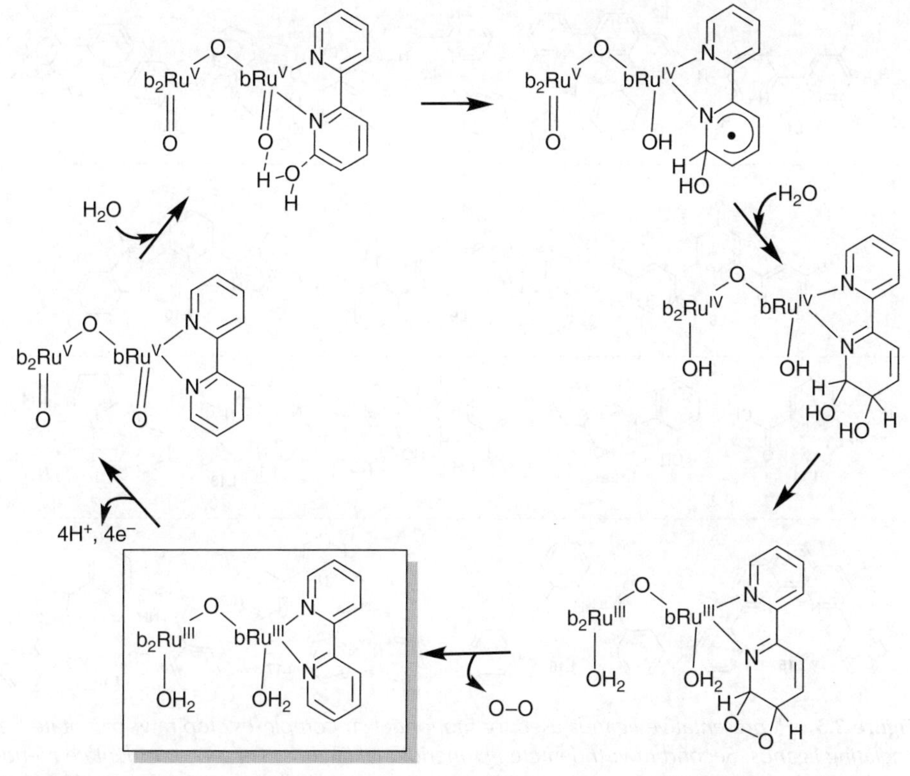

Scheme 3.3 *Ligand-based mechanism proposed for the dinuclear WOC 3 ("b" represents the bpy ligand)*

formation takes place, generating a μ-1, 2-peroxo species, $\{Ru^{III}\text{-}(OO)\text{-}Ru^{III}O\}^{3+}$, which is later followed by the formation of a hydroperoxidic intermediate that finally evolves oxygen [47], as depicted in Scheme 3.4. This intramolecular mechanistic proposal is further supported by a thorough theoretical analysis of intermediates and transition states based on density functional theory (DFT) and CASPT2 calculations [46]. Moreover, [18]O labeling data, together with the first-order kinetics observed for the formation of the intermediate, discard the potential bimolecular nature of the process, as well as the WNA mechanism.

Related tetranuclear Ru–aqua complexes based on connecting two Hbpp units with a xylilic group showed water oxidation activity coupled to ligand degradation involving the benzylic methylene group, thus generating both O_2 and CO_2 [48], a fact that has also been observed in other Ru complexes [49].

The strategy of bridging two Ru metal centers with a rigid bridging ligand was also used by the groups of Thummel, Sun and Tanaka (Figure 3.4). A series of symmetrical complexes of general formula *trans, trans*-[Ru$_2$(μ-Cl)(μ-binapypyr)(4-R-py)$_4$]$^{3+}$ **(5)** (binapypyr = 3, 6-bis(6-(1, 8-naphthyridin-2-yl)pyridin-2-yl)pyridazine; R = H, Me, NH$_2$, OCH$_3$, etc.) were reported, using the binapypyr scaffold as bridging ligand. The complex with R = –OCH$_3$ exhibited the best catalytic performance, with TON values of 689 [50, 51]. Higher TON values, up to 1690, were later obtained by Sun and coworkers

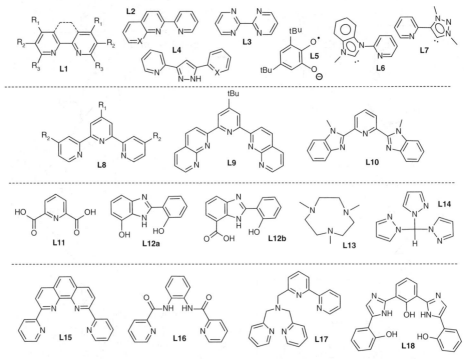

Figure 3.5 *Representative ligands used for mononuclear complexes. Top raw: bidentate N2 chelating ligands. Second raw: tridentate N3 meridional ligands. Third raw: tridentate meridional NO2 ligands and N3 facial ligands. Bottom raw: tetra and pentadentate ligands. X = N or CH and R = H, Me, F, OMe, tBu, or COOH*

using 6, 6′-(pyridazine-3, 6-diyl)dipicolinato (pdd^{2-}) as bridging ligand, where the terminal naphthyridyl moieties of binapypyr were replaced by anionic carboxylates [52]. Additional information on Sun's complexes can be found in Chapter 4. Similarly, Tanaka's group has reported a dinuclear Ru–Cl complex {[Ru(Cl)(bpy)]$_2$(μ-btpyan)}$^{2+}$ (**6**) ((btpyan = 1, 8-bis(2, 2′ : 6′2″)-terpyridylanthracene; see Figure 3.4) and has managed to obtain a crystal structure [53]. Upon treatment with Ce(IV) in acidic solution, complex **6** generates molecular oxygen up to 400 TON. It is proposed that the O–O bond formation takes places in an intramolecular fashion.

The first POM complex capable of oxidizing water to dioxygen electrochemically, with a molecular formula of [Zn$_2$Ru$_2$(OH$_2$)$_2$(ZnW$_9$O$_{34}$)$_2$]$^{14-}$ (**7**), was reported by Shannon and coworkers [54], who showed that for this particular complex the replacement of the Ru metals by Zn totally inhibited the water oxidation capacity of the molecule. Bonchio and Hill groups independently reported in 2008 the synthesis of a tetranuclear Ru complex containing a dinucleating tetradentate polyoxometalate ligand, [Ru$_4$O$_4$(OH)$_2$(H$_2$O)$_4$(γ-SiW$_{10}$O$_{36}$)$_2$]$^{10-}$ (**8**) (POM = γ-SiW$_{10}$O$_{36}$$^{8-}$; Figure 3.4) [55, 56]. The geometry of **8** consists in a central adamantane {Ru$_4$O$_6$} unit in which the four Ru atoms are situated in a tetrahedral disposition alternating with O atoms, which are situated in turn in the vertex of a hypothetical octahedron. The POM units then act as tetradentate

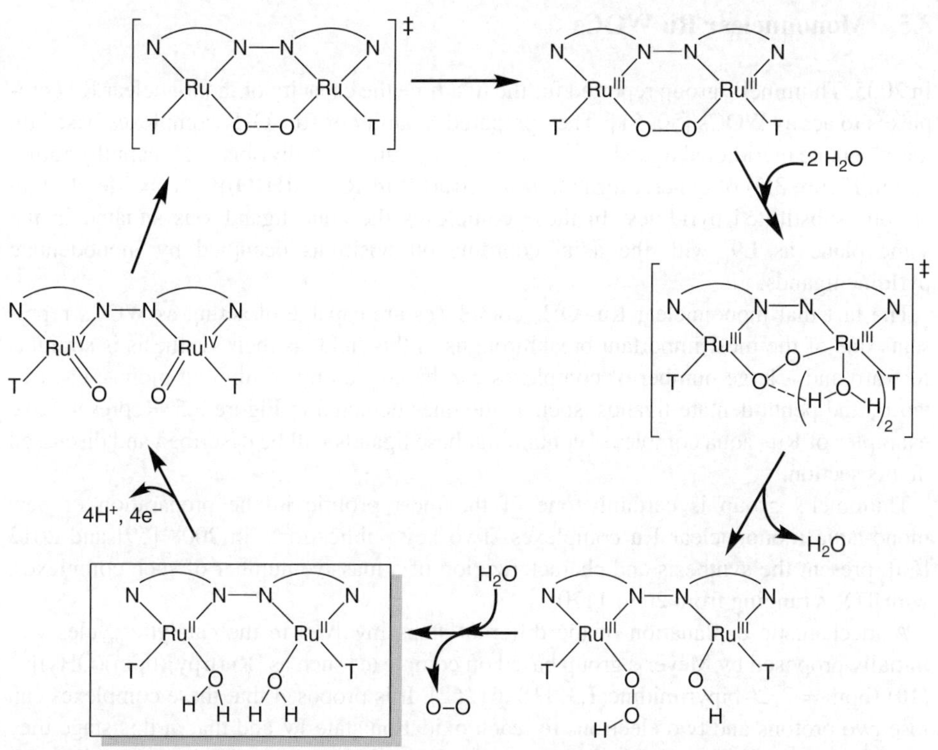

Scheme 3.4 *I2M-type mechanism proposed for dinuclear complex* **4** *("T" represents the trpy ligand and the 4N atoms connected with arcs the bpp⁻ ligand)*

bridging ligands between two Ru centers each. Finally, each Ru atom completes its octahedral coordination with a terminal aqua ligand. This catalyst is very stable against degradation and it is reported that the TON is limited by the availability of Ce(IV). Its TOF value is 7.5 min⁻¹. The high stability of this complex is attributed to the absence of organic ligands that avoid bimolecular deactivation pathways. The mechanistic pathway of **8** was reported in 2009; the report included a DFT study and proposed that the Ru₄ core is sequentially oxidized by 4e⁻ and 4H⁺:

$$[Ru^{IV}_4(OH_2)_4]^{10-} - 4H^+ - 4e^- \rightarrow [Ru^V_4(OH)_4]^{10-} \tag{3.9}$$

Once the four Ru atoms have reached formal oxidation state V, one of the Ru^V-OH groups undergoes a WNA and generates a $Ru^{III}-OOH$ species, which evolves dioxygen [57]. Recently, ab initio DFT calculations by Piccinin and Fabris [58] have shown that the free-energy difference between the initial and final oxidation states is significantly lower than the thermodynamic limit for water oxidation. This might suggest either that the DFT model does not adequately describe the electronic properties of this complex or that oxidation states higher than Ru^V or oxidized intermediates such as $Ru^{IV}-OO$ are involved. Additional information on Hill's complexes can be found in Chapter 9.

3.5 Mononuclear Ru WOCs

In 2005, Thummel's group reported for the first time the capacity of mononuclear Ru complexes to act as WOCs [50, 51]. They prepared a family of Ru–OH_2 complexes based on the tridentate meridional ligand 2, 2'-(4-(t-butyl)pyridine-2, 6-diyl)bis-(1, 8-naphthyridine) (L9 in Figure 3.5) of general formula *trans*-[Ru(L9)(4-R-py)$_2$(H$_2$O)]$^{2+}$ (R = Me, **9**) with various substituted pyridines. In these complexes the aqua ligand was situated in the same plane as L9, with the axial coordination positions occupied by monodentate pyridine ligands.

The fact that mononuclear Ru–OH_2 complexes are capable of acting as WOCs represents one of the most important breakthroughs in this field, as their synthesis is straight-forward and a large number of complexes can be created by combining mono-, bi-, tri-, tetra-, and penta-dentate ligands, such as the ones depicted in Figure 3.5. Representative examples of Ru–aqua complexes containing these ligands will be described and discussed in this section.

Thummel's group is certainly one of the most prolific in the preparation of new mono-aqua mononuclear Ru complexes. Two key publications, in 2008 [59] and 2012 [60], present the synthesis and characterization of a massive number of such complexes, with TONs ranging from 20 to 1170.

A mechanistic explanation of the different steps involved in the catalytic cycles was initially proposed by Meyer's group based on complexes such as [Ru(trpy)(bpm)(OH$_2$)]$^{2+}$ (**10**) (bpm = 2, 2'-bipyrimidine; L3) [18, 61, 62]. It is proposed that these complexes can lose two protons and two electrons to reach oxidation state IV and that at this stage they can further lose one more electron to reach the highly reactive oxidation state V. Ru(V) then suffers a nucleophilic attack of a solvent water molecule, forming a {Ru^{III}N$_5$(OOH)} intermediate which then undergoes a rapid one-electron oxidation accompanied by a proton loss to form a {Ru^{IV}N$_5$(OO)} derivative, as depicted in Scheme 3.5. Finally, this Ru(IV) peroxo complex evolves O$_2$ to regenerate the initial Ru(II) complex. Alternatively, the {Ru^{IV}N$_5$(OO)} intermediate can be further oxidized to {Ru^{V}N$_5$(OO)}, which then releases O$_2$ and generates a {Ru^{III}–OH} species. A thorough mechanistic description of the water oxidation pathways, along similar lines but more complete, has been reported recently for complex **9** by Polyanski and Fujita [63]. Additionally, Berlinguette and coworkers [64] have also lately reported on the water oxidation mechanism of mononuclear [Ru(trpy)(bpy)(OH$_2$)]$^{2+}$ derivatives (**11**), highlighting the influence of electronic effects using substituted bpy ligands (L1) with electron-donating and electron-withdrawing groups [65]. Sakai an coworkers also carried out a related mechanistic work using tridentate facial ligands such as 1,4,7-trimethyl-1,4,7-triazacyclononane (L13) and trispyrazolyl-methane (tpm; L14) of formula [Ru(T)(bpy)(OH$_2$)]$^{2+}$ (**12** and **13**), for T = L13 and L14, respectively [20]. A dinuclear Ru complex {[Ru^{III}(L18)(4-Me-py)$_2$]$_2$(μ-I)}$^+$ (**14**) containing the pentadentate ligand L18 acting as tridentate has been described in which the two metal centers are bridged by an iodido ligand [66]. This most likely acts as a mononuclear complex in water and can carry out water oxidation catalysis either using Ce(IV) at pH 1.0, with TON values ranging from 600 to 1000 and oxidative efficiencies in the range of 1.3–21.5% after 10 hours, or using [Ru(bpy)$_3$]$^{3+}$ at pH 7.2, with TON of 100 after 14 minutes. It has also been shown that the main deactivation pathway is caused by ligand oxidation to form CO$_2$.

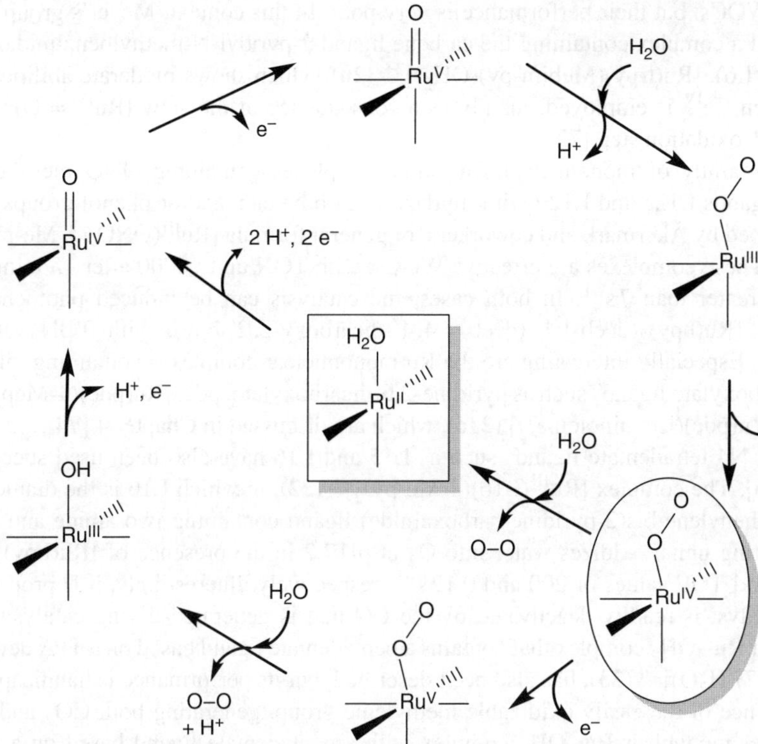

Scheme 3.5 *General WNA mechanism proposed for mononuclear Ru complexes*

New mononuclear isomeric Ru complexes *in-* and *out-*[Ru(trpy)(L4)(OH$_2$)]$^{2+}$ (*in-***15** and *out-***15**), in which the L4 ligand acts in a chelate manner, have been prepared and shown to display differentiated catalytic water oxidation activity [67]. In this line, the isomeric effects of placing fluoro substituents at different positions on the bpy ligands have been evaluated for [Ru(trpy)(5, 5′-F$_2$-bpy)(OH$_2$)]$^{2+}$ (**16**) and [Ru(trpy)(6, 6′-F$_2$-bpy)(OH$_2$)]$^{2+}$ (**17**). In the latter complex, a through space interaction can take place with the F atom and the Ru–O group, where the O–O bond formation must take place [68]. Another example of the subtle influence of electronic and steric effects over the water oxidation capacity of Ru catalysts has recently been reported by Yagi and Thummel [69, 70], who have prepared and isolated the two geometrical isomers of the complex [Ru(trpy)(pnaph)(OH$_2$)]$^{2+}$ (**18**) (pnaph = 2-(pyridin-2-yl)-1, 8-naphthyridine; L2, X = N). Surprisingly, while the isomer that has the L2-naphthyridine moiety *trans* to the Ru–OH$_2$ group is highly active (more than 50 TON), the one with the L2-pyridyl moiety *trans* to Ru–OH$_2$ is practically inactive.

Carbenes have also been used as potential ligands for the first coordination sphere of the Ru center. A series of mononuclear Ru complexes with triazolylidene carbenes (L7), [Ru(L7)(MeCN)$_4$]$^{2+}$ (**19**), where the carbene is coordinated in an abnormal manner, can carry out the oxidation of water to oxygen extremely quickly, at a TOF close to 120 min^{-1} [71]. Interestingly, related Ru complexes with carbenes coordinating in a normal fashion also act as WOCs, but their performance is very poor. In this context, Meyer's group has also reported a complex containing the carbene ligand 2-pyridyl-N-methylbenzimidazole (Mebim-py; L6), $[Ru(trpy)(Mebim-py)(OH_2)]^{2+}$ (**20**), which shows moderate abilities as a WOC when Ce^{IV} is employed, mainly as a consequence of the slow $[Ru^{IV} = O]^{2+}$ to $[Ru^{V} = O]^{3+}$ oxidation step [72].

Further, a family of mononuclear Ru–aqua complexes containing ONO meridional trianionic ligands L12a and L12b with imidazole, carboxylate, and/or phenol groups has been developed by Åkermark and coworkers, of general formula $[Ru^{III}(ONO)(4\text{-Me-py})_3]$ (**21a**) [73]. These complexes are effective WOCs, with TON up to 4000 after 15 minutes and TOF_i greater than $7 s^{-1}$. In both cases, the catalysis can be induced photochemically using $[Ru(bpy)_2(deeb)]^{2+}$ (deeb = 4, 4'-dicarboxy-2, 2'-bpy), with TON values close to 200. Especially interesting are the Ru mononuclear complexes containing mixed pyridyl–carboxylato ligands such as pyridine-2,6-dicarboxylato, pdc, $[Ru(pdc)(4\text{-Mepy})_2]$ (**21b**) and $[Ru(pdc)(isoquinoleine)_2]$ (**21c**), which are discussed in Chapter 4 [74].

Equatorial N4 tetradentate ligands such as L15 and L16 have also been used successfully [75, 76]. The complex $[Ru^{III}(L16)(4\text{-Me-py})_2]^+$ (**22**), in which L16 is the dianionic N, N' -1,2-phenylene-bis(2-pyridine-carboxamide) ligand containing two amide and two py coordinating units, oxidizes water into O_2 at pH 7.2 in the presence of $[Ru(bpy)_3]^{3+}$ with TON and TOF values of 200 and $0.12 s^{-1}$, respectively. Interestingly, it is proposed that this catalyst is readily deactivated by the CO that is generated during catalysis. A mononuclear Ru–OH$_2$ complex that contains a pentadentate ligand based on a TPA derivative, $[Ru(L17)(H_2O)]^{2+}$ (**23**), has also been described, but its performance is handicapped by the presence of the easily oxidizable methylenic group, generating both CO_2 and O_2 [49]. Another monuclear Ru–OH$_2$ complex with a pentadentate ligand based on a heteroundecatungstate containing additional Si and Ge, $[Ru^{III}(H_2O)(SiW_{11}O_{39})]^{5-}$ (**24**) and $[Ru^{III}(H_2O)(GeW_{11}O_{39})]^{5-}$ (**25**), has also been reported to act as a WOC. Mechanistic investigations suggest that the O–O bond formation takes places through a WNA pathway [77].

Finally, the water oxidation activity and mechanistic pathway of the *cis*-$[Ru(bpy)_2(OH_2)_2]^{2+}$ (**26**) (Figure 3.4) [78] has been studied. ^{18}O -labelling experiments combined with DFT analysis (MO6-L DFT and CASSCF/CASPT2 calculations) have provided evidence for the existence of a WNA mechanism operating in this case [79, 80].

3.6   Anchored Molecular Ru WOCs

A number of WOCs have been anchored on to solid surfaces with the objective of increasing their stability and/or enhancing their reactivity. The different strategies used to support the homogeneous WOCs are described in this section.

Direct deposition of the dinuclear complex $\{[Ru(OH)(3, 5\text{-tBu}_2sq)]_2(\mu\text{-btpyan})\}^{2+}$ (**27**) $(3, 5\text{-tBu}_2sq$ = the semiquinone ligand L5) into indium tin oxide (ITO), adsorbed mainly through van der Waals interactions, was described by Tanaka and coworkers more than 10 years ago [81, 82]. This hybrid material was able to generate molecular oxygen with an outstanding TON of 33 500 (phosphate buffer, pH 4) at an applied potential of 1.7 V versus Ag/AgCl across 40 hours.

Electrostatic interactions between the catalyst and the surface can also be used for catalyst immobilization. Bonchio's group took catalyst **8**, which is deca-anionic, and immobilized it on carbon nanotubes (CNTs) functionalized with polyamidoamines dendrites [83]. The performance of this new material in water oxidation is spectacular and reaches TOF values comparable to those of the homogeneous counterpart.

Another strategy consists in taking advantage of $\pi-\pi$ stacking interactions between the surface and the catalyst [84].

Finally, the most robust strategy is based on covalently anchoring the molecular catalyst into a surface without modifying the intrinsic properties of the homogeneous catalyst. For surfaces based on oxides, the preferred anchoring groups are phosphonates or carboxylates. Meyer's group followed this procedure in order to anchor their catalysts into TiO_2, ZrO_2, fluorine-doped tin oxide (FTO), ITO, or nanoITO. One example that illustrates this strategy is the complex $[Ru(Mebimpy)(4,4'-(CH_2PO_3)_2bpy)(OH_2)]^{2+}$ (**28**) (Mebimpy = 2,6-bis(1-methylbenzimidazol-2-yl)pyridine, L10; $4,4'-(CH_2PO_3)_2$bpy = ([2,2'-bipyridine]-4,4'-diylbis(methylene))diphosphonate; see Figure 3.6) [85]. The phosphonate–bpy group can be attached at the surface of the electrode and thus acts as a bridge between the electrode and the Ru metal center [86, 87], as depicted in Figure 3.6, to generate the new hybrid material **FTO–28**. With an applied potential of 1.85 V versus standard hydrogen electrode (SHE) at pH 5, the **FTO-28** derivative is reported to sustain electrocatalytic oxidation of water for at least 8 hours, reaching TON values of around 11 000 with a TOF of $21.6\,min^{-1}$. Additionally, the same group has also reported a mononuclear Ru–aqua complex containing a Ru-based redox mediator linked by a bridging bpm ligand (L3) of formula $[(4,4'-(CH_2P(O)(OH)_2)_2bpy)_2Ru(\mu-bpm)Ru(Mebimpy)(OH_2)]^{4+}$ (**29**). When attached to the surface of conductive solid supports such as ITO, FTO, or FTO–TiO_2 (Figure 3.6), this generates **ITO-29**. This modified electrode shows constant catalytic currents, giving a TON of at least 28 000, with TOF of $36\,min^{-1}$ and a Faradaic efficiency of around 98% after 8 hours of continuous O_2 evolution at an applied potential of 1.8 V versus SHE [88].

The RuHbpp catalyst **4** has also been anchored in solid surfaces. A pyrrole-substituted derivative $[Ru_2(\mu-bpp)(\mu-OAc)(t-trpy)_2]^{2+}$ (**4a**) (t-trpy = 4'-(p-pyrrolylmethylphenyl)-2,2' : 6',2''-terpyridine; see Figure 3.6) was prepared and anodically electropolymerized in the surface of vitreous carbon sponge (VCS), forming **VCS/poly-4a** [89]. The new hybrid materials was shown to maintain the intrinsic electronic properties of its molecular analog **4** and drastically improved the catalytic performance by minimizing catalyst–catalyst deactivation pathways, giving a TON of 250. Nano-TiO_2 was also used as an oxidatively rugged solid support to anchor **4** analogs. In this case, the carboxylate derivative *in,in*- $\{[Ru^{II}(trpy)(H_2O)]_2(\mu-bpp-R_a)\}^{3+}$ (**4b**) (Hbpp-R_a = 4 -((3,5-di(pyridine-2-yl)-1H-pyrazol-4-yl)methyl)benzoic acid) was synthesized and anchored into TiO_2-rutile in MeCN, forming **TiO_2-4b**. Upon activation of this species with Ce(IV) at pH 1.0, coevolution of O_2 and CO_2 was observed, together with the leaching of the catalyst from the solid support [13].

3.7 Light-Induced Ru WOCs

Three main components are needed in order to carry out light-driven water oxidation in homogeneous phase: a photosensitizier capable of harvesting sunlight energy, a WOC and a sacrificial electron acceptor. $[Ru(bpy)_3]^{2+}$ and its derivatives are the most commonly used photosensitizers due to their strong absorbance in the visible spectrum,

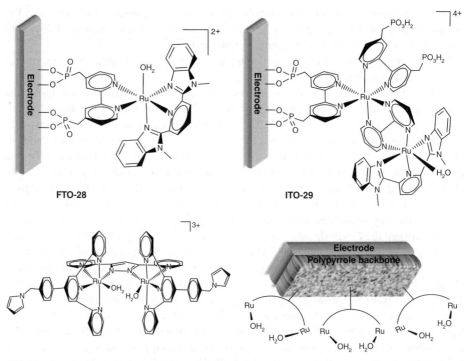

Figure 3.6 *Top left: monuclear Ru complex heterogenized at the surface of an electrode via phosphonate functionalization (**FTO-28**). Top right: as top left but with the addition of a redox mediator (**ITO-29**). Bottom left: structure of the Ru–bis–aqua complex **4a** containing pyrrole functionalities. Bottom right: schematic drawing of the polymeric material **VCS/poly-4a***

microsecond excited-state lifetimes at room temperature, and high redox potentials. When light with adequate wavelength is shined over $[Ru(bpy)_3]^{2+}$, an excited state is generated, $[Ru(bpy)_3]^{2+*}$, that is capable of transfering an electron to a sacrificial electron acceptor such as $[Co^{III}(NH_3)_5Cl]^{2+}$ or $Na_2S_2O_8$. The former generates a Co(II) complex, which decomposes, and $[Ru(bpy)_3]^{3+}$, that is a powerful oxidizer that can activate a WOC from its low oxidation state to a higher one (WOC$_A$; see Scheme 3.6), which in turn oxidizes water to dioxygen.

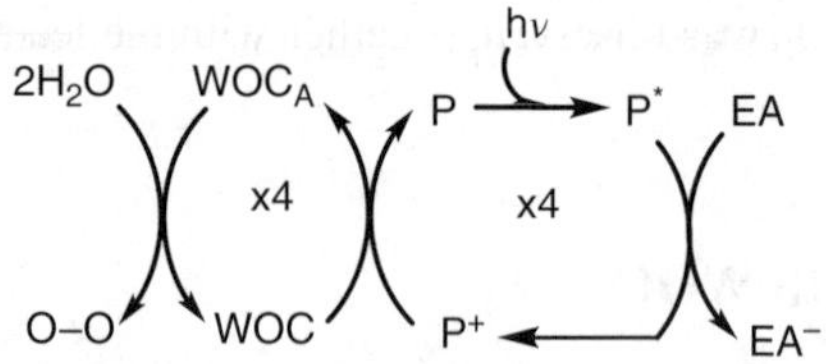

$$2H_2O \quad WOC_A \qquad h\nu$$
$$\qquad\qquad P \longrightarrow P^* \quad EA$$
$$\qquad x4 \qquad\qquad x4$$
$$O\text{–}O \qquad WOC \quad P^+ \longleftarrow EA^-$$

Scheme 3.6 *Combination of productive reactions involved in light-induced water oxidation. WOC, catalyst in a nonactive oxidation state; WOC$_A$, catalyst in its active oxidation state; P, photosensitizier; P*, photosensitizer in its excited state; EA, sacrificial electron acceptor*

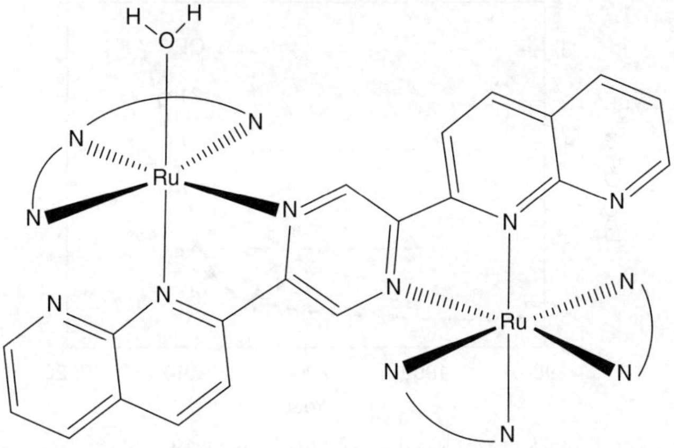

Figure 3.7 *Structure of dinuclear dyad complex **30**. The three N connected with arcs represent tBu-terpyridine, while the two connected with an arc represent the bpy ligand*

Several mono- and polynuclear Ru WOCs have been tested following this light-driven strategy in homogeneous phase with modest performances due to the complexity of the reactions involved [90, 91], as is also the case for *in-* and *out-***15** [67]. Another strategy is to anchor both the photosensitizer and the catalyst on the surface of an n-type semiconductor, as was done with **8**, which was adsorbed on a sensitized nanocrystalline TiO_2 surface [92]. Another approach to this system was published by Meyer's group, who attached the photosensitizer to the ITO and modified the WOC with long alkyl chains to allow anchoring to the surface over the dye [93]. This system proved unstable when cycling from 0.2 to 1.6 V.

The building of a chromophore–catalyst "dyad" molecule is another potential strategy for light-driven water oxidation. In this system, the two ruthenium centers are connected by a bridging ligand having different roles: one acts as the light-harvesting antenna (RuP) and the other as a WOC (RuC). There are a number of examples of this strategy in the literature for organic substrate oxidation [94–98], but very few for water oxidation [99, 100]. An illustrative example capable of water oxidation has been described recently by Thummel group, in which an N6 hexadentate bridging ligand binds to a $\{Ru^{II}(tBu_3\text{-trpy})(H_2O)\}$ fragment as the Ru–Cat moiety, generating an analog of **10**. A $\{Ru^{II}(bpy)_2\}$ moiety coordinates to the other side of the bridging ligand, generating the Ru–P moiety as an analog of $[Ru(bpy)_3]^{2+}$, as can be seen in Figure 3.7. Shining light into this dyad (**30**) in the presence of $S_2O_8^{2-}$ as sacrificial electron accepter generates O_2, reaching a TON of 134 over 6 hours [101].

3.8 Conclusion

The urgent need for a clean and renewable energy source to replace fossil fuels, which are becoming exhausted and are environmentally contaminating, has prompted intense research into light-driven water splitting. Within this context, the water oxidation process

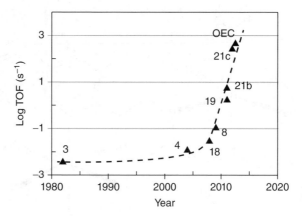

Figure 3.8 *Plot of log of initial TOF for the catalytic water oxidation reaction versus time (year reported) for a series of representative Ru WOCs*

has traditionally been recognized as one of the bottlenecks hampering development. In the last 7 years a tremendous advance has been made in this field; from the single catalyst available in 1982 (the blue dimer), literally dozens of active WOCs with increasing TONs and TOFs have been described recently. Figure 3.8 shows the log of TOF of a select number of molecular WOCs, including the OEC-PSII, plotted against the year in which they were reported. It is impressive to see that in the last 4 years the reported TOF has increased by more than four orders of magnitude and that we are quickly approaching the values of nature. Impressive advancements have also been taking place at a mechanistic level, and a few O–O bond formation pathways have been nicely elucidated. As a consequence of all these efforts, the construction of robust photoelectrochemical cells (PECs) now seems feasible and a few examples have already been described. The main challenge that still lies ahead concerns the creation of a PEC for water splitting using only sunlight, in which all the reactions occur in a harmonious manner, generating hydrogen and oxygen.

Acknowledgments

Support from Generalitat de Catalunya (2009 SGR-69), MINECO (CTQ-2010-21497, CTQ2011-26440, PRI-PIBIN-2011-1278), and "La Caixa" Foundation is gratefully acknowledged.

References

[1] Armaroli, N., Balzani, V. (2007) Angew. Chem. Int. Ed., 46: 52–66.
[2] Raven, P. H., Evert, R. F., Eichhorn, S. E. (2005) In Biology of Plants 7th ed. W.H. Freeman and Company Publishers, New York, NY, p. 124.
[3] Nugent, J. H. A., Rich, A. M., Evans, M. C. W. (2001) Biochim. Biophys. Acta, 1503: 138–146.
[4] Renger, G. (2001) Biochim. Biophys. Acta, 1503: 210–228.

[5] Nugent, J. H. A., Ball, R. J., Evans, M. C. W. (2004) Biochim. Biophys. Acta, 1655: 217–221.

[6] Umena, Y., Kawakami, K., Shen, J. R., Kamiya, N. (2011) Nature, 473: 55–60.

[7] Vrettos, J. S., Stone, D. A., Brudvig, G. W. (2001) Biochemistry, 40: 7937–7945.

[8] Meyer, T. J., Huynh, M. H. V., Thorp, H. H. (2007) Angew. Chem. Int. Ed., 46: 5284–5304.

[9] Brudvig, G. W. (2008) Philo. T. Roy. Soc. B., 363, 1211–1219.

[10] Kanady, J. S., Tsui, E. Y., Day, M. W., Agapie, T. (2011) Science, 333, 733–736.

[11] Zuccaccia, C., Bellachioma, G., Bolaño, S., Rocchigiani, L., Savini, A., Macchioni, A. (2012) Eur. J. Inorg. Chem., 9: 1462–1468.

[12] Tsai, M.-K., Rochford, J., Polyansky, D. E., Wada, T., Tanaka, K., Fujita, E., Muckerman, J. T. (2009) Inorg. Chem., 48: 4372–4383.

[13] Francas, L., Sala, X., Benet-Buchholz, J., Escriche, L., Llobet, A. (2009) Chem. Sus. Chem., 2: 321–329.

[14] Li, F., Jiang, Y., Huang, F., Li, Yanqing, Zhang, B., Sun, L. (2011) Chem. Commun., 47: 8949–8951.

[15] Barber, J. (2009) Chem. Soc. Rev., 38: 185–196.

[16] Meyer, T. J., Huynh, M. H. V. (2003) Inorg. Chem., 42: 8140–8160.

[17] Huynh, M. H. V., Meyer, T. J. (2007) Chem. Rev., 107: 5004–5064.

[18] Concepcion, J. J., Jurss, J. W., Templeton, J. L., Meyer, T. J. (2008) J. Am. Chem. Soc., 130: 16 462–16 463.

[19] Llobet, A. (1994) Inorg. Chim. Acta, 221: 125.

[20] Yoshida, M., Masaoka, S., Abe, J., Sakai, K. (2010) Chem. Asian J., 5: 2369–2378.

[21] Siegbahn, P. (2011) J. Photochem. Photobiol., 104, 94–99.

[22] Sproviero, E. M., Gascon, J. A., McEvoy, J. P., Batista, V. (2008) J. Am. Chem. Soc., 130: 3428.

[23] Zheng, J., Wang, D., Thiel, W., Shaik, S. (2006) J. Am. Chem. Soc., 128, 13 204–13 215.

[24] Gao, Y., Akermark, T., Liu, J., Sun, L., Akermark, B. (2009) J. Am. Chem. Soc., 131: 8726–8727.

[25] Gestern, S. W., Samuels, G. J., Meyer, T. J. (1982) J. Am. Chem. Soc., 104, 4029–4030.

[26] Nagoshi, K., Yamashita, S., Yagi, M., Kaneko, M. (1999) J. Mol. Catal. A: Chem., 144: 71–76.

[27] Collin, J. P., Sauvage, J. P. (1986) Inorg. Chem., 25: 135–141.

[28] Lebeau, E. L., Adeyemi, S. A., Meyer, T. J. (1998) Inorg. Chem., 37: 6476–6484.

[29] Geselowitz, D., Meyer, T. J. (1990) Inorg. Chem., 29: 3894–3896.

[30] Chronister, C. W., Binstead, R. A., Ni, J., Meyer, T. J. (1997) Inorg. Chem., 36: 3814–3815.

[31] Binstead, R. A., Chronister, C. W., Ni, J., Hartshorn, C. M., Meyer, T. J. (2000) J. Am. Chem. Soc., 122: 8464–8473.

[32] Hurst, J. K., Zhou, J., Lei, Y. (1992) Inorg. Chem., 31, 1010–1017.

[33] Lei, Y., Hurst, J. K. (1994) Inorg. Chem., 33: 4460–4467.

[34] Lei, Y., Hurst, J. K. (1994) Inorg. Chim. Acta, 226: 179–185.

[35] Yamada, H., Hurst, J. K. (2000) J. Am. Chem. Soc., 122: 5303–5311.

[36] Yamada, H., Koike, T., Hurst, J. K. (2001) J. Am. Chem. Soc., 123, 12 775–12 780.

[37] Yamada, H., Siems, W. F., Koike, T., Hurst, J. K. (2004) J. Am. Chem. Soc., 126: 9786–9795.

[38] Liu, F., Concepcion, J. J., Jurss, J. W., Cardolaccia, T., Templeton, J. L., Meyer, T. J. (2008) Inorg. Chem., 47: 1727–1752.

[39] Concepcion, J. J., Jurss, J. W., Templeton, J. L., Meyer, T. J. (2008) Proc. Nat. Acad. Sci. U. S. A., 105: 17 632–17 635.

[40] Concepcion, J. J., Jurss, J. W., Brennaman, M. K., Hoertz, P. G. Patrocinio, A. O. T., Murakami, N. Y., Templeton, J. L., Meyer, T. J. (2009) Acc. Chem. Res., 42: 1954–1965.

[41] Cape, J. L., Hurst, J. K. (2008) J. Am. Chem. Soc., 130: 827–829.

[42] Sens, C., Romero, I., Rodríguez, M., Llobet, A., Parella, T., Benet-Buchholz, J. (2004) J. Am. Chem. Soc., 126: 7798–7799.

[43] Planas, N., Christian, J. G., Mas-Marza, E., Sala, X., Fontrodona, X., Maseras, F., Llobet, A. (2010) Chem. Eur. J., 16: 7965–7968.

[44] Sala, X., Romero, I., Rodríguez, M., Escriche, L., Llobet, A. (2009) Angew. Chem. Int. Ed., 48: 2842–2852.

[45] Romero, I., Rodríguez, M., Sens, C., Mola, J., Kollipara, M. R., Francàs, L., Mas-Marza, E., Escriche, L., Llobet, A. (2008) Inorg. Chem., 47: 1824–1834.

[46] Bozoglian, F., Romain, S., Ertem, M. Z., Todorova, T. K., Sens, C., Mola, J., Rodriguez, M., Romero, I., Benet-Bucholz, J., Fontrodona, X., Cramer, C. J., Gagliardi, L., Llobet, A. (2009) J. Am. Chem. Soc., 131: 15 176–15 187.

[47] Romain, S., Bozoglian, F., Sala, X., Llobet, A. (2009) J. Am. Chem. Soc., 131: 2768–2769.

[48] Francas, L., Sala, X., Escudero-Adan, E., Benet-Buchholz, J., Escriche, L., Llobet, A. (2011) Inorg. Chem., 50: 2771–2781.

[49] Radaram, B., Ivie, J. A., Singh, W. M., Grudzien, R. M., Reibenspies, J. H., Webster, C. E., Zhao, X. (2011) Inorg. Chem., 50: 10 564.

[50] Zong, R., Thummel, R. P. (2005) J. Am. Chem. Soc., 127: 12 802–12 803.

[51] Deng, Z., Tseng, H.-T., Zong, R., Wang, D., Thummel, R. (2008) Inorg. Chem., 47: 1835–1848.

[52] Xu, Y., Åkermark, T., Gyollai, V., Zou, D., Eriksson, L., Duan, L., Zhang, R., Åkermark, B., Sun, L. (2009) Inorg. Chem., 48: 2717–2719.

[53] Wada,T., Ohtsu, H., Tanaka, K. (2012) Chem. Eur. J., 18: 2374–2381.

[54] Howells, A. R., Sankarraj, A., Shannon, C. (2004) J. Am. Chem. Soc., 126: 12 258–12 259.

[55] Geletii, Y. V., Botar, B., Kögerler, P., Hillesheim, D. A., Musaev, D. G., Hill, C. L. (2008) Angew. Chem., Int. Ed., 47, 3896–3899.

[56] Sartorel, A., Carraro, M., Scorrano, G., Zorzi, R. D., Geremia, S., McDaniel, N. D., Bernhard, S., Bonchio, M. (2008) J. Am. Chem. Soc., 130: 5006–5007.

[57] Sartorel, A., Miro, P., Salvadori, E., Romain, S., Carraro, M., Scorrano, G., Valentin, M. D., Llobet, A., Bo, C., Bonchio, M. (2009) J. Am. Chem. Soc., 131: 16 051–16 053.

[58] Piccinin, S., Fabris, S. (2011) Phys. Chem. Chem. Phys., 13: 7666–7674.

[59] Tseng, H. W., Zong, R., Muckerman, J. T., Thummel, R. (2008) Inorg. Chem., 47: 11 763–11 773.

[60] Kaveevivitchai, N., Zong, R., Tseng, H.-W., Chitta, R., Thummel, R. (2012) Inorg. Chem., 51: 2930–2939.

[61] Concepcion, J. J., Tsai, M.-K., Muckerman, J. T., Meyer, T. J. (2010) J. Am. Chem. Soc., 132: 1545–1557.

[62] Concepcion, J. J., Jurss, J. W., Norris, M. R., Chen, Z., Templeton, J. L., Meyer, T. J. (2010) Inorg. Chem., 49: 1277–1279.

[63] Polyansky, D. E., Muckerman, J. T., Rochford, J., Zong, R., Thummel, R. P., Fujita, E. (2011) J. Am. Chem. Soc., 133: 14 649.

[64] Wasylenko, D. J., Ganesamoorthy, C., Koivisto, B. D., Henderson, M. A., Berlinguette, C. P. (2010) Inorg. Chem., 49: 2202–2209.

[65] Wasylenko, D. J., Ganesamoorthy, C., Henderson, M. A., Koivisto, B. D., Osthoff, H. D., Berlinguette, C. P. (2010) J. Am. Chem. Soc., 132: 16 094–16 106.

[66] Kärkäs, M. D., Johnston, E. V., Karlsson, E. A., Lee, B.-L., Åkermark, T., Shariat-gorji, M., Ilag, L., Hansson, O., Bäckvall, J.-E., Åkermark, B. (2011) Chem. Eur. J., 17: 7953–7959.

[67] Roeser, S., Farràs, P., Bozoglian, F., Martínez-Belmonte, M., Benet-Buchholz, J., Llobet, A. (2011) ChemSusChem, 4: 197–207.

[68] Maji, S., Lopez, I., Bozoglian, F., Benet-Buchholz, J., Llobet A. (2013) Inorg. Chem., 52: 3591–3593.

[69] Yamazaki, H., Hakamata, T., Komi, M., Yagi, M. (2011) J. Am. Chem. Soc., 133: 8846.

[70] Boyer, J. L., Polyansky, D. E., Szalda, D. J., Zong, R., Thummel, R., Fujita, E. (2011) Angew. Chem. Int. Ed., 50: 12 600–12 604.

[71] Bernet, L., Lalrempuia, R., Ghattas, W., Mueller-Bunz, H., Vigara, L., Llobet, A., Albrecht, M. (2011) Chem. Commun., 47: 8058–8060.

[72] Norris, M. R., Concepcion, J. J., Harrison, D. P., Binstead, R. A., Ashford, D. L., Fang, Z., Templeton, J. L., Meyer, T. J. (2013) J. Am. Chem. Soc., 135: 2080–2083.

[73] Kärkäs, M. D., Åkermark, T., Johnston, E. V., Shams, R. K., Laine, T. M., Lee, B.-L., Åkermark, T., Privalov, T., Åkermark, B. (2012) Angew. Chem. Int. Ed., 51: 11 589–11 593.

[74] Duan, L., Xu, Y., Gorlov, M., Tong, L., Andersson, S., Sun, L. (2010) Chem. Eur. J., 16: 4659–4668.

[75] Zhang, G., Zong, R., Tseng, H.-W., Thummel, R. P. (2008) Inorg. Chem., 47: 990–998.

[76] Kärkäs, M. D., Åkermark, T., Chen, H., Sun, J., Åkermark, B. (2013) Angew. Chem. Int. Ed., 52: 4189–4193.

[77] Murakami, M., Hong, D., Suenobu, T., Yamaguchi, S., Ogura, T., Fukuzumi, S. (2011) J. Am. Chem. Soc., 133: 11 605.

[78] Dobson, J. C., Meyer, T. J. (1988) Inorg. Chem., 27: 3283–3291.

[79] Sala, X., Ertem, M. Z., Vigara, L., Todorova, T. K., Chen, W., Rocha, R. C., Aquilante, F., Cramer, C. J., Gagliardi, L., Llobet, A. (2010) Angew. Chem. Int. Ed., 122: 10 911–10 913.

[80] Planas, N., Vigara, L., Cady, C., Miro, P., Huang, P., Hammarstrom, L., Styring, S., Leiden, N., Dau, H., Haumann, M., Gagliardi, L., Cramer, C. J., Llobet, A. (2011) Inorg. Chem., 50: 11 134–11 142.

[81] Wada, T., Tsuge, K., Tanaka, K. (2000) Angew. Chem. Int. Ed., 39: 1479–1482.

[82] Wada, T., Tsuge, K., Tanaka, K. (2001) Inorg. Chem., 40: 329–337.

[83] Toma, F. M., Sartorel, A., Iurlo, M., Carraro, M., Parisse, P., Maccato, C., Rapino, S., Gonzalez, B. R., Amenitsch, H., Da Ros, T., Casalis, L., Goldoni, A., Marcaccio, M., Scorrano, G., Scoles, G., Paolucci, F., Prato, M., Bonchio, M. (2010) Nat. Chem., 2: 826–831.

[84] Li, F., Zhang, B., Li, X., Jiang, Y., Chen, L., Li, Y., Sun, L. (2011) Angew. Chem. Int. Ed., 50: 12 276–12 279.

[85] Liu, F., Cardolaccia, T., Hornstein, B. J., Schoonover, J. R., Meyer, T. J. (2007) J. Am. Chem. Soc., 129: 2446–2447.

[86] Chen, Z., Concepcion, J. J, Jurss, J. W., Meyer, T. J. (2009) J. Am. Chem. Soc., 131: 15 580–15 581.

[87] Chen, Z., Concepcion, J. J., Hull, J. F., Hoertz, P. G., Meyer, T. J. (2010) Dalton Trans., 39: 6950–6952.

[88] Concepcion, J. J., Jurss, J. W., Hoertz, P. G., Meyer T. J. (2009) Angew. Chem. Int. Ed., 48: 9473–9476.

[89] Mola, J., Mas-Marzà, E., Sala, X., Romero, M., Rodríguez, I., Viñas, C., Llobet, A. (2008) Angew. Chem. Int. Ed., 47: 5830–5832.

[90] Geletii, Y. V., Huang, Z., Hou, Y., Musaev, D. G., Lian, T., Hill, C. L. (2009) J. Am. Chem. Soc., 131: 7522–7523.

[91] Puntoriero, F., La Ganga, G., Sartorel, A., Carraro, M., Scorrano, G., Bonchio, M., Campagna, S. (2010) Chem. Commun., 46: 4725–4727.

[92] Orlandi, M., Argazzi, R., Sartorel, A., Carraro, M., Scorrano, G., Bonchio, M., Scandola, S. (2010) Chem. Commun., 46: 3152–3154.

[93] Glasson, C. R. K., Song, W., Ashford, D. L., Vannucci, A., Chen, Z., Concepcion, J. J., Holland, P. L., Meyer, T. J. (2012) Inorg. Chem., 51: 8637–8639.

[94] Chen, W., Rein, F. N., Rocha, R. C. (2009) Angew. Chem., 121: 9852–9855; Angew. Chem. Int. Ed., 48: 9672–9675.

[95] Chen, W., Rein, F. N., Scott, B. L., Rocha, R. C. (2011) Chem. Eur. J., 17: 5595–5604.

[96] Hamelin, O., Guillo, P., Loiseau, F., Boissonnet, M. F., nage, S. M. (2011) Inorg. Chem., 50: 7952–7954.

[97] Guillo, P., Hamelin, O., Batat, G., Jonusauskas, G., McClenaghan, N. D., Nage, S. M. (2012) Inorg. Chem., 51: 2222–2230.

[98] Farràs, P., Maji, S., Benet-Buchholz, J., Llobet, A. (2013) Chem. Eur. J., 19: 7162–7172.

[99] Li, F., Jiang, Y., Zhang, B., Huang, F., Gao, Y., Sun, L. (2012) Angew. Chem. Int. Ed., 51: 2417–2420.

[100] Ashford, D. L., Stewart, D. J., Glasson, C. R., Binstead, R. A., Harrison, D. P., Norris, M. R. M. R, Concepcion, J. J., Fang, Z., Templeton, J. L., Meyer, T. J. (2012) Inorg. Chem., 51: 6428–6430.

[101] Kaveevivitchai, N., Chitta, R., Zong, R., El Ojaimi, M., Thummel, R. P. (2012) J. Am. Chem. Soc., 134: 10 721–10 724.

4

Towards the Visible Light-Driven Water Splitting Device: Ruthenium Water Oxidation Catalysts with Carboxylate-Containing Ligands

Lele Duan[1], Lianpeng Tong[1], and Licheng Sun[1,2]

[1]*Department of Chemistry, School of Chemical Science and Engineering, KTH Royal Institute of Technology, Stockholm, Sweden*
[2]*State Key Lab of Fine Chemicals, DUT-KTH Joint Education and Research Center on Molecular Devices, Dalian University of Technology, Dalian, China*

4.1 Introduction

Water oxidation involves the transfer of multiple electrons and protons and the formation of an O–O bond. Therefore, it is a highly energy-demanding reaction that is difficult to catalyze (great overpotential is required). In the natural photosynthesis system II, water oxidation is carried out at the oxygen-evolving complex (OEC), a μ-oxo-bridged Mn_4O_5Ca cluster surrounded by carboxylate- and imidazole-containing residues [1] with an amazingly small overpotential (η) of a mere 160 mV at pH 6.5 [2–4]. The presence of oxo and carboxylate ligands is crucial to decreasing the redox potentials of the OEC. It has been proved that negatively charged ligands can stabilize the high oxidation states of various transition metal-based complexes and lower their oxidation potentials. With the view of developing artificial water oxidation catalysts (WOCs) with small overpotentials, we have focused in this chapter on the complexation of transition metals (primarily ruthenium) with carboxylate-containing ligands.

Molecular Water Oxidation Catalysis: A Key Topic for New Sustainable Energy Conversion Schemes,
First Edition. Edited by Antoni Llobet.
© 2014 John Wiley & Sons, Ltd. Published 2014 by John Wiley & Sons, Ltd.

4.2　Binuclear Ru Complexes

Before moving to the carboxylate-containing ligands, we would like to introduce our effort to increase the electron-donating ability of the 3, 6-di-(6′-[1″, 8″-naphthyrid-2″-yl]-pyridin-2′-yl)pyrazine ligand (the ancillary ligand of Thummel's complex **1** [5]; Figure 4.1) by removing two electron-withdrawing pyridyl rings from the naphthyridinyl motifs and introducing electron-donating alkyl groups. Accordingly, we synthesized two binuclear Ru complexes **2** and **3** bearing 3, 6-bis-(4, 4′-dimethyl-(2, 2′-bipyridin)-6-yl)pyridazine and 3, 6-bis((2, 2′-bipyridin)-6-yl)-4, 5-cyclopenteno-1, 2-diazine, respectively (Figure 4.1) [6]. Unfortunately, the oxidation potentials of complexes **2** and **3** are not significantly shifted towards the anodic direction (compared to those of complex **1**) as we expected. What is intriguing is that these two complexes displayed considerably higher activities towards Ce^{IV}-driven ($Ce^{IV} = Ce(NH_4)_2(NO_3)_6$) water oxidation than complex **1**. Under optimized catalytic conditions, 3500 turnovers for **2** and 4500 turnovers for **3** were achieved. According to the mass spectrometry analysis of the catalytic reaction mixture, H_2O/4-picoline (pic) ligand exchange occurred for **3** during catalysis, and the resulting

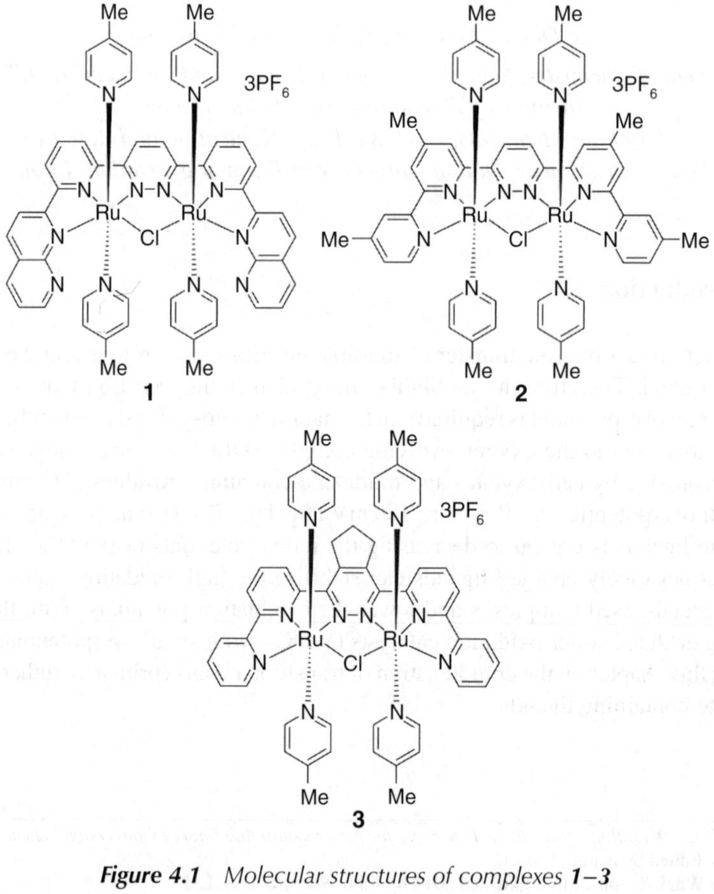

Figure 4.1　Molecular structures of complexes 1–3

$[M - pic + OH_2]^{3+}$ and $[M - 2pic + 2OH_2]^{3+}$ species were proposed to be the real catalysts driving water oxidation.

In 2009, Xu and coworkers published our first binuclear Ru WOC $[Ru_2(cppd)(pic)_6](PF_6)$ ($H_3cppd = 3,6$-bis(6'-COOH-pyrid-2'-yl)pyridazine, **4**; Figure 4.2), carrying a carboxylate-containing ligand $cppd^{3-}$. It exhibited a significant improvement in reactivity towards Ce^{IV}-driven water oxidation in comparison with other binuclear WOCs, such as complex **1** [7].

The X-ray crystal structure of complex **4** is depicted in Figure 4.2. It is composed of two Ru^{II} ions, a triply charged anionic ligand $cppd^{3-}$, and six ancillary 4-picoline ligands. Noticeably, these two Ru ions are in anti-positions with each other (at opposite sides of the pyridazine ring): one coordinates to a nitrogen site on the central pyridazine and the other to a carbon site. The μ-bridged chloride, as observed in Thummel's *cis*-Ru$_2$ complex **1**, is not present in complex **4**.

The cyclic voltametry (CV) curve of complex **4** exhibits two reversible waves at $E_{1/2} = 0.294$ and 0.797 V versus saturated calomel electrode (SCE), respectively corresponding to its $Ru_2^{II,III/II,II}$ and $Ru_2^{III,III/III,II}$ couples. One irreversible oxidation wave at ca. 1.83 V versus SCE is observed, most likely due to the oxidation of $Ru_2^{III,III}$ to $Ru_2^{III,IV}$ complex. Compared to complex **1** ($E_{1/2}^{ox.} = 1.25$ and 1.66 V versus SCE) [5], the oxidation potentials of **4** are drastically decreased: by 0.96 V for the $Ru_2^{II,III/II,II}$ couple and 0.86 V for the $Ru_2^{III,III/III,II}$ couple. As expected, introduction of the negatively charged carboxylate ligand lowers the oxidation potentials of complex **4** significantly.

The catalytic activity of **4** was evaluated in typical Ce^{IV}-driven water oxidation conditions (Equation 4.1). At a high level of Ce^{IV} concentration ($[Ce^{IV}] = 330$ mM), the turnover number (TON) of water oxidation by **4** is ca. 1700 over a reaction time of 15 hours. This value is more than threefold that for **1** (TON = 538). It was also found that the reactivity of **4** is dependent on the concentration of Ce^{IV} in the experimental medium. When the Ce^{IV} concentration was at a low level (5 mM), a TON of 4700 was obtained by **4**, with an initial

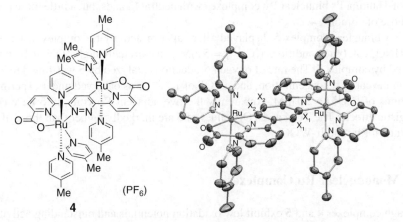

Figure 4.2 *Molecular structure of **4** and its Oak Ridge Thermal Ellipsoid Plot Program (ORTEP) view with thermal ellipsoids at 50% probability. H atoms and the PF_6 counter ion are omitted for clarity. The X_1 and X_2 are either C and N or N and C, respectively, with 50% probability. Adapted with permission from [7]. Copyright © 2009, American Chemical Society*

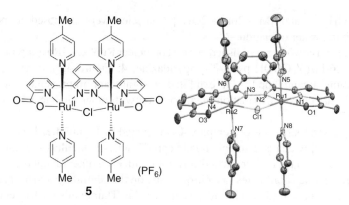

Figure 4.3 *Molecular structure of 5 and its ORTEP view with thermal ellipsoids at 30% probability. H atoms, solvate H_2O, and the $(PF_6)^{-1}$ counter ion are omitted for clarity [8]. Reproduced with permission from [8]. Copyright © 2010, WILEY-VCH Verlag GmbH & Co. KGaA, Weinheim*

turnover frequency (TOF) of $0.28\,s^{-1}$ [8].

$$2H_2O + 4Ce^{IV} \xrightarrow{\text{catalyst}} O_2 + 4H^+ + 4Ce^{III} \qquad (4.1)$$

In order to keep the two ruthenium cores of the binuclear complex in a *cis*-position, the ligand 1, 4-bis(6′-COOH-pyrid-2′-yl)phthalazine (H_2bcpp) was designed and synthesized by replacing the central pyridazine moiety of H_3cppd with a phthalazine unit. Complexation between the Ru ion and bcpp^{2-} avoids formation of the antibinuclear Ru product but affords a *cis*-binuclear Ru complex **5** (Figure 4.3) [8].

Complex **5** in acetonitrile displays two reversible redox couples at $E_{1/2} = 0.903$ and 1.396 V versus normal hydrogen electrode (NHE), respectively, assigned to the corresponding $Ru_2^{II,III/II,II}$ and $Ru_2^{III,III/III,II}$ processes. These potentials are notably more negative than those of Thummel's binuclear Ru complexes with neutral ligands, but a little more positive than those of complex **4**.

The *cis*-binuclear complex **5** displayed a far superior activity to complex **4**. Under the Ce^{IV}-HNO_3 catalytic conditions ($[Ce^{IV}] = 5\,mM$), an extremely high TON of 10 400 was obtained by complex **5**. The rate of oxygen production catalyzed by **5** was found to be first order in the catalyst concentration, and a TOF of $1.2\,s^{-1}$ was observed under experimental conditions of $[Ce^{IV}] = 20\,mM$. Apparently the two adjacent Ru cores in **5** have certain synergistic effects. The TON and TOF values of **5** are markedly higher than those of other well-known Ru- and Ir-WOCs.

4.3   Mononuclear Ru Complexes

Although complexes **4** and **5** exhibit low oxidation potentials and outstanding activities in Ce^{IV}-driven water oxidation, their complexity prevents us from understanding the details of their catalytic paths. Nevertheless, the positive result of introducing carboxylate-containing ligands in the construction of WOCs has greatly encouraged us to further explore mononuclear WOCs using the same concept.

4.3.1 Ru–O₂N–N₃ Analogs

4.3.1.1 Ru–pdc Complexes

The first series of mononuclear WOCs we designed and synthesized is a family of mononuclear Ru complexes **6a–d** (Figure 4.4) containing a tridentate pdc²⁻ (H₂pdc = 2, 6-pyridinedicarboxylic acid) backbone ligand that features the tridentate binding ability of H₃cppd but has a relatively simple structure [9, 10].

The catalytic and electrochemical data for **6a–d**, together with those for [Ru(tpy)(bpy)(OH₂)]²⁺ (employed as a benchmark complex; tpy = 2, 2′ : 6′, 2″-terpyridine; bpy = bpy = 2, 2′-bipyridine), are listed in Table 4.1 [9–12]. Complexes **6a–c** contain electron-donating pyridyl ligands and are able to catalyze water oxidation with considerably high rates (up to $0.29\,\text{s}^{-1}$) and moderate TONs (up to 560), while complex **6d**, which has electron-deficient pyrazine ligands, shows a negligible catalytic activity (trivial TOF and 50 turnovers). Electron-donating groups thus enhance the activity of [Ru(pdc)L₃]-type WOCs. For water oxidation catalyzed by **8a–b** (Figure 4.4) containing a strong electron-withdrawing dimethyl sulfoxide (dmso) ligand, no oxygen production is observed.

From a closer look at the details of water oxidation by **6b**, we can see that its equatorial 4-picoline ligand is labile under acidic conditions, leading to the formation of *trans-*

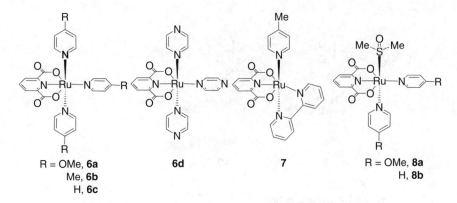

R = OMe, **6a**
Me, **6b**
H, **6c**

6d

7

R = OMe, **8a**
H, **8b**

Figure 4.4 *Structures of complexes **6a–d**, **7**, and **8a–b***

Table 4.1 *Electrochemistry and catalytic data for **6a–d** and [Ru(tpy)(bpy)(OH₂)]²⁺*

Complex	E^{ox}(V)	TON	$TOF_{initial}(s^{-1})$
6a	0.38, 1.21	560	0.29
6b	0.53, 1.22	550	0.23
6c	0.58, 1.24	460	0.09
6d	0.83, 1.30	50	ca. 0
[Ru(tpy)(bpy)(OH₂)]²⁺	1.04, 1.23	310	0.029

[Ru(pdc)(pic)$_2$(sol)] (sol = solvent; see more discussions in the next section). When the labile equatorial pic ligand was changed to a bpy ligand (one of the axial pic ligands was also replaced by one pyridine moiety of the bpy ligand), the resulting complex **7** showed a negligible activity towards water oxidation, implying that the equatorial site of the Ru center is the active site. In addition, mass spectrometry studies suggest that **6b** plays the role of a pre-catalyst and that the aqua complexes *trans*-[Ru(pdc)(pic)$_2$(OH$_2$)]$^+$ and/or *cis*-[Ru(pdc)(pic)(OH$_2$)$_2$]$^+$ are responsible for the activity of **6b**.

4.3.1.2   The Ru–hqc Complex

It is inferred that the anionic carboxylate donor facilitates dissociation of the equatorial 4-picoline ligand from complex **6b** and generation of the authentic aqua-Ru WOC in the acidic Ce^{IV} medium. Complex [Ru(hqc)(pic)$_3$] (**9**, H$_2$hqc = 8-hydroxyquinoline-2-carboxylic acid; Figure 4.5) was prepared and compared with **6b** and [Ru(tpy)(pic)$_3$]$^{2+}$ (**10**; Figure 4.5) so as to explore the effect of anionic ligands on water oxidation by Ru catalysts [13]. Both pdc^{2-} and hqc^{2-} ligands contain negatively charged oxygen groups ($-$O$^-$), which bear extra lone pairs and are able to play the role of π-donor in the $p\pi-d\pi$ interaction between oxygen and ruthenium of the coordination bond. In principle, the phenolate group in hqc^{2-} is a better π donor than the carboxylate group of pdc^{2-}.

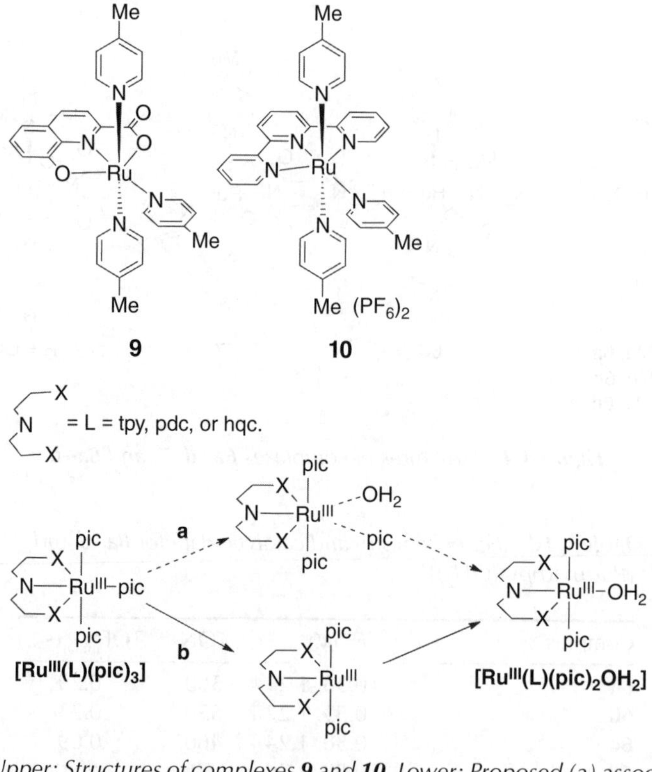

Figure 4.5   *Upper: Structures of complexes* **9** *and* **10**. *Lower: Proposed (a) associative and (b) dissociative ligand-exchange pathways for complexes* **6b**, **9** *and* **10**. *Adapted with permission from [13]. Copyright © 2012, American Chemical Society*

Mass spectrometry experiments indicated that one of the pic ligands of **9** was replaced by a water molecule or a coordinative organic solvent molecule (such as acetone) when **9** was oxidized to its trivalent state. In contrast, a similar pic–water or pic–solvent ligand exchange was not observed in the mass spectra of **10**. We therefore proposed that, compared to the neutral tridentate tpy ligand, the anionic tridentate hqc ligand is capable of enhancing the rate of pic–water exchange under the conditions of homogeneous catalysis of water oxidation. Two pic–water ligand exchange pathway scenarios were suggested and were examined by density functional theory (DFT) calculations upon trivalent complexes **6b**, **9** and **10**. One is the concerted associative pathway (Figure 4.5a), in which coordination of a water molecule to the Ru^{III} core and departure of a 4-picoline ligand occur simultaneously (via a seven-coordinated Ru^{III} transition state). The other pathway is the dissociative pathway (Figure 4.5b), in which one 4-picoline ligand dissociates from the Ru^{III} center first, giving a five-coordinated Ru^{III} intermediate, and then a water molecule comes into its coordination sphere. The results of theoretical simulations show that the dissociative pathway is favored for all three complexes. Moreover, the required energy of activation for 4-picoline dissociation from **6b** or **9** (12.7 and 12.2 kcal/mol, respectively) is dramatically lower than that for **10** (22.8 kcal/mol). The effect of anionic ligands over 4-picoline dissociation is primarily ascribed to stabilization of the electron-deficient Ru^{III} center by the oxygen donors – phenolate and carboxylate particularly – in the intuitive five-coordinated Ru^{III} intermediate.

Compared to **10**, complex **6b** and **9** have higher highest occupied molecular orbitals (HOMOs). The Ru^{II}/Ru^{III} redox potential of **9** (0.23 V) and **6b** (0.5 V) are much more negative than that of **10** (1.5 V) in organic electrolytes. The catalytic behavior of **9** towards water oxidation is similar to that of **6b**. As soon as **9** was injected into the HNO_3 solution of largely excess Ce^{IV}, instant O_2 evolution was detected, and the initial O_2-evolving rate was first-order in the concentration of the catalyst. An initial TOF of $0.32\,s^{-1}$ was achieved for **9**.

4.3.2 Ru–O_2N_2–N_2 Analogs

The successful development of binuclear Ru catalyst **4** and mononuclear Ru catalyst **6b** in early 2009 motivated us to synthesize mononuclear Ru complexes containing tetradentate O^N^N^O ligands, such as H_2bda (2, 2′-bipyridine-6, 6′-dicarboxylic acid) and H_2pda (1,10-phenanthroline-2,9-dicarboxylic acid), which are educed from H_3cppd by removing the central pyridazine moiety. This leads us to the isolation of an extraordinary seven-coordinate intermediate, the discovery of a family of highly active WOCs with TOFs up to $300\,s^{-1}$, and an understanding of the structure–activity relationships of Ru-bda and Ru-pda complexes [14–18].

4.3.2.1 Ru–bda Water Oxidation Catalysts

The Catalytic Mechanism. The first example of Ru–bda series catalysts is [Ru(bda)(pic)$_2$] (**11**) [14], whose X-ray single crystal structure is depicted in Figure 4.6, revealing that the Ru center locates in a six-coordinate octahedral configuration with large distortion. The cleft of O2–Ru1–O3 is 122.99°, much higher than the ideal 90° of an octahedral

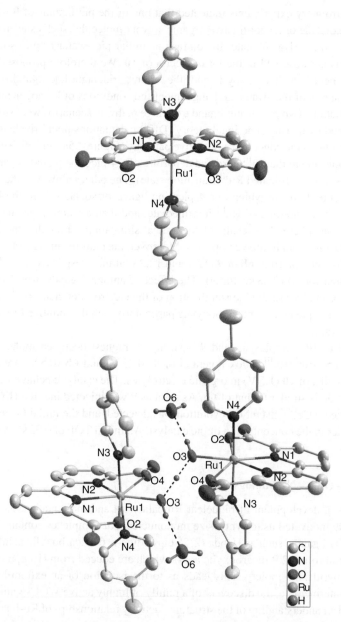

Figure 4.6 *Crystal structures of **11** (upper) and **12** (lower) with thermal ellipsoids at 50% probability. Hydrogen atoms are omitted for clarity, except for the H—O type. Adapted with permission from [14]. Copyright © 2009, American Chemical Society*

configuration. Further studies have revealed that this large angle is essential to the complex accepting water as a seventh ligand.

Through electrochemical study, we noticed that complex **11** exhibits very low oxidation potentials, and that the Ru^{IV} state is essentially thermodynamically stable at pH 1.0 conditions. This allows us to isolate a tetravalent ruthenium intermediate from the Ce^{IV}-driven water oxidation reaction catalyzed by **11**.

The isolated Ru^{IV} species was determined by single-crystal diffraction as a seven-coordinate dimeric complex μ-(HOHOH)[Ru^{IV}(bda)(pic)$_2$]$_2$[PF$_6$]$_3$ (**12**, Figure 4.6) with two solvated water molecules. The bridging ligand [HOHOH]$^-$ coordinates, via two O atoms, with two Ru cores through the O–Ru–O clefts. Each Ru core is seven-coordinated and exists in a highly distorted pentagonal bipyramidal configuration (the O2–N1–N2–O4 dihedral angle, 23.13°), while angles in the pentagonal plane are all close to the ideal value of 72°. Of note is that the hydrogen bond of H–O⋯H⋯O–H in the [HOHOH]$^-$ bridge is ca. 0.2 Å shorter than that of O3–H⋯O6, manifesting a strong hydrogen bonding network. Both solvated water molecules are bound with the [HOHOH]$^-$ bridge, implying a purported proton transfer path from the reaction center to the solvation shell during water oxidation.

A water molecule could coordinate to the Ru center of complex **11** in an aqueous solution, forming an Ru^{II}–OH$_2$ complex. Currently, it is not clear which configuration, six- or seven-coordination, this Ru^{II}–OH$_2$ complex possesses. Its Pourbaix diagram is depicted in Figure 4.7 [15]. The $Ru^{III/II}$ redox process is coupled with one proton transfer in the pH range 5.5–12.9, as evidenced by a diagonal line of 59 mV/pH. In the pH region below 5.5 and above 12.9, only electron transfer is observed in the Ru^{II}–Ru^{III} process. The pKa values of Ru^{III}–OH$_2$ and Ru^{II}–OH$_2$ are 5.5 and 12.9, respectively. Over the whole pH range from 1.0 to 13.5, the oxidation of Ru^{III} to Ru^{IV} is a one-electron–one-proton proton-coupled electron transfer (PCET) process, forming Ru^{IV}–OH at pH < 5.5 and Ru^{IV}=O at pH > 5.5. The Ru^{V}=O species, which is proposed to trigger water oxidation, is generated via one-electron oxidation of Ru^{IV}=O or via one-electron oxidation of Ru^{IV}–OH

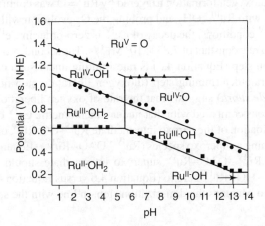

Figure 4.7 *Pourbaix diagram of complex **11**. The zones of stability of the different species as a function of pH and E are indicated by the oxidation state of the Ru metal and the degree of protonation of the aqua group [15]. Reproduced with permission from [15]. Copyright © 2012, Rights Managed by Nature Publishing Group*

coupled with one-proton transfer. In the typical acidic conditions of Ce^{IV}-driven water oxidation (pH < 2), the proton/electron transfer sequence for **11** is as follows: $Ru^{II}-OH_2 \rightarrow Ru^{III}-OH_2 \rightarrow Ru^{IV}-OH \rightarrow Ru^{V}=O$.

The kinetics of sequential oxidation of **11** was studied by mixing the complex with stoichiometric amounts of Ce^{IV} at pH 1.0 and continually monitoring the ultraviolet/visible (UV/vis) absorption of the sample using the stopped-flow technique [15]. Under aerobic conditions, $Ru^{II}-OH_2$ is readily oxidized to $Ru^{III}-OH_2$ by aerial oxygen, so $Ru^{II}-OH_2 + Ce^{IV} \rightarrow Ru^{III}-OH_2 + Ce^{III}$ was not investigated. In addition, the consequent oxidation step $Ru^{III}-OH_2 + Ce^{IV} \rightarrow Ru^{IV}-OH + H^+ + Ce^{III}$ was too fast to measure. When $Ru^{III}-OH_2$ was mixed with two equivalents of Ce^{IV}, the measured initial electron transfer process corresponded to the $Ru^{IV} \rightarrow Ru^{V}$ oxidation (Equation 4.2), which is first-order in both Ce^{IV} and the catalyst. The rate constant (k_{ET}) of this process is calculated as $k_{ET} = 2.3 \times 10^5 \, M^{-1}/s$ at 35 °C. Once $Ru^{V}=O$ is reached, two rapidly couple, forming a peroxo-bridged binuclear Ru intermediate (Equation 4.3). This dimerization step is independent of the Ce^{IV} concentration and is second-order in catalyst with a rate constant $k_D(35\,°C) = 1.1 \times 10^5 \, M^{-1}/s$. The last step involves the decomposition of $Ru^{IV}-OO-Ru^{IV}$ and is responsible for the oxygen liberation. This step is first-order with regard to catalyst, with a rate constant $k_{O2}(35\,°C) = 5.8 \, s^{-1}$ (Equation 4.4).

$$Ru^{IV}-OH + Ce^{IV} \xrightarrow{k_{ET1}} Ru^{V}=O + H^+ + Ce^{III} \qquad (4.2)$$

$$2Ru^{V}=O \xrightarrow{k_D} Ru^{IV}-OO-Ru^{IV} \qquad (4.3)$$

$$Ru^{IV}-OO-Ru^{IV} \xrightarrow{k_{O2}} 2Ru^{III} + O_2 \qquad (4.4)$$

The Gibbs free energy of activation ($\Delta G^{\ddagger}$) for Equations 4.2, 4.3, and 4.4 at room temperature (295 K) was calculated to be 10.3 ± 0.5, 10.8 ± 1.4, and $16.5 \pm 1.0 \, kcal/mol$, respectively. Under stoichiometric Ce^{IV} conditions, oxygen liberation is the rate-limiting step. Additionally, dioxygen formation triggered by $Ru^{V}=O$ was confirmed by mixing two equivalents of Ce^{IV} with $Ru^{III}-OH_2$ and probing the O_2 evolution with a Clark electrode.

Under excess Ce^{IV} conditions, the decay of Ce^{IV} is zero-order in Ce^{IV} and second-order in catalyst **11**, with a rate constant of $7.83 \times 10^5 \, M^{-1}/s$. The second-order rate law indicates that the dimerization step (Equation 4.3) is rate-limiting under such catalytic conditions. The switch of the rate-determining step (dioxygen liberation step under stoichiometric Ce^{IV} conditions, *vide supra*) suggests that the rate of oxygen liberation is elevated under excess Ce^{IV} conditions compared with that under stoichiometric Ce^{IV} conditions. We proposed that a fast oxidation of the peroxo intermediate Ru^{IV}-OO-Ru^{IV} occurred (calculated $E^{ox} = 1.03$ V), forming a superoxo species ($Ru^{IV}-O\dot{-}O-Ru^{IV}$) (Equation 4.5). The higher oxidation states of $Ru^{IV}-O\dot{-}O-Ru^{IV}$ superoxo intermediate should generate O_2 much faster than the $Ru^{IV}-OO-Ru^{IV}$ peroxo (Equation 4.6 versus Equation 4.4). Consequently, the rate-limiting step becomes the dimerization step, in line with the second-order dependence.

$$Ru^{IV}-OO-Ru^{IV} + Ce^{IV} \xrightarrow{fast} Ru^{IV}-O\dot{-}O-Ru^{IV} + Ce^{III} \qquad (4.5)$$

$$Ru^{IV}-O\dot{-}O-Ru^{IV} \xrightarrow{fast} Ru^{III} + Ru^{IV} + O_2 \qquad (4.6)$$

Tuning the Activity of Ru–bda Water Oxidation Catalysts: The [Ru(bda)(isoq)$_2$] Complex. In natural photosystem II (PSII), water oxidation is catalyzed by the OEC with markedly high reaction rates: $100 - 400 \, \mathrm{s}^{-1}$. In comparison, synthetic WOCs are still much slower than the OEC. Aiming at further improvement of Ru–bda catalysts, we introduced iso-quinolines to replace the axial 4-picoline ligands of **11** and discovered a superior fast cata-lyst [Ru(bda)(isoq)$_2$] (**13**, isoq = isoquinoline; Figure 4.8) [15]. Oxygen-evolution-versus-time plots of **11** and **13** are depicted in Figure 4.8. Under the given conditions, a remarkable increment of O$_2$-evolving rate was observed when complex **13** was employed as a catalyst instead of **11**. TOFs of $303 \pm 9.6 \, \mathrm{s}^{-1}$ for **13** versus $32 \, \mathrm{s}^{-1}$ for **11** were attained. At relatively low levels of catalyst concentration, large TONs were obtained for both **11** (2010 ± 57 at $[\mathbf{11}] = 5.88 \times 10^{-5} \, \mathrm{M}$) and **13** ($8360 \pm 91$ at $[\mathbf{13}] = 1.50 \times 10^{-5} \, \mathrm{M}$). CV curves of **11** and **13** (Figure 4.8) showed almost identical redox wave positions, revealing negligible elec-tronic effect differences between 4-picoline and isoquinoline insofar as the redox potentials

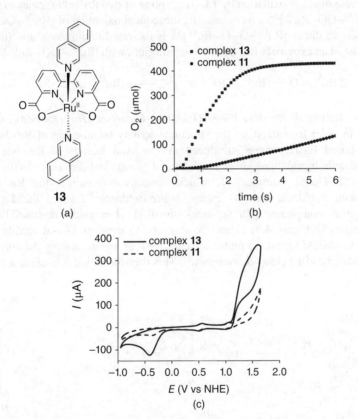

Figure 4.8 *(a) Molecular structure of* **13**. *(b) Oxygen evolution versus time for com-plexes* **11** *and* **13** *(conditions: CF$_3$SO$_3$H aqueous solutions (3.7 mL) containing Ce^{IV} (0.48 M, 1.79 × 10^{-3} mol) and catalyst (2.16 × 10^{-4} M, 8 × 10^{-7} mol)). (c) CVs of* **11** *and* **13** *(conditions: [catalyst] = 1 mM; solvent = mixed CF$_3$CH$_2$OH/pH 1.0 (v/v = 1 : 2); scanning rate = 100 mV/s; working electrode = pyrolytic graphite electrode (basal plane)). Reproduced with permission from [15]. Copyright © 2012, Rights Managed by Nature Pub-lishing Group*

of Ru–bda complexes are concerned. The enhanced reactivity of **13** in compassion with **11** should not primarily arise from the electronic effect. The O–O bond formation step via coupling of two Ru^V=O species is regarded as rate-limiting when **11** is employed as the WOC under typical acidic Ce^{IV} catalytic conditions. Noncovalent attractions between individual catalysts via isoquinoline ligands, such as $\pi - \pi$ stacking and hydrophobic effects, play an essential role in lowering the energy barrier of this coupling step and thus increasing the formal rate of the whole catalytic cycle. DFT calculations provide insight into the transition state of the O–O bond formation step in the case of water oxidation by **13** and disclose how $\pi - \pi$ stacking of isoquinolines contributes to the stabilization of the transition state.

Kinetic studies under excess Ce^{IV} conditions revealed that Ce^{IV} consumption is second-order in catalyst **13** and first-order in Ce^{IV}. Due to the noncovalent attractive interactions, the O–O bond formation step is significantly accelerated, making it fast enough to compete with the next oxidation step. According to this formal rate law, we propose that, in the catalytic cycle of water oxidation by **13**, (i) coupling of two Ru^V=O species to the peroxo dimer $[Ru^{IV}-OO-Ru^{IV}]^{2+}$ is very fast, (ii) subsequent oxidation of $[Ru^{IV}-OO-Ru^{IV}]^{2+}$ to the superoxo dimer $[Ru^{IV}-O\dot{-}O-Ru^{IV}]^{3+}$ is the rate-limiting step, and (iii) Ru^V=O species exist in an extremely fast dynamic equilibrium with $[Ru^{IV}-OO-Ru^{IV}]^{2+}$:

$$2\,Ru^V = O \rightleftharpoons [Ru^{IV}-OO-Ru^{IV}]^{2+} \xrightarrow{-e^-} [Ru^{IV}-O\dot{-}O-Ru^{IV}]^{3+} \quad (4.7)$$

Tuning the Activity of Ru–bda Water Oxidation Catalysts: The [Ru(bda)(Imd)(dmso)] Analogs. In order to understand the structure–activity relationship of Ru–bda WOCs, imidazole-based ligands were introduced as the axial ligands of Ru–bda catalysts. These imidazole ligands have diverse substituent groups and different electron-donating abilities [18]. Unlike complex **11**, which contains 4-piconline, the Ru complexes (**14a–d**) with imidazole ligands, except 5-nitroimidazole, favored the formation of C_s symmetric complexes $[Ru^{II}(\kappa^3\text{-bda})(dmso)L_2]$ (L = imidazole-based ligand; see their structures in Figure 4.9) in aprotic solvents. Complexes **14a–d** readily lose their equatorial imidazole ligands in protic solvents, such as water, forming the corresponding $[Ru^{II}(\kappa^4\text{-bda})(dmso)L]$ (**15a–d**) complexes. This transformation was observed occurring

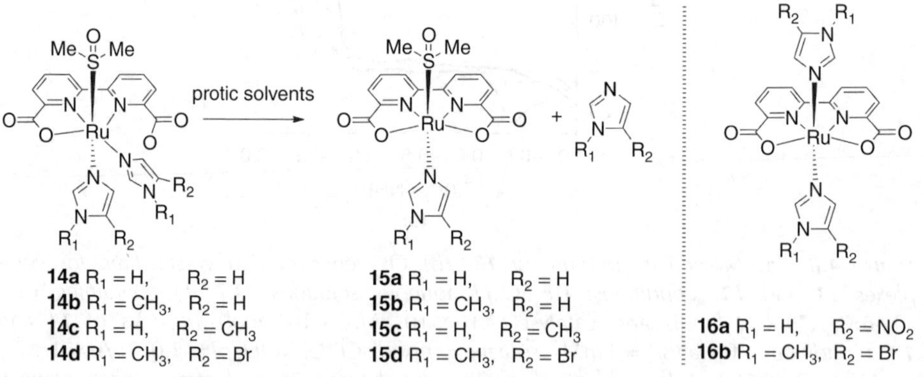

14a R_1 = H, R_2 = H
14b R_1 = CH_3, R_2 = H
14c R_1 = H, R_2 = CH_3
14d R_1 = CH_3, R_2 = Br

15a R_1 = H, R_2 = H
15b R_1 = CH_3, R_2 = H
15c R_1 = H, R_2 = CH_3
15d R_1 = CH_3, R_2 = Br

16a R_1 = H, R_2 = NO_2
16b R_1 = CH_3, R_2 = Br

Figure 4.9 *Structures of complexes **14a–d**, **15a–d**, and **16a–b**, as well as the schematic conversion of **14a–d** to **15a–d** under protic solvents*

spontaneously in acidic aqueous solutions. When 5-nitroimidazole was employed as the axial ligand, $[Ru^{II}(\kappa^4\text{-bda})(5\text{-nitroimidazole})_2]$ (**16a**) was the favored product, rather than $[Ru^{II}(\kappa^3\text{-bda})(dmso)(5\text{-nitroimidazole})_2]$. When $RuCl_3 \cdot 3H_2O$ was used instead of *cis*-$[Ru(dmso)_4Cl_2]$ as a Ru precursor, $[Ru^{II}(\kappa^4\text{-bda})(5\text{-bromoimidazole})_2]$ (**16b**) was able to be prepared in a dmso-free environment.

Single crystals suitable for X-ray crystallographic analysis have been obtained from a solution of **14d**. Nevertheless, the resolved X-ray crystal structure turns out to be **15d** formed *in situ* from **14d** (Figure 4.10) during crystallization. The configuration of the Ru center of **15d** is a highly distorted octahedron, involving one bda^{2-} as the equatorial ligand and one imidazole and one S-bound dmso as the axial ligands. The O2–Ru1–O3 angle of **15d** is slightly greater than that of complex **11** by 1.7°. This bigger bite angle should make access to the coordination sphere of complex **15d** easier for water.

Electrochemistry and Ce^{IV}-driven water oxidation studies revealed that a catalytic mechanism similar to complex **11** could be applied to complexes **15a–d** and **16a–b**. Their electrochemical and catalytic data are represented in Table 4.2. Complexes **15a–d**, generated *in situ* after **14a–d**, were dissolved in pH 1.0 aqueous solutions. Their oxidation potentials are barely different from each other, < 30 mV, no matter which redox couples ($Ru^{III/II}$, $Ru^{IV/III}$, or $Ru^{V/IV}$) are concerned. In contrast, **15a–d** display much more positive oxidation potentials than complexes **16a–b**, reflecting the far lower electron-donating ability of the dmso ligand compared to the imidazole ligands. Interestingly, complexes

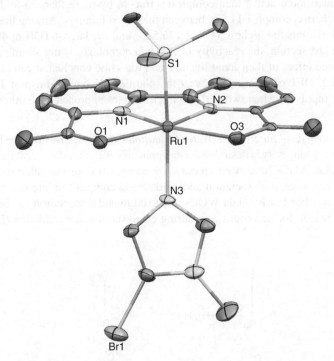

Figure 4.10 *X-ray crystal structure of complex **15d** with thermal ellipsoids at 50% probability. Hydrogen atoms and solvated molecules are omitted for clarity. Adapted with permission from [18]. Copyright © 2012, American Chemical Society*

Table 4.2 *Electrochemistry and catalytic data from selected Ru complexes. Adapted with permission from [18]. Copyright © 2012, American Chemical Society.*

Complex	$E_{1/2}^{ox.}$ (V versus NHE[a])			TON[c]	TOF[d] (s^{-1})
	Ru$^{III/II}$	Ru$^{IV/III}$	Ru$^{V/IV}$		
15^a	0.90	1.15	1.39	2032	137.6
15^b	0.89	1.13	1.37	2365	146.2
15^c	0.92	1.14	1.38	550	150.8
15^d	0.91	1.14	1.40	4050	176.5
16^a	0.68	1.05	-	1094	3.4
16^b	0.70	1.08	-	1150	4.5
11	0.63	1.11	-	2000	41.2

Note: Measured in pH solution (0.1 M CF$_3$SO$_3$H), scan rate 0.1 V/s
^{a}[Ru(bpy)$_3$]$^{2+}$ was used as a reference with $E_{1/2}$ = 1.26 V versus NHE;
^{b}Onset potential of the catalytic curve;
^{c}Conditions: catalyst (1.20 × 10^{-5} M, 3.60 × 10^{-8} mol for **15a–d**; 2.99 × 10^{-5} M, 1.01 × 10^{-7} mol for **16a** and **16b**) and Ce^{IV} (0.4 M) in 3.3 mL CF$_3$SO$_3$H aqueous solutions;
^{d}Conditions: catalyst (2.16 × 10^{-4} M, 7.99 × 10^{-7} mol for all catalysts) and Ce^{IV} (0.4 M) in 3.7 mL CF$_3$SO$_3$H aqueous solutions

15a–d are much more active than complexes **16a–b**, by more than 30-fold in terms of their TOFs, whereas complex **11** lies between these two families. Among these catalysts, complex **15d** exhibits the highest TOF at 176.5 s^{-1} and the largest TON at 4050. As discussed in the last section, the reactivity of Ru–bda complexes is not significantly related to the electronic effect of their axial ligands, and the same conclusion can also be drawn from Table 4.2. DFT calculations suggest that the flexible bound dmso of **15a–d** avoid serious steric repulsion when two Ru^V=O intermediates approach each other and form a O–O bond.

Tuning the Longevity of the Ru–bda Water Oxidation Catalysts. Besides the high reactivity, extraordinary longevity is desirable for an applicable water oxidation catalyst. Although many molecular WOCs have been prepared, decomposition of the catalysts is always a general problem. Ligand dissociation and oxidative decomposition are two major degradation pathways. For fast Ru–bda WOCs, the axial ligand dissociation has been found to be the major reason for decomposition during catalysis of water oxidation (Figure 4.11).

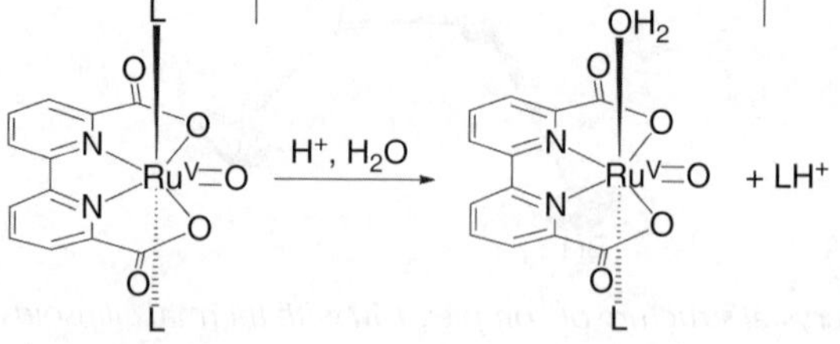

Figure 4.11 *Illustration of the ligand-exchange reaction studied in DFT calculations [16]*

According to DFT calculations, ligand exchange is more pronounced in the Ru^V state than in lower-valency states of the complex. With a view to axial ligands with stronger binding ability than 4-picoline and isoquinoline (axial ligands of **11** and **13** respectively), we screened a series of nitrogen heterocyclic compounds as axial ligands, including 4-picoline, isoquinoline, pyrimidine, pyrazine, pyridazine, cinnoline, phthalazine, and 4,5-dimethoxypyridazine, and compared their dissociation energies via the pathway shown in Figure 4.11 [16].

We inspected the electronic structure, structural properties, and pK$_a$ values of various systems and finally were able to roughly correlate the HOMO energy of the ligand in the gas phase with the Gibbs free energy of the ligand exchange process (Figure 4.12). It can be deduced that Ru−bda complexes containing ligands in the top-right corner of Figure 4.12, such as pyridazine, cinnoline, 4,5-dimethoxypyridazine, and phthalazine, are considerably more resistant to aquation than those with 4-picoline, isoquinoline, pyrimidine, or pyrazine as axial ligands (bottom-left).

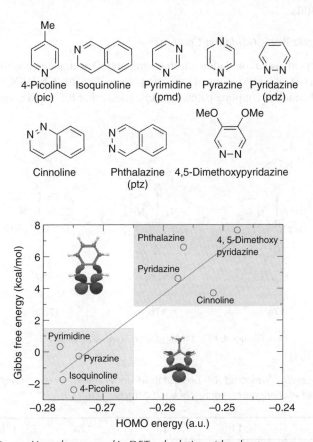

Figure 4.12 Upper: Ligands screened in DFT calculations (the short names are given in parentheses). Lower: Gibbs free energy of reaction in pH 0 aqueous solution at 298 K as a function of HOMO energy of the ligand in vacuum (in the inset are the HOMO of 4-picoline and the HOMO−1 of phthalazine, calculated with DFT) [16]

Three complexes [Ru(bda)L$_2$] (L = pyrimidine (pmd), **17a**; pyridazine (pdz), **17b**; phthalazine (ptz), **17c**; Figure 4.13) were selectively synthesized and compared with complexes **11** and **13** (Table 4.3). As we predicted from Figure 4.12, the lifetimes of complexes **17b** and **17c**, under catalytic conditions, are notably longer than those of complexes **11**, **13**, and **17a**. Meanwhile, the longevity of Ru–bda complexes increase linearly with the elevation of the HOMO energy of their corresponding axial ligands, with the exception of complex **11**. This good correlation paves the way to development of robust Ru–bda WOCs through simple calculation of the HOMO energy of the axial ligand. Thus, prediction of the longevity of Ru–bda-type catalysts becomes very convenient through the avoidance of time-consuming DFT calculations on the whole-ligand dissociation pathway and synthesis/testing of the targeting complex.

The TON of the most robust catalyst **17c** has reached 55 419 ± 959, while complexes **17a** and **17b** give 1702 ± 107 and 4563 ± 172, respectively, under specific reaction conditions. Under high [catalyst] conditions, their TOFs are 70 ± 12 s^{-1} for **17a**, 31 ± 3 s^{-1} for **17b**, and 286 ± 21 s^{-1} for **17c**. It is worth noting that **17c** is the second example that reaches a TOF close to 300 s^{-1}.

4.3.2.2 *Ru–pda Water Oxidation Catalysts*

Two well-accepted water oxidation mechanisms are (i) water nucleophilic attack on a metal–oxo and (ii) radical coupling of two metal–oxos. The Ru–bda WOCs catalyze water oxidation via the radical coupling pathway, while most other Ru–aqua complexes catalyze

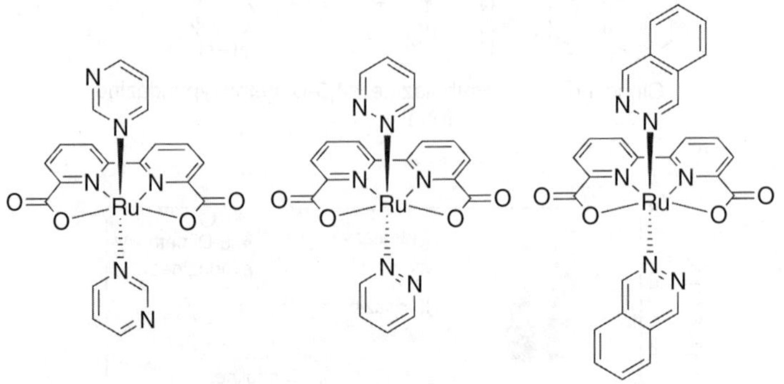

17a   17b   17c

Figure 4.13 *Molecular structures of **17a–c***

Table 4.3 *Longevity of selected Ru–bda catalysts under the given conditions and the calculated ligand-exchange free energy of each complex [16]*

Complex	11	13	17a	17b	17c
ΔG (kcal/mol)	2.38	1.75	0.33	4.61	6.59
Longevity (hours)[a]	0.29	0.064	0.332	1.07	1.37

[a]The longevity reported in this table is defined as the time until the oxygen production rate is 5% of the initial rate

water oxidation via the water nucleophilic attack pathway. This always raises the question, what is the critical structural factor that controls the O–O bond formation mechanism? For Ru–bda catalysts, the water substrate coordinates to the Ru center as the seventh ligand through the big cleft of O–Ru–O. At first we thought it was the seven-coordinate configuration that made Ru–bda catalysts so different from others. However, it turns out that the ligand reorganization is more critical to their catalytic behavior. The resulting seven-coordinate Ru–bda complex requires ligand reorganization to accommodate the pentagonal bipyramidal configuration. Consequently, the bipyridyl backbone is bent towards the Ru side and distorted with a dihedral angle of O2–N1–N2–O4, being 23.13°. With a view to examining the ligand reorganization effect on WOC, a rigid tetradentate anionic pda^{2-} ligand was employed instead of the flexible bda^{2-} ligand and Ru complexes [Ru(pda)L$_2$] (**18a–c**; Figure 4.14) were synthesized [17].

In the Ru^{II} state, the coordination of carboxylate is labile. In a mixed acetonitrile/water solvent, one of the carboxylate ligands could be replaced by an acetonitrile molecule, generating [Ru(κ^3-pda)L$_2$] species.

In a case study, after addition of four equivalents of Ce^{IV} in the acetonitrile/water solution of **18a**, two Ru-containing species were resolved by mass spectrometry: a seven-coordinate Ru^{IV} complex, [Ru^{IV}(pda)(pic)$_2$(OH)]$^+$, and a six-coordinate Ru^{III} species, [Ru^{III}(pda)(pic)$_2$]$^+$. DFT calculations favor the formation of a seven-coordinate Ru^{III}–aqua complex [Ru^{III}(pda)(pic)$_2$(OH$_2$)]$^+$ in protic aqueous solutions. Therefore, the Ru–pda complexes retain the seven-coordination feature, as Ru–bda complexes have.

All three complexes are capable of catalyzing Ce^{IV}-driven water oxidation, with TONs of 336 for **18a**, 310 for **18b**, and 190 for **18c** over a period of 6 hours. The initial TOFs of complexes **18a** (0.092 s^{-1}) and **18b** (0.102 s^{-1}) are close to each other and are more than two times greater than that of complex **18c** (0.040 s^{-1}). Electron-donating groups thus conclusively increase the catalytic activity of Ru–pda complexes.

It is interesting that the rate of catalytic O$_2$-evolving reaction is first-order in catalysts **18a–c** rather than second-order in the case of **11**. Accordingly, a mononuclear catalytic pathway involving water nucleophilic attack on the metal–oxo step (O–O bond formation step) is proposed for **18a–c** (Figure 4.15). Oxidation of Ru^{II}–OH$_2$ gives Ru^{III}–OH$_2$

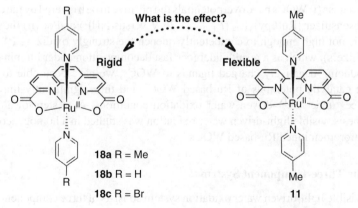

Figure 4.14 *Molecular structures of complexes **18a–c** and **11**, highlighting their structural differences*

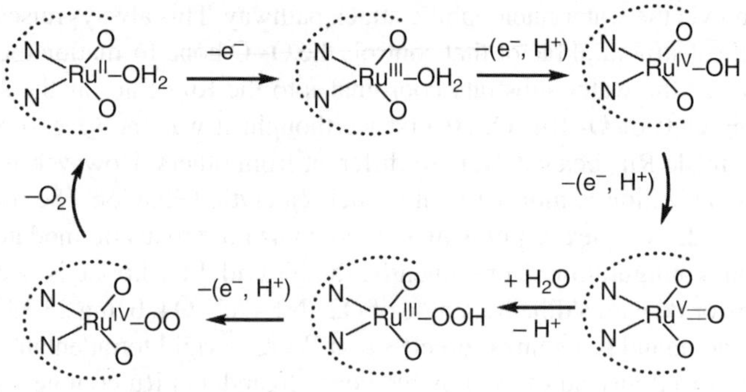

Figure 4.15 *Proposed catalytic cycle for complexes **18a–c***

species; oxidation from Ru^{III} to Ru^{IV} and further to Ru^V are both PCET processes, yielding the $Ru^V=O$ intermediate; water nucleophilic attack on $Ru^V=O$ affords a hydroperoxide $Ru^{III}-OOH$ complex, which undergoes one PCET oxidation and forms $Ru^{IV}-OO$; eventually dioxygen is released from this Ru^{IV} peroxide. At the transition state of the water attack step, the Ru cation shifts towards one side of the pda^{2-} ligand. According to a comparative simulation, the preorganized pda^{2-} ligand, rather than the flexible bda^{2-}, provides the Ru cation a bigger and more rigid cavity, which favors the water nucleophilic attack pathway over the coupling of two $Ru^V=O$ units. It is also found that the relatively weakly bound carboxylate group acts as a proton reshuffle in the catalytic process.

4.4   Homogeneous Light-Driven Water Oxidation

The goal we are pursuing is of course to apply highly active and robust WOCs in artificial photosynthesis devices that convert photo energy to chemical energy. Although many WOCs have been reported to be capable of catalyzing Ce^{IV}-driven water oxidation, visible light-driven water oxidation in a homogeneous system with a high performance is rarely achieved because (i) WOCs have overpotentials too positive to be triggered by the commonly used photosensitizer $[Ru(bpy)_3]^{2+}$ ($E_{1/2} = 1.26$ V versus NHE) and/or (ii) the efficiency of WOCs is not high enough to competently quench the strongly oxidizing $\mathbf{P}^+$ (oxidized photosensitizers), which as a result undergoes fast decomposition under illumination.

By introducing negatively charged ligands to WOCs, we have been able to markedly reduce the catalytic potentials of Ru-based WOCs and thus significantly improve their performances under Ce^{IV}-driven water oxidation conditions. As a step toward artificial photosynthesis, visible light-driven water oxidation was studied in a homogeneous system using the aforementioned Ru-based WOCs.

4.4.1   The Three-Component System

A typical visible light-driven water oxidation system consists of three components: a WOC, a photosensitizer, and a sacrificial electron acceptor. There are two well-known sacrificial electron acceptors that are coupled with $[Ru(bpy)_3]^{2+}$-type photosensitizers: $S_2O_8^{2-}$ and

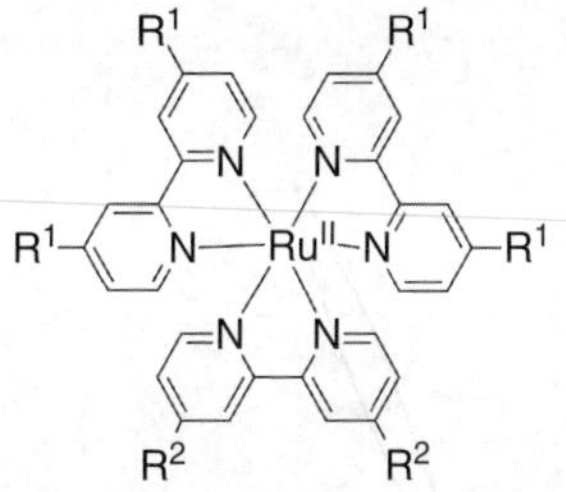

P1: R^1, R^2 = Me; $E_{1/2}(Ru^{III/II})$ = 1.10 V
P2: R^1, R^2 = H; $E_{1/2}(Ru^{III/II})$ = 1.26 V
P3: R^1 = H, R^2 = CO_2Et; $E_{1/2}(Ru^{III/II})$ = 1.40 V
P4: R^1 = CO_2Et, R^2 = Me; $E_{1/2}(Ru^{III/II})$ = 1.49 V
P5: R^1 = CO_2Et, R^2 = H; $E_{1/2}(Ru^{III/II})$ = 1.54 V

Figure 4.16 *Structures of Ru–bpy photosensitizers* **P1–P5** *and their oxidation potentials*

$[Co(NH_3)_5Cl]^{2+}$. Since different WOCs show diverse onset catalytic potentials, photosensitizers **P1–P5** with oxidation potentials spanning from 1.10 to 1.54 V are used in our study (Figure 4.16). A number of WOCs developed in our group are capable of catalyzing water oxidation driven by visible light [8, 9, 13, 17, 19–21]. Light-driven water oxidation by **4** and **11** is described in this section as a representative example [19, 21].

Figure 4.17 shows the CV curves of complex **11** and **P2** in pH 7.0 phosphate buffer solutions. The onset potential of **11** is around 0.98 V, which is 0.28 V less than the $E_{1/2}(Ru^{III/II})$ of **P2**, indicating that a light-driven water oxidation system that includes **11** as a catalyst and **P2** as a photosensitizer is thermodynamically favored under neutral conditions.

One of three-component homogeneous systems we constructed consists of **P2** as a photosensitizer, **11** as a WOC, and $[Co(NH_3)_5Cl]Cl_2$ as a sacrificial electron acceptor in pH 7.0 phosphate buffer. Figure 4.17 exhibits the oxygen formation versus time plot monitored by a Clark-type oxygen electrode. Oxygen evolves from the previously mentioned reaction system upon irradiation, while the reaction ceases once light irradiation is switched off. A TON of 100 and a TOF of $0.06\,s^{-1}$ were achieved. Explicit control experiments revealed that all three components are necessary to achieve light-driven water oxidation.

Electrochemical study of dinuclear Ru complex **4** suggested that its catalytic potential is about 1.3 V. Therefore, only photosensitizers with $E_{1/2} > 1.3$ V are qualified to drive light-driven water oxidation in a three-component homogeneous system that includes **4** as a WOC. Three photosensitizers **P3–P5** containing electron-withdrawing substituents met the potential criterion and therefore were employed to couple with **4** in our investigation. The results showed that the photosensitizer with the more positive oxidation potential led to oxygen evolution at a faster rate and with a greater TON. Worth noting is that the catalyst was still active when oxygen evolution ceased for the first run. The system could be revived by addition of fresh sensitizer, base, and sacrificial electron acceptor $S_2O_8^{2-}$. The $4-P5-S_2O_8^{2-}$ system finally achieved 1270 turnovers after four reviviscences.

4.4.2 The Supramolecular Assembly Approach

Insofar as the efficiency of the whole photosynthesis system is concerned, there are several drawbacks of the three-component light-driven water oxidation system: (i) the rate of electron transfer from catalysts to photosensitizers is controlled by molecular diffusion and therefore has an upper-bound limitation; (ii) this leads to a slow quench of the oxidized photosensitizers and their fast decomposition; and (iii) when catalysts and photosensitizers are immobilized on the electrode surface in order to construct a photo-anode,

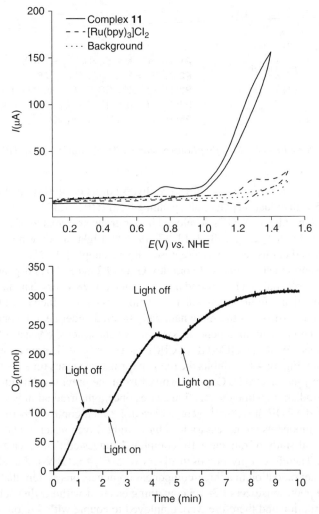

Figure 4.17   *Upper: Cyclic voltammograms of **11** (1.0 mM) and [Ru(bpy)₃]Cl₂ (1.0 mM), plus background in phosphate buffer (pH 7.0, 50 mM) solution containing 10% acetonitrile. Lower: Light-control experiment of photochemical water oxidation in 2 mL phosphate buffer (initial pH 7.0, 50 mM) solution of [Co(NH₃)₅Cl]Cl₂ (2.9 × 10⁻² M), [Ru(bpy)₃]Cl₂ (6.7 × 10⁻⁵ M) and **11** (5.5 × 10⁻⁶ M). Adapted with permission from [21]. Copyright © 2010, American Chemical Society*

the relatively large separation of these two components results in a serious intermolecular charge-transfer problem. A supramolecular assembly, in which the sensitizer and catalyst are covalently linked, is of practical advantage in solving those problems encountered in the three-component system. However, this will also create other problems, such as fast charge recombination between the excited sensitizer and the highly oxidizing catalyst.

We designed two supramolecular assemblies **19a** and **19b**, both of which are based on an Ru–bda motif and tethered with two Ru–polypyridyl units (Figure 4.18). More

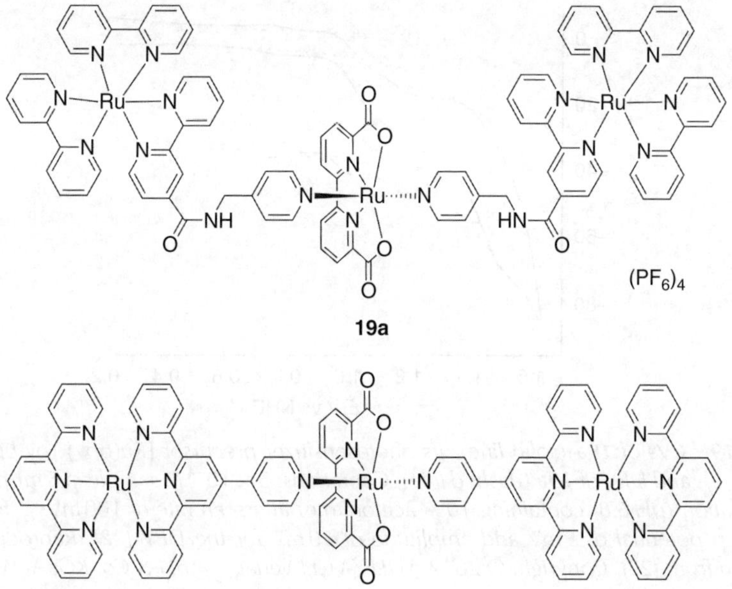

19a

(PF$_6$)$_4$

19b   (PF$_6$)$_4$

Figure 4.18   *Structures of Ru–based molecular assemblies **19a** and **19b***

importantly, **19a** is the first example of functional molecular assembly towards light-driven water oxidation [22]. This project is led by Sun and Li at DUT-KTH Joint Education and Research Center on Molecular Devices, Dalian University of Technology (DUT), China.

Figure 4.19 depicts the redox waves of complex **19a** and its photosensitizer precursor [Ru(bpy)$_2$(py-bpy)]$^{2+}$ (py-bpy = 2-(4-methyl-2-pyridyl)-N-(4-pyridylmethyl)isonicotinamide). From the DPV of **19a**, three oxidation peaks related to the Ru–bda core were observed at 0.74, 1.00, and 1.19 V, assigned to the respective Ru$^{III/II}$, Ru$^{IV/III}$, and Ru$^{V/IV}$ processes. A forth peak at 1.42 V was assigned to the Ru$^{III/II}$ process of the Ru photosensitizer motifs of the assembly, while the free photosensitizer [Ru(bpy)$_2$(py-bpy)]$^{2+}$ alone displayed a redox wave at ca. 1.32 V. For the molecular assembly **19b**, four redox waves were also observed at 0.79, 1.04, 1.21, and 1.43 V: the first three were assigned to the Ru$^{III/II}$, Ru$^{IV/III}$, and Ru$^{V/IV}$ processes of the Ru–bda core and the fourth to the Ru$^{III/II}$ process of the [Ru(tpy)$_2$]$^{2+}$ motifs. In addition, the electrochemical water oxidation current was observed at $E \approx 1$ V for both molecular assemblies under neutral conditions. These data indicate that the photo-generated, oxidized photosensitizer could thermodynamically oxidize the Ru–bda core to its Ru^V state, which triggers water oxidation.

However, many other factors, such as the lifetime of the sensitizer motif, influence the activity of **19a–b** towards light-driven water oxidation. The photochemistry was performed in phosphate buffer solutions containing assemblies and Na$_2$S$_2$O$_8$ as the sacrificial electron acceptor. The combination of **19a**/Na$_2$S$_2$O$_8$ is capable of catalyzing water oxidation under light radiation. Control experiments showed that all three components – light, assembly, and sacrificial electron acceptor – are necessary for water oxidation. Under optimized conditions, 38 turnovers were achieved by **19a** after 70 minutes of illumination. In contrast,

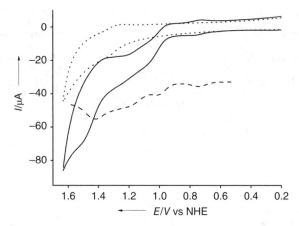

Figure 4.19 *CVs of **19a** (solid line), its photosensitizer precursor [Ru(bpy)$_2$(py-bpy)](PF$_6$)$_2$ (dotted line), and DPV of **19a** (dashed line). Conditions: 5 × 10^{-4} M sample in 5 mL phosphate buffer solution (pH 6.8, containing 10% acetonitrile) at a scan rate of 100 mVs^{-1} for the CV, with a step potential of 5 mV and amplitude of 50 mV for the DPV [22]. Reproduced with permission from [22]. Copyright © 2012, WILEY-VCH Verlag GmbH & Co. KGaA, Weinheim*

the **11**−[Ru(bpy)$_3$]$^{2+}$−Na$_2$S$_2$O$_8$ three-component system resulted in only eight turnovers based on catalyst under the same conditions (**11** : [Ru(bpy)$_3$]$^{2+}$ = 1 : 2). On the other hand, the **19b**−Na$_2$S$_2$O$_8$ system is not active due to a short excited [Ru(tpy)$_2$]$^{2+}$ motif lifetime, leading to kinetic problems of electron transfer from the catalyst to the photosensitizer.

4.5   Water Oxidation Device

The goal of our work is to create a water-splitting device that can convert solar energy to chemical energy without using any sacrificial reagents. In principle, such a device would be composed of three parts: a visible light-harvesting antenna with charge-separation function, a WOC integrated on the anode that donates electrons, and a catalyst on the cathode that carries out reduction reactions and accepts protons and electrons. The anode could be electrochemically active or photo-active, depending on whether the light-harvesting antenna is a separate system, such as a solar panel, or an integrated layer of the photo-anode.

4.5.1   Electrochemical Water Oxidation Anode

Recently, Sun and coworkers at DUT-KTH Joint Education and Research Center on Molecular Devices reported an efficient water oxidation anode modified by a molecular Ru−bda catalyst immobilized on multiwall carbon nanotubes (MWCNTs) [23]. MWCNTs were coated on an indium tin oxide (ITO) glass electrode as a solid support that possesses excellent chemical stability and electronic property. The Ru−bda catalyst bearing pyrene substituents and the MWCNTs were combined via noncovalent $\pi - \pi$ stacking interactions (Figure 4.20, upper). When a potential of 1.4 V versus NHE was applied on this electrode, a catalytic current density of 220 μA/cm^2 was obtained. A turnover number of 11 000 was

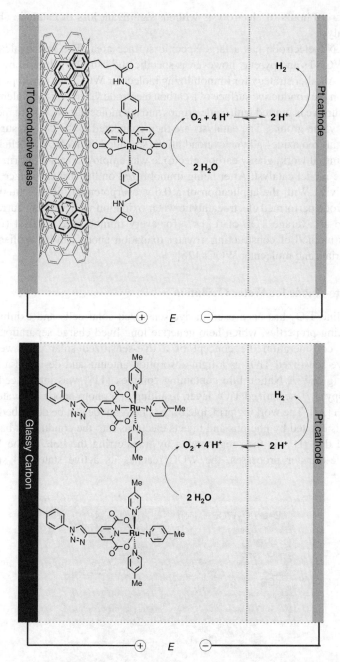

Figure 4.20 *Schematic illustration of water-splitting electrochemical cells using a modified MWCNT electrode (upper) or a functionalized glassy carbon electrode (lower) as an anode*

achieved over 10 hours of electrolysis without significant loss of catalytic ability of the Ru−bad catalyst.

The MWCNT electrode has a large specific surface area. The noncovalent attraction between MWCNTs and pyrene, however, is specific and can't be achieved by other kinds of electrode. Another strategy for immobilizing molecular WOCs we developed is to tether the catalyst on the conductive surface of a carbon electrode via a robust covalent bond [24]. The carbon surface is grafted with azide groups and the molecular WOCs are modified with terminal acetylene groups. The catalysts are then connected to the carbon surface via the copper(I)-catalyzed azide−alkyne cycloaddition (so-called "CuAAC" or "click") reaction [25]. In our initial work, glassy carbon electrode was employed as a platform and **6b** was chosen as the model catalyst. After being immobilized on the carbon surface, **6b** kept its catalytic activity. With the application of a 0.6 V overpotential, **6b**-functionlized glassy carbon electrode performed electrocatalytic water oxidation with a salient current that corresponded to an average TOF of $0.71\,s^{-1}$ for every immobilized catalyst (Figure 4.20, lower). This method of constructing a water oxidation anode can be applied to various carbon materials and molecular WOCs [26].

4.5.2   Photo-Anode for Water Oxidation

Nanocrystalline TiO_2 has been used in dye-sensitized solar cells and exhibits excellent semiconducting properties, which help generate long-lived charge separation dye states. Taking into consideration the concept of a dye-sensitized solar cell, we chose the Ru-based dye-sensitized TiO_2 as a light-absorption antenna and designed a light-driven water-splitting cell. A Nafion film containing complex $[11]^+$ was introduced on the top of the $[Ru(bpy)_3]^{2+}$-sensitized TiO_2 layer, resulting in a photo-anode for water oxidation (Figure 4.21) [27]. The working principle of the photo-anode can be described as follows: the dye (D) is excited by photons and injects electrons into the conduction band of TiO_2; the oxidized dye (D^+) is then regenerated by transferring the hole to the WOC; after multiple hole-transfer processes, the WOC reaches its active state and subsequently

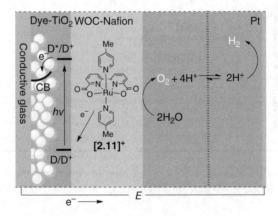

Figure 4.21   *Illustration of the working principle of a photoelectrochemical cell towards artificial water splitting. Reproduced with permission from [28], The Royal Society of Chemistry*

oxidizes water to dioxygen. Electrons generated by excitation flow from the anode to the Pt cathode through the external circuit of the cell. At the Pt cathode, the electrons are consumed by proton reduction reaction. Under illumination of visible light, the photoelectrochemical cell gives 16 turnovers over 60 minutes, with an initial turnover frequency of $27\,h^{-1}$ [12].

4.6 Conclusion

The anionic carboxylate possesses strong electron-donating ability thanks to the $p\pi-d\pi$ interaction between the bound oxygen and the Ru center. Carboxylate-containing ligands have been shown to reduce the oxidation potential of Ru complexes and to stabilize their high valent states. For this reason, a seven-coordinate Ru^{IV} species has been isolated as a key intermediate of Ru-catalyzed water oxidation.

Ru-WOCs with carboxylate-containing ligands exhibit remarkably high activity towards water oxidation, with TOFs of up to $300\,s^{-1}$. Their ancillary ligands have drastic effects on the activity.

The strong electron-donating ligand promotes the ligand exchange between the solvent molecule and the bound ligand of a Ru complex. Accordingly, carboxylate-containing ligands may significantly enhance the rate of the oxygen-releasing step.

A DFT-directed development of robust Ru-WOCs has been demonstrated, showing one of the advantages of molecular WOCs.

Visible light-driven water oxidation has been achieved using both a three-component system and a supramolecular assembly.

On the basis of all the advantages of Ru-WOCs with carboxylate ligation, the electrochemical active anode and the photo-active water oxidation anode have been realized.

References

[1] Umena, Y., Kawakami, K., Shen, J.-R., Kamiya, N. (2011) Nature, 473: 55.
[2] Metz, J. G., Nixon, P. J., Rogner, M., Brudvig, G. W., Diner, B. A. (1989) Biochem., 28: 6960.
[3] Vass, I., Styring, S. (1991) Biochem., 30: 830.
[4] Geijer, P., Morvaridi, F., Styring, S. (2001) Biochem., 40: 10 881.
[5] Zong, R., Thummel, R. P. (2005) J. Am. Chem. Soc., 127: 12 802.
[6] Xu, Y., Duan, L., Åkermark, T., Tong, L., Lee, B.-L., Zhang, R., Åkermark, B., Sun, L. (2011) Chem. Eur. J., 17: 9520.
[7] Xu, Y., Åkermark, T., Gyollai, V., Zou, D., Eriksson, L., Duan, L., Zhang, R., Åkermark, B., Sun, L. (2009) Inorg. Chem., 48: 2717.
[8] Xu, Y., Fischer, A., Duan, L., Tong, L., Gabrielsson, E., Åkermark, B., Sun, L. (2010) Angew. Chem. Int. Ed., 49: 8934.
[9] Duan, L., Xu, Y., Gorlov, M., Tong, L., Andersson, S., Sun, L. (2010) Chem. Eur. J., 16: 4659.
[10] An, J., Duan, L., Sun, L. (2012) Faraday Discuss., 155: 267.
[11] Yoshida, M., Masaoka, S., Sakai, K. (2009) Chem. Lett., 38: 702.
[12] Wasylenko, D. J., Ganesamoorthy, C., Henderson, M. A., Koivisto, B. D., Osthoff, H. D., Berlinguette, C. P. (2010) J. Am. Chem. Soc., 132: 16 094.

[13] Tong, L., Wang, Y., Duan, L., Xu, Y., Cheng, X., Fischer, A., Ahlquist, M., Sun, L. (2012) Inorg. Chem., 51: 3388.

[14] Duan, L., Fischer, A., Xu, Y., Sun, L. (2009) J. Am. Chem. Soc., 131: 10 397.

[15] Duan, L., Bozoglian, F., Mandal, S., Stewart, B., Privalov, T., Llobet, A., Sun, L. (2012) Nat. Chem., 4: 418.

[16] Duan, L., Araujo, C. M., Ahlquist, M. S. G., Sun, L., (2012) P. Natl. Acad. Sci. U. S. A., 109: 15 584.

[17] Tong, L., Duan, L., Xu, Y., Privalov, T., Sun, L. (2011) Angew. Chem. Int. Ed., 50: 445.

[18] Wang, L., Duan, L., Stewart, B., Pu, M., Liu, J., Privalov, T., Sun, L. (2012) J. Am. Chem. Soc., 134: 18 868.

[19] Xu, Y., Duan, L., Tong, L., Akermark, B., Sun, L. (2010) Chem. Commun., 46: 6506.

[20] Duan, L., Xu, Y., Tong, L., Sun, L. (2011) ChemSusChem, 4: 238.

[21] Duan, L., Xu, Y., Zhang, P., Wang, M., Sun, L. (2010) Inorg. Chem., 49: 209.

[22] McConnell, I. L., Grigoryants, V. M., Scholes, C. P., Myers, W. K., Chen, P.-Y., Whittaker, J. W., Brudvig, G. W. (2012) J. Am. Chem. Soc., 134: 1504.

[23] Li, F., Zhang, B., Li, X., Jiang, Y., Chen, L., Li, Y., Sun, L. (2011) Angew. Chem. Int. Ed., 50: 12 276.

[24] Tong, L., Göthelid, M., Sun, L. (2012) Chem. Commun., 48: 10 025.

[25] Meldal, M., Tornee, C. W. (2008) Chem. Rev., 108: 2952.

[26] Pinson, J., Podvorica, F. (2005) Chem. Soc. Rev., 34: 429.

[27] Li, L., Duan, L., Xu, Y., Gorlov, M., Hagfeldt, A., Sun, L. (2010) Chem. Commun., 46 : 7307.

[28] Duan, L., Tong, L., Xu, Y., Sun, L. (2011) Energy Environ. Sci., 4: 3296.

5

Water Oxidation by Ruthenium Catalysts with Non-Innocent Ligands

Tohru Wada[1], Koji Tanaka[2], James T. Muckerman[3], and Etsuko Fujita[3]

[1]*Department of Chemistry, College of Science and Research Center for Smart Molecules, Rikkyo University, Toshima, Tokyo, Japan*
[2]*Institute for Integrated Cell-Material Sciences, Kyoto University, Kyoto, Japan*
[3]*Chemistry Department, Brookhaven National Laboratory, Upton, NY, USA*

5.1 Introduction

Since the first report on the catalytic oxidation of water by Meyer and coworkers, in which a homogeneous catalyst known as the ruthenium blue dimer (i.e. $[(OH_2)(bpy)_2Ru-O-Ru(bpy)_2(OH_2)]^{4+}$; bpy = 2,2′-bipyridine) was used [1–12], a variety of Mn [13–15], Ru [16–26], and Ir [27, 28] metal-cluster compounds have proven to be active towards the oxidation of water. On the basis of these results, the participation of two or more metal centers was considered to be an essential component for the activation of two water molecules to form an O−O bond. However, mononuclear Ru [29–36], Ir [37–41], Fe [42], and Co [43] complexes (and in some cases their decomposition products) have recently been shown to exhibit catalytic activity towards the oxidation of water. The possible reaction path for O−O bond formation has been a point of contention among investigators of water oxidation, but there is a growing consensus acknowledging two main pathways [44, 45]: (1) a nucleophilic attack of water on a high-valent metal−oxo species, which affords end-on (η^1) hydroperoxide complexes (Scheme 5.1a) [31, 46, 47]; and (2) a coupling reaction of two metal−oxo species to give μ-peroxide intermediates (Scheme 5.1b) [48, 49].

In the case of mononuclear ruthenium catalysts, high-valent $Ru^V=O$ species undergo a nucleophilic attack by water due to the low reactivity of the $Ru^{IV}=O$ moiety and the weak nucleophilicity of neutral and acidic water [31]. However, some $Ru^{IV}=O$ species can react with water to produce an $Ru^{III}-OOH$ species under neutral and basic conditions [50].

Molecular Water Oxidation Catalysis: A Key Topic for New Sustainable Energy Conversion Schemes,
First Edition. Edited by Antoni Llobet.
© 2014 John Wiley & Sons, Ltd. Published 2014 by John Wiley & Sons, Ltd.

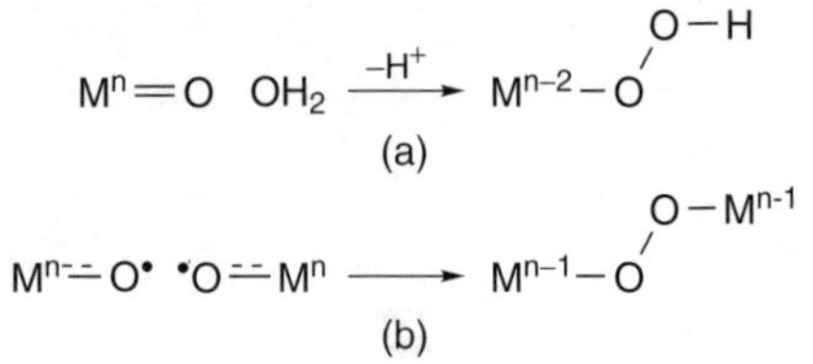

$$\text{M}^n = \text{O} \quad \text{OH}_2 \xrightarrow{-\text{H}^+} \text{M}^{n-2} - \text{O}^{\diagup\text{O}-\text{H}}$$

(a)

$$\text{M}^{n} \doteq \text{O}^\bullet \quad {}^\bullet\text{O} \doteq \text{M}^n \longrightarrow \text{M}^{n-1} - \text{O}^{\diagup\text{O}-\text{M}^{n-1}}$$

(b)

Scheme 5.1   *Proposed pathways of water oxidation*

While nucleophilic attack by water on the $\text{Ru}^V = \text{O}$ moiety in the blue dimer may be an important step in O–O bond formation, dinuclear complexes in which two metals are fixed by a bridging ligand at an appropriate position to accommodate an O–O bond are suitable for water oxidation via an intramolecular coupling (Scheme 5.1b). Furthermore, dinuclear complexes may have an advantage in oxidizing water at a less positive potential because intramolecular coupling of Ru=O may not need the high-valent oxidation state required for nucleophilic attack of water on $\text{Ru}^V = \text{O}$, as investigated at pH 1 [44]. However, control of the electronic density on the oxo ligand is necessary for smooth O–O bond formation via the intramolecular coupling mechanism (Scheme 5.1b) because electrostatic repulsion between O^{2-} ligands of the Ru=O moieties inhibits the intramolecular coupling of the oxo groups.

Dioxolenes are known as "non-innocent" ligands (NILs). As shown in the bottom row of Scheme 5.2, the dioxolene ligand can take on three oxidation states: quinone (q), semiquinone (sq), and catecholate (cat) [51]. The term "non-innocent" is used when the ligands in a complex have strong electrochemical interactions with the central metal [52]. Metal complexes with NILs are characterized by particular combinations of metals and ligands, rather than by redox-active ligands alone. Ambiguous or non-integer oxidation-state assignments of the metal and NIL fragments occur when there is a large degree of mixing between the metal and NIL orbitals. The possible "oxidation states" for the Ru–NIL systems and their spin multiplicities are depicted in Scheme 5.2. Among various metal complexes bearing NILs, such as dioxolenes, dithiolenes, and benzoquinone diimines, Ru–dioxolene complexes are particularly interesting because of the close alignment of energy levels between the d-orbitals of the metal and the π^*-orbitals of the dioxolene ligand [53–65]. As a result, there are several possible electronic structures for Ru–dioxolene complexes (Scheme 5.2) in the absence of an additional redox-active ligand or substrate such as a water molecule. The presence of a substrate water molecule complicates this simple picture greatly, especially when coupled with proton removal from the H_2O. When a dioxolene ligand is introduced in the *trans* position to the aqua ligand of an Ru–aqua complex, the electronic state of the Ru–oxo moiety generated by deprotonation and/or oxidation of the complex is strongly influenced by the dioxolene ligand through d-orbitals of the Ru center (Figure 5.1).

More than a decade ago, Wada and Tanaka reported a novel dinuclear Ru complex known as the Tanaka catalyst, $[\text{Ru}_2(\text{OH})_2(3,6\text{-Bu}_2\text{q})_2(\text{btpyan})](\text{SbF}_6)_2$ ($[\mathbf{1}(\text{OH})_2(3,6\text{-Bu}_2\text{q})_2](\text{SbF}_6)_2$, $\mathbf{1} = \text{Ru}_2(\text{btpyan})$, $3,6\text{-Bu}_2\text{q} = 3,6$-di-*tert*-butyl-1,2-benzoquinone, btpyan $= 1,8$-bis($2,2':6',2''$-terpyrid-$4'$-yl)anthracene; Scheme 5.3), which contains non-innocent quinone ligands and has excellent electrocatalytic activity for water oxidation when immobilized on an indium tin oxide (ITO) electrode [66, 67]. As shown in the top-left drawings

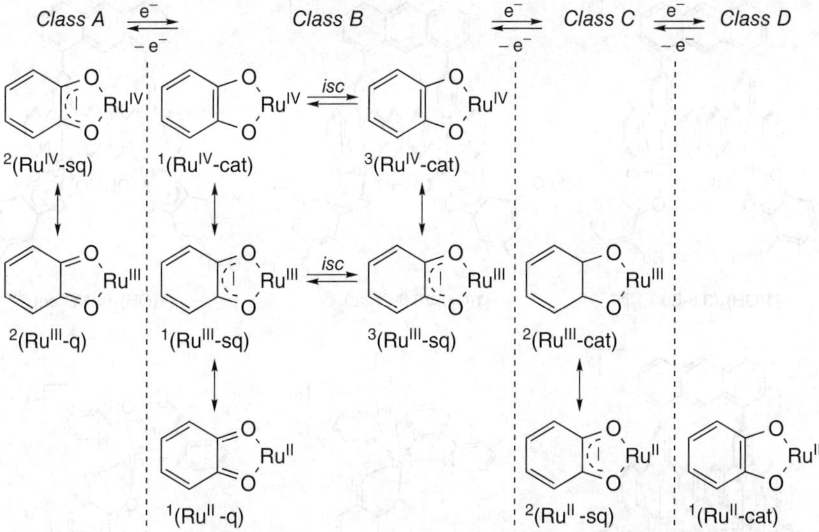

Scheme 5.2 *Oxidation states in the various resonance structures for Ru–dioxolene-type complexes. The various redox states have been divided into four classes. Spin multiplicities are indicated as leading superscripts in the oxidation state designations. Possible intersystem crossing between singlet and triplet states is indicated by "isc"*

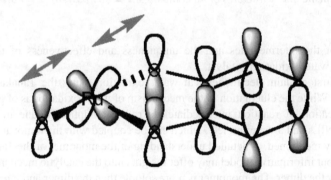

Figure 5.1 *Interaction between oxo and dioxolene through an Ru dπ-orbital*

in Scheme 5.3, two [Ru(OH)(3,6-Bu$_2$q)]$^+$ units are complexed by anthracene-bridged bis-terpyridine ligands with a geometry in which the two OH groups on the Ru centers are in close proximity, allowing creation of an O–O bond. The novel features of the dinuclear and related mononuclear Ru species with quinone ligands, and comparison of their properties with those of the Ru analogs in which the bpy ligand (bpy = 2,2′-bipyridine) replaces quinone, are summarized in this chapter. While insight into the mechanism of water oxidation, including the identity and oxidation state of key intermediates with the Tanaka catalyst and its related mononuclear species, is still evolving, we believe it is important to present the current status of our knowledge due to the intriguing electronic

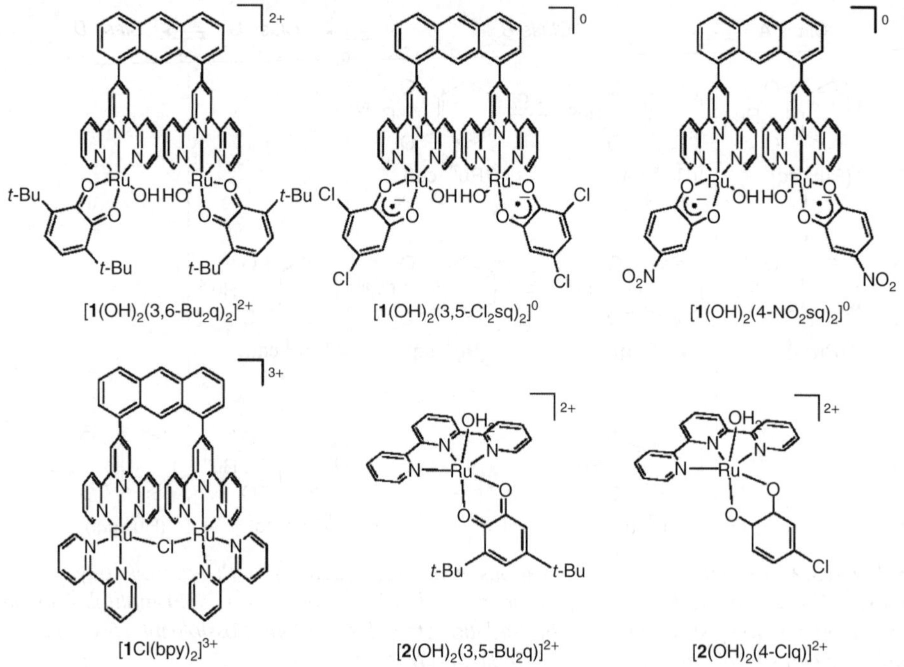

Scheme 5.3   *Structures of Ru complexes: 1 = Ru$_2$(btpyan), 2 = Ru(tpy)*

structures of the intermediates and the uniqueness and effectiveness of the dinuclear species as a water oxidation catalyst.

We will first explain electrochemical water oxidation using the Tanaka catalyst, in Section 5.2. While the elucidation of the mechanism of water oxidation is of great interest, the identification of reaction intermediates is rather complicated due to the catalyst containing NILs and a water moiety, which can be coupled with the removal of proton(s), as previously mentioned. A detailed understanding of the monomer of the Tanaka catalyst as a simple but informative model may offer insights into the catalytic mechanism of water oxidation by the dimer. The monomer is more soluble than the dimer and more convenient for electrochemical measurements; its size is also more amenable to detailed electronic structure theory studies. Therefore, we first describe electrocatalytic water oxidation by the Tanaka catalyst and a related [Ru$_2$(μ-Cl)(bpy)$_2$(btpyan)]$^{3+}$ ([1Cl(bpy)$_2$]$^{3+}$), which is a precatalyst of the water oxidation catalyst [Ru^{III}$_2$(OH)$_2$(bpy)$_2$(btpyan)]$^{4+}$ ([1(OH)$_2$(bpy)$_2$]$^{4+}$), then explain the acid–base equilibrium and redox behavior of mononuclear Ru–aqua complexes with a variety of dioxolene ligands, since a dioxolene ligand on the ruthenium has substantial effects on the electronic state and catalytic activity of the ruthenium complexes. We will also discuss the more complicated redox and catalytic properties of the Tanaka catalyst and its related dinuclear complexes. We now have clear evidence of O–O bond formation via intramolecular coupling (Scheme 5.1b), detected by resonance Raman spectra measurements of a dinuclear ruthenium complex with bpy ligands ([1(μ-O$_2$)(bpy)$_2$]$^{4+}$), probably because its catalytic activity for water oxidation is inferior to that of the dioxolene complex $[1(OH)_2(3,6\text{-}Bu_2q)_2]^{2+}$. We hope to convey the striking features of both the dinuclear and the mononuclear species, gleaned from theoretical and experimental investigations.

5.2 Water Oxidation Catalyzed by Dinuclear Ruthenium Complexes with NILs

The arrangement of two Ru–oxyl radical species is extremely important for the formation of an O–O bond via intramolecular coupling (Scheme 5.1b). We have developed the btpyan ligand as a novel bridging ligand for water oxidation catalysts and solved the X-ray crystal structure as shown in Figure 5.2 [66, 68]. The two terpyridyl groups linked to the 1,8-positions of anthracene are located face to face and are parallel. The btpyan ligand has a rotation axis that includes C atoms at the 9- and 10-positions of anthracene, and the two terpyridyl groups are crystallographically equivalent in the crystal structure. The dihedral angles between anthracene and the terpyridines are 69.06°. Free rotation of the two terpyridyl groups must be inhibited because of these steric repulsions. The short distance of 4.22 Å between the two central N atoms of the two terpyridyl groups, compared with the distance of 4.88 Å between the two C atoms at the 1- and 8-positions of anthracene, is explained by the $\pi-\pi$ stacking of the two terpyridyl groups. The distance between the two metals is therefore estimated to be in the range 4–5 Å when btpyan forms a dinuclear complex.

Using btpyan, we prepared complexes $[1(OH)_2(3,6\text{-}Bu_2q)_2]^{2+}$ [66], $[1(OH)_2(3,5\text{-}Cl_2sq)_2]^0$, and $[1(OH)_2(4\text{-}NO_2sq)_2]^0$ [68] and the analogs $[1Cl_2(bpy)_2]^{2+}$ and $[1Cl(bpy)_2]^{3+}$ [69]. The two-electron-reduced species $[1(OH)_2(3,6\text{-}Bu_2sq)_2]^0$ can be

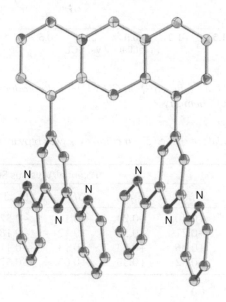

Figure 5.2 *Crystal structure of btpyan*

readily obtained by chemical reduction with $Na_2S_2O_3$ [68]. The cyclic voltammetries (CVs) of these complexes were measured up to 1.0 V versus saturated calomel electrode (SCE) in CH_2Cl_2. The CV of $[1(OH)_2(3,6-Bu_2sq)_2]^0$ is shown in Figure 5.3 and the redox potentials of dinuclear Ru complexes containing btpyan and NIL are summarized in Table 5.1.

The CVs of $[1(OH)_2(3,6-Bu_2q)_2]^{2+}$ were measured up to +1.8 V in 2,2,2-trifluoroethanol/ether (50/50 v/v) to detect the metal-centered Ru^{II}/Ru^{III} redox couples of these complexes. A broad, irreversible anodic wave appeared at $E_p = +1.39$ V, in addition to waves for the reversible $[Ru^{II}(q), Ru^{II}(q)]^{2+}/[Ru^{II}(q), Ru^{II}(sq)]^+$ and $[Ru^{II}(q), Ru^{II}(sq)]^+/[Ru^{II}(sq), Ru^{II}(sq)]^0$ redox couples at $E_{1/2} = +0.31$ and +0.20 V, respectively. The area of the broad anodic wave at +1.39 V was almost the same as the total area of the two anodic waves at $E_{1/2} = +0.31$ and +0.20 V. In addition, controlled-potential electrolysis of

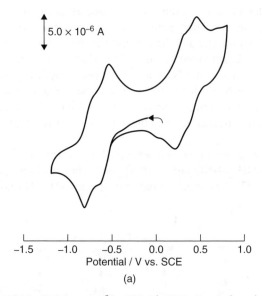

5.0×10^{-6} A

-1.5	-1.0	-0.5	0.0	0.5	1.0

Potential / V vs. SCE

(a)

Figure 5.3   *CV of $[1(OH)_2(3,6-Bu_2sq)_2]^0$ in CH_2Cl_2 [68]. .Reproduced with permission from [68], The Royal Society of Chemistry*

Table 5.1   *Redox potentials of dinuclear Ru complexes with btpyan and NIL in CH_2Cl_2*

Complex	Potential/V versus SCE					
	$E_{1/2}(1)$	$E_{1/2}(2)$	ΔE^{oxa}	$E_{1/2}(3)$	$E_{1/2}(4)$	ΔE^{reb}
$[1(OH)_2(3,6-Bu_2sq)_2]^0$	+0.40	+0.24	+0.16	-0.57	-0.80	+0.23
$[1(OH)_2(3,5-Cl_2sq)_2]^0$	+0.71	+0.59	+0.12	-0.18	-0.30	+0.12
$[1(OH)_2(4-NO_2sq)_2]^0$	+0.86	+0.73	+0.13	-0.11	-0.18	+0.07
$[1(OAc)(3,6-Bu_2sq)_2]^+$	+0.42	+0.35	+0.07	-0.57	-0.70	+0.13

$^a\Delta E^{ox} = E_{1/2}(1) - E_{1/2}(2)$;
$^b\Delta E^{re} = E_{1/2}(3) - E_{1/2}(4)$
Note: Reproduced by permission of the Royal Society of Chemistry [68]

$[\mathbf{1}(OH)_2(3,6\text{-}Bu_2q)_2]^{2+}$ at +1.65 V consumed two electrons per $[\mathbf{1}(OH)_2(3,6\text{-}Bu_2q)_2]^{2+}$ complex in 2,2,2-trifluoroethanol/ether.

Addition of H_2O to the solution (10%) caused strong catalytic currents at potentials more positive than +1.0 V (Figure 5.4), suggesting that $[\mathbf{1}(OH)_2(3,6\text{-}Bu_2q)_2]^{2+}$ has the ability to catalyze water oxidation. In fact, the controlled-potential electrolysis of $[\mathbf{1}(OH)_2(3,6\text{-}Bu_2q)_2](SbF_6)_2$ (1.5 μmol) in trifluoroethanol containing water (10%) at +1.75 V evolved 0.69 mL of dioxygen with a current efficiency of 91% (21 turnovers). During the progress of water oxidation, the presence of $[\mathbf{1}(OH)_2(3,6\text{-}Bu_2q)_2]^{2+}$ in the solution was evidenced in the electrospray ionization coupled to mass spectroscopy (ESI-MS) and electronic spectra. On the other hand, $[\mathbf{1}(OH)_2(3,6\text{-}Bu_2q)_2]^{2+}$ was not confirmed in the ESI-MS and electronic spectra of the solution after the catalytic currents had completely stopped. Consistent with this result, no anodic current flowed in the absence of $[\mathbf{1}(OH)_2(3,6\text{-}Bu_2q)_2]^{2+}$ under similar reaction conditions. Thus, it is concluded that the oxidized form of $[\mathbf{1}(OH)_2(3,6\text{-}Bu_2q)_2]^{2+}$ is the active species for water oxidation.

The CV of $[\mathbf{1}(OH)_2(3,6\text{-}Bu_2q)_2]^{2+}$ physically immobilized on an ITO electrode $(1.2 \times 10^{-8}\ mol/2.0\ cm^2)$ in water (pH 4.0) exhibits a broad redox wave centered at +0.37 V, an irreversible anodic wave at +1.24 V and a strong anodic current at potentials more positive than +1.65 V (Figure 5.5a). The controlled-potential electrolysis of $[\mathbf{1}(OH)_2(3,6\text{-}Bu_2q)_2]^{2+}$ modified on an ITO electrode was therefore conducted at +1.75 V in water (pH 4.0), and 1.5 mL of O_2 was generated after 27.5C passed in the electrolysis. The current efficiency for O_2 evolution was 95%, based on gas chromatographic (GC)

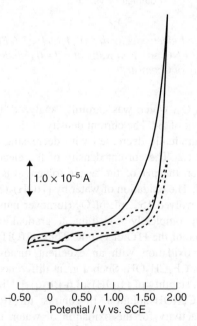

1.0×10^{-5} A

−0.50   0   0.50   1.00   1.50   2.00
Potential / V vs. SCE

Figure 5.4 *CVs of $[\mathbf{1}(OH)_2(3,6\text{-}Bu_2q)_2]^{2+}$ in 2,2,2-trifluoroethanol/ether (1/1 v/v) in the absence (dotted line) and presence (10%, solid line) of water. Reproduced with permission from [66], The Royal Society of Chemistry*

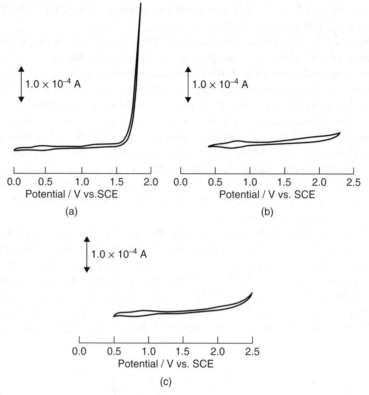

Figure 5.5 *CVs of modified electrodes with (a) [1(OH)$_2$(3,6-Bu$_2$sq)$_2$]^0, (b) [1(OH)$_2$(3,5-Cl$_2$sq)$_2$]^0, and (c) [1(OH)$_2$(4-NO$_2$sq)$_2$]^0 in water at pH 4.0 [68]. Reproduced with permission from [68], The Royal Society of Chemistry*

analysis of the generated O$_2$, which was carefully analyzed by monitoring an N$_2$ peak caused by a very small leak of air. The current density of the electrode was 0.12 mA/cm^2 in the initial stage; this gradually decreased with decreasing pH in the aqueous phase and almost ceased at pH 1.2. The current density of the electrode for the oxidation of water was recovered when the pH of the water solvent was readjusted to 4.0 by the addition of aqueous KOH. The oxidation of water by [1(OH)$_2$(3,6-Bu$_2$q)$_2$]$^{2+}$ modified on an ITO electrode finally evolved 15.2 mL of O$_2$ (turnover number: 33 500) in 40 hours, and then the O$_2$ evolution completely ceased due to gradual exfoliation of [1(OH)$_2$(3,6-Bu$_2$q)$_2$]$^{2+}$ from the surface of the ITO electrode. Thus, [1(OH)$_2$(3,6-Bu$_2$q)$_2$]$^{2+}$ modified on ITO-catalyzed water oxidation with an excellent turnover number compared to [1(OH)$_2$(3,6-Bu$_2$q)$_2$]$^{2+}$ in CF$_3$CH$_2$OH. Such a great difference in catalytic ability probably results from the high stability of [1(OH)$_2$(3,6-Bu$_2$q)$_2$]$^{2+}$ in the solid state compared with its stability in CF$_3$CH$_2$OH solution. In contrast to [1(OH)$_2$(3,6-Bu$_2$q)$_2$](SbF$_6$)$_2$, which showed catalytic activity for the oxidation of water, the analogous monomeric complex [2(OH$_2$)(3,6-Bu$_2$q)](ClO$_4$)$_2$ (2 = Ru(tpy), tpy = 2,2′,2″-terpyridine) showed no such activity in H$_2$O under similar conditions (see Section 5.4). The two Ru–hydroxo units of [1(OH)$_2$(3,6-Bu$_2$q)$_2$]$^{2+}$ therefore play the key role in the oxidation of water.

Does an electron-withdrawing group on the quinone enhance the catalytic activity for water oxidation? We have carried out CVs in water (pH 4.0) using an ITO electrode modified with $[1(OH)_2(3,5-Cl_2sq)_2]^0$ and $[1(OH)_2(4-NO_2sq)_2]^0$. The results shown in Figure 5.5b,c indicate that the Cl- and NO_2-substituted complexes do not show any catalytic activity upon the application of up to 2.0 V versus Ag/AgCl. This result is in sharp contrast to that with the t-Bu_2-substituted complex. Since the electron-withdrawing groups shift all potentials (including q/sq and sq/cat) to more positive values, as well as affecting the relative electronegativities of the ligands, the catalyst may not behave in the same way. A delicate balance of the charge distribution over the Q–Ru–OH framework may control the catalytic activity for water oxidation. We will therefore present the detailed properties of mononuclear Ru species with non-innocent dioxolene ligands, investigated by experimental and theoretical methods, in Section 5.4. However, in order to allow an understanding of the intramolecular O–O coupling mechanism for water oxidation, we must first describe the systems in which non-innocent dioxalene ligands are replaced by the bpy ligand, as given in the next section.

5.3 Water Oxidation by Intramolecular O–O Coupling with $[Ru^{II}_2(\mu\text{-Cl})(bpy)_2(btpyan)]^{3+}$

A dinuclear complex, $[1(OH)_2(3,6-Bu_2q)_2]^{2+}$, shows catalytic activity towards the four-electron oxidation of water. In this process, the q/sq redox reaction is coupled with the deprotonation of the Ru–OH groups. It may play a key role in the catalysis. Despite our many efforts to detect O–O bond formation in the catalytic cycle, no direct evidence has been obtained to date. We prepared the $[Ru^{II}_2(\mu\text{-Cl})(bpy)_2(btpyan)]^{3+}$ ($[1Cl(bpy)_2]^{3+}$) ion as an analog of the quinone complex $[1(OH)_2(3,6-Bu_2q)_2]^{2+}$ ion to investigate the intermediates of water oxidation [69]. $[1Cl_2(bpy)_2](SbF_6)_2$ was prepared by the reaction of $[Ru_2Cl_6(btpyan)]$ with two equivalents of the bpy ligand in the presence of NEt_3 and purified using chromatographic techniques (Scheme 5.4). X-ray crystallographic analysis of $[1Cl_2(bpy)_2](SbF_6)_2$ revealed that the Cl ligands are bound to the Ru atoms in a monodentate manner; that is, one ligand is placed in the cavity formed by the btpyan ligand and two Ru atoms and the other is located outside the cavity (Figure 5.6a).

The tpy unit of the {$Ru_2(bpy)(tpy)$} moiety therefore effectively relieves the steric repulsion between the Cl2 ligand and the {Ru1Cl(bpy)(trpy)} moiety. Both structures of the two {RuCl(bpy)(trpy)} moieties are distorted octahedra. The Ru1–Cl1 bond is slightly longer than the Ru2–Cl2 bond. In fact, the Cl2 ligand was easily removed by treatment of $[1Cl_2(bpy)_2]^{2+}$ ions with $AgBF_4$ to give the chloride-bridged dimer $[Ru_2(\mu\text{-Cl})(bpy)_2(btpyan)](BF_4)_3$ ($[1Cl(bpy)_2](BF_4)_3$).

$[1Cl(bpy)_2](BF_4)_3$ displayed two reversible redox couples at $E_{1/2}(1) = +0.66$ V and $E_{1/2}(2) = +0.85$ V and an irreversible anodic peak at around $E = +1.2$ V in the cyclic voltammogram in water at pH 1.0 (red line in Figure 5.7). The two reversible redox reactions at $E_{1/2}(1)$ and $E_{1/2}(2)$ are assigned to the $[Ru^{II}, Ru^{II}]^{3+}/[Ru^{II}, Ru^{III}]^{4+}$, and $[Ru^{II}, Ru^{III}]^{4+}/[Ru^{III}, Ru^{III}]^{5+}$ couples on the basis of the rest potential of the solution ($E_{rest} = +0.30$ V) and the consumption of 1.0 F/mol of electrons in the controlled potential electrolysis of $[1Cl(bpy)_2]^{3+}$ ions at $E = +0.70$ V. The $E_{1/2}(1)$ value is independent of pH changes. On the other hand, the redox potential of $E_{1/2}(2)$ shifts to negative potentials with

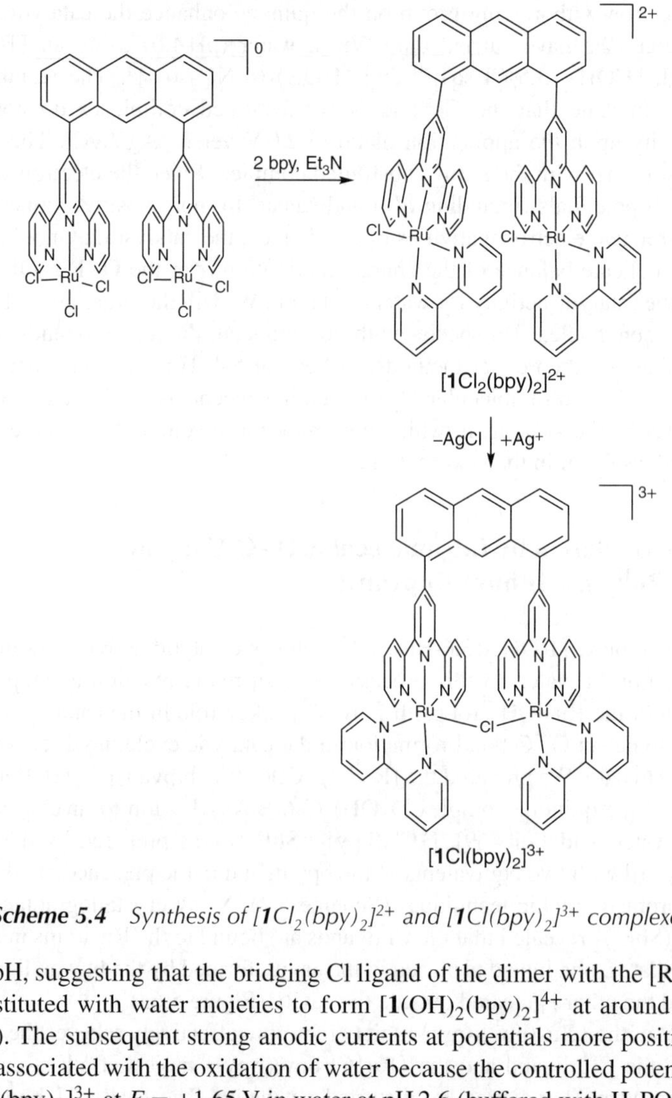

Scheme 5.4   *Synthesis of $[1Cl_2(bpy)_2]^{2+}$ and $[1Cl(bpy)_2]^{3+}$ complexes*

increasing pH, suggesting that the bridging Cl ligand of the dimer with the $[Ru^{III}, Ru^{III}]^{5+}$ core is substituted with water moieties to form $[1(OH)_2(bpy)_2]^{4+}$ at around $E = +0.8$ V (Figure 5.7). The subsequent strong anodic currents at potentials more positive than $E = +1.2$ V are associated with the oxidation of water because the controlled potential electrolysis of $[1Cl(bpy)_2]^{3+}$ at $E = +1.65$ V in water at pH 2.6 (buffered with H_3PO_4/NaH_2PO_4) catalytically evolves dioxygen. However, the reaction rate for water oxidation catalyzed by this complex is slower than that of the quinone complex $[1(OH)_2(3,6-Bu_2q)]^{2+}$ [69].

Furthermore, the addition of $[1Cl(bpy)_2](BF_4)_3$ (1.0 μmol) in CH_3CN (100 μL) to an aqueous solution of $(NH_4)_2Ce(NO_3)_6$ (10 mL, 2.5 mmol) at pH 1.0 (adjusted with HNO_3) causes the evolution of 414 μmol of O_2 (Figure 5.8). Thus, chemical oxidation with a Ce^{IV} species and electrochemical oxidation at $E = +1.65$ V of $[1Cl(bpy)_2]^{3+}$ ions enables the catalysis of the four-electron oxidation of H_2O. The initial rate of oxygen evolution in the presence of excess amounts of the Ce^{IV} species is proportional to the concentration of $[1Cl(bpy)_2]^{3+}$ ions in a range of 1.0–6.0 μM (Figure 5.8, insert).

The complex ion $[1Cl(bpy)_2]^{3+}$, with a $\{Ru^{II}, Ru^{II}\}$ core, gives rise to a strong absorption band at $\lambda_{max} = 468$ nm that is assignable to a metal–ligand charge transfer (MLCT) from

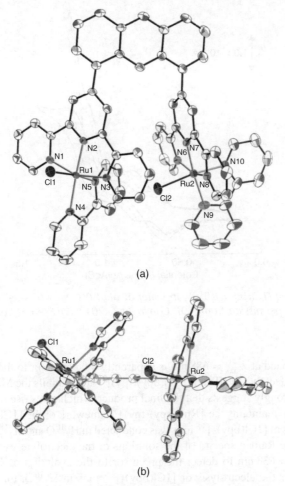

Figure 5.6 *Crystal structure of the [1Cl₂(bpy)₂]²⁺ ion (50% probability). (a) Side view. (b) Top view. The hydrogen atoms have been omitted for clarity*

the Ru^{II} center to the bpy or btpyan ligand in water (pH 1.0) [69]. Electrochemical oxidation of the complex at $E = +0.90$ V results in a decrease in the absorbance of the band. Further oxidation of the solution at $E = +1.40$ V results in the disappearance of the band at $\lambda_{max} = 468$ nm (Figure 5.9a) and bleaching of the green solution. Instead, a new band emerges at $\lambda_{max} = 688$ nm during the electrolysis reaction (Figure 5.9b). When the absorbance becomes constant, the electrolysis is stopped. The resultant solution slowly evolves O₂ when left without the application of potentials. The evolution of O₂ completely stops within 10 hours, giving an amount consistent with that of the [1Cl(bpy)₂]³⁺ ions (Figure 5.10). During the evolution of an equivalent of O₂ based on the amount of [1Cl(bpy)₂]³⁺ from the green solution (Figure 5.9), another new band appears at $\lambda_{max} = 475$ nm, at the expense of the band at $\lambda_{max} = 688$ nm (Figure 5.9c). The peak position of the band at $\lambda_{max} = 475$ nm in the final solution is close to the MLCT band of the [1Cl(bpy)₂]³⁺ ions (i.e. $\lambda_{max} = 468$ nm), but the patterns of the two bands are different from each other. The transient species that

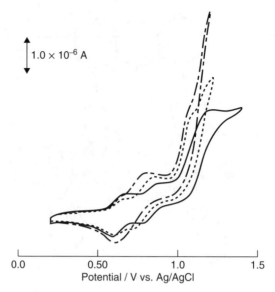

Figure 5.7 CVs of [1Cl(bpy)₂](BF₄)₃ in water at pH 1.0 (–), 2.0 (–––), and 3.0 (–·–·–). Reproduced with permission from [69]. Copyright © 2012 WILEY-VCH Verlag GmbH & Co. KGaA, Weinheim

gives rise to the band at λ_{max} = 688 nm is apparently the precursor to the evolution of O_2. Furthermore, the fact that the [$Ru^{II}(OH_2)(bpy)(tpy)$]$^{2+}$ ion exhibits the MLCT band at λ_{max} = 477 nm [70] strongly suggests that the final product, which gives rise to the MLCT band at λ_{max} = 475 nm, maintains the {$Ru^{II}(bpy)(tpy)$} framework of the [1Cl(bpy)₂]$^{3+}$ ion.

The electrolysis of [1Cl(bpy)₂]$^{3+}$ ions was conducted in $H_2{}^{16}O$ and $H_2{}^{18}O$ at $E = +1.45$ V and the resonance Raman spectra of the solutions of the electrolyte were measured with irradiation at $\lambda = 633$ nm to detect the precursor to the evolution of O_2 (Figure 5.11). Immediately after the electrolysis of [1Cl(bpy)₂]$^{3+}$ ions in $H_2{}^{16}O$, the electrolyte solution displayed two Raman bands at $v = 442$ and 824 cm^{-1}, which shifted to $v = 426$ and 780 cm^{-1}, respectively, under similar measurement conditions in $H_2{}^{18}O$. These bands completely disappeared within 1 hour after stopping the electrolysis. The isotope frequency shifts in both solutions ($\Delta v = 16$ and 44 cm^{-1}) were very consistent with the calculated values of the Ru–O and O–O stretching modes, respectively. The transient Raman bands observed from the electrochemical oxidation of [1Cl(bpy)₂]$^{3+}$ ions also appeared from the chemical oxidation with a Ce^{IV} species. The addition of 10 equivalents of the Ce^{IV} species to an aqueous solution of [1Cl(bpy)₂]$^{3+}$ ions at pH 1 (adjusted with HNO₃) resulted in the disappearance of the band at $\lambda = 468$ nm that arises from the complex. The resultant colorless solution slowly evolved O_2 for 1 hour and became green. The electronic absorption spectrum of the green solution showed a band at $\lambda = 688$ nm, which was the same as that in the electrochemical oxidation of [1Cl(bpy)₂]$^{3+}$ ions at $E = 1.40$ V. The green solution also displayed Raman bands at $v = 442$ and 824 cm^{-1}, which were assignable to the v(Ru–O) and v(O–O) modes so long as the absorption band at $\lambda = 688$ nm remained. The Raman spectrum was completely consistent with that observed for the electrochemical two-electron oxidation of [1Cl(bpy)₂]$^{3+}$ plus a replacement of Cl$^-$ by 2OH$^-$ to form the Ru(III)

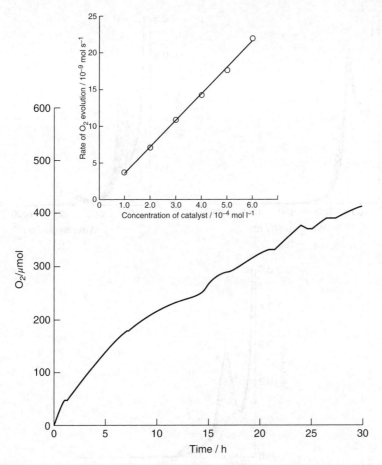

Figure 5.8 *Evolution of oxygen catalyzed by [**1**Cl(bpy)$_2$](BF$_4$)$_3$ (1.0 mmol), using (NH$_4$)$_2$Ce(NO$_3$)$_4$ (2.5 mmol) as an oxidant in water (10 mL) at pH 1.0. Inset: the initial rates of evolution of O$_2$ versus concentrations of the catalyst. Reproduced with permission from [69]. Copyright © 2012 WILEY-VCH Verlag GmbH & Co. KGaA, Weinheim*

complex, $[\mathbf{1}(OH)_2(bpy)_2]^{4+}$, followed by further oxidation to form $[\mathbf{1}(\mu\text{-}O_2)(bpy)_2]^{4+}$ in H$_2$O. The Raman spectra of dinuclear metal complexes with an η^1 or η^2 μ-O$_2$ group have been documented, and the $v(O-O)$ bands of μ-peroxide and μ-superoxide modes are generally observed in the ranges $v = 740-950$ and $1000-1200\,cm^{-1}$, respectively [71]. Lippard and Suzuki isolated μ-peroxide dinuclear Fe complexes that exhibited the $v(O-O)$ band around $v = 900\,cm^{-1}$ [72, 73]. Dinuclear Ru complexes with a μ-O$_2$ group are rare, and $[\{Ru^{IV}(OH)(edta)_2\}_2(\mu\text{-}O_2)]$ (edta = ethylenediaminetetraacetato) is the only known example. It showed a $v(O-O)$ band at $v = 890\,cm^{-1}$, which was assigned as the end-on μ-peroxide unit [74]. The precursor for the evolution of O$_2$ in the present study must be the dinuclear complex with a $\{Ru_2(\mu\text{-}O_2)(bpy)_2(tpy)_2\}$ framework.

A possible mechanism for O$_2$ generation by $[\mathbf{1}Cl(bpy)_2]^{3+}$ is shown in Scheme 5.5. The bulk electrolysis $[\mathbf{1}Cl(bpy)_2]^{3+}$ at $+1.65\,V$ and pH 1 with four-electron oxidation has

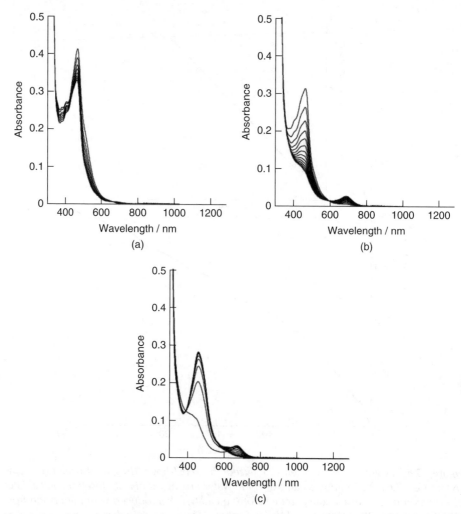

Figure 5.9　*Ultraviolet/visible/near-infrared spectral changes of [1Cl(bpy)$_2$]$^{3+}$ ions in water during electrolysis at E = (a) +0.90 and (b) +1.45 V and (c) after electrolysis has stopped. Reproduced with permission from [69]. Copyright © 2012 WILEY-VCH Verlag GmbH & Co. KGaA, Weinheim*

been demonstrated to produce [1(μ-O$_2$)(bpy)$_2$]$^{4+}$, which slowly reacts with water to generate O$_2$. Taking into account that pentacoordinate Ru^{II}–polypyridyl complexes, such as [Ru^{II}(tpy)(bpy)]$^{2+}$, are not considered to be stable under normal conditions, the stepwise addition of two water molecules to the Ru centers might be carried out prior to the evolution of O$_2$ from [1(μ-O$_2$)(bpy)$_2$]$^{4+}$ [69]. In the absence of an applied potential, the mechanism shown here for the sequential displacement of the bridging peroxide ligand by water molecules is probably correct, but will be slow. There could be (and probably is) another pathway under an applied potential: one that couples an oxidation step to the attack by each water molecule.

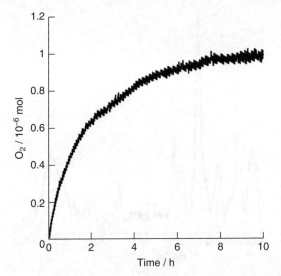

Figure 5.10 *Slow O_2 evolution after electrolysis at +1.65 V. [$1Cl(bpy)_2$](BF_4)$_3$ (1.45 mg, 1.0 μmol) was used as a catalyst of water oxidation. After 2.6 F/mol of electrons (based on the amount of catalyst) were consumed, the electrolysis was stopped. Reproduced with permission from [69]. Copyright © 2012 WILEY-VCH Verlag GmbH & Co. KGaA, Weinheim*

These results represent the first direct evidence for O–O bond formation through the intramolecular coupling of two oxo groups derived from two water molecules activated on two metal ions [69]. Therefore, btpyan is quite suitable as a ligand for stabilization of the Ru–O–O–Ru structure. However, the reaction rate catalyzed by bipyridine [$1Cl(bpy)_2$]$^{3+}$ is slower than that catalyzed by the dioxolene complex [$1(OH)_2(3,6\text{-}Bu_2q)_2$]$^{2+}$ [66]. Further innovations in the acceleration of the elimination of the O_2 molecule from the Ru–O–O–Ru intermediate without high overpotential would improve water oxidation and lead to faster reaction rates.

5.4 Mononuclear Ru–Aqua Complexes with a Dioxolene Ligand

5.4.1 Structural Characterization

We prepared the Ru–aqua complexes [$2(OH_2)(3,5\text{-}Bu_2q)$]($ClO_4$)$_2$ and [$2(OH_2)(4\text{-}Clq)$](ClO_4)$_2$ and solved the X-ray structure of [$2(OH_2)(3,5\text{-}Bu_2q)$]($ClO_4$)$_2$ (Figure 5.12) [75]. The formal electronic state of [$2(OH_2)(3,5\text{-}Bu_2q)$]$^{2+}$ is Ru^{II}(q), Ru^{III}(sq), or Ru^{IV}(cat), as can be seen from class B in Scheme 5.2.

A variety of spectroscopic measurements, such as ultraviolet/visible (UV/vis), electron paramagnetic resonance (EPR), X-ray photoelectron spectra (XPS), infrared (IR), and Raman, along with electrochemical experiments, X-ray crystallography, and theoretical studies can assist in elucidating the electron distribution within a given system. The C–C and C–O distances within the NIL are often employed as a diagnostic tool for determining the NIL's oxidation state. Bhattacharya and coworkers proposed the C–O

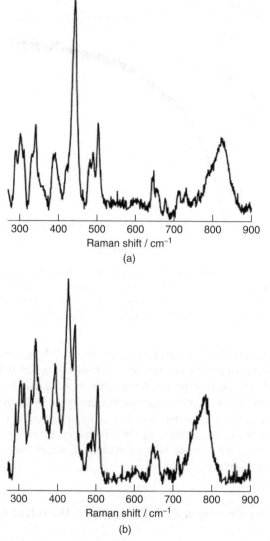

Figure 5.11 *Resonance Raman spectra of the electrochemically oxidized form of the* $[1Cl(bpy)_2]^{3+}$ *ions at E = +1.4 V in (a) $H_2{}^{16}O$ and (b) $H_2{}^{18}O$. Reproduced with permission from [69]. Copyright © 2012 WILEY-VCH Verlag GmbH & Co. KGaA, Weinheim*

and C–C bond lengths (± 0.01 Å) for the three oxidation states of the dioxolene ligands: q 1.22 and 1.48 Å, sq 1.30 and 1.43 Å, and cat 1.34 and 1.42 Å [76]. The C–O bond lengths of ruthenium complexes with 3,5-di-*tert*-butyl-substituted dioxolenes are summarized in Table 5.2. The two C–O bond distances, C1–O1 and C2–O2, of the dioxolene ligands of $[2(OH_2)(3,5\text{-}Bu_2q)]^{2+}$ are 1.293(5) and 1.280(5) Å, respectively. The C–O bond lengths of Ru(sq) complexes are between 1.26 and 1.35 Å (Table 5.2). An electron-acceptor ligand in the *trans* position strongly affects the C–O bond length of the dioxolene. However, the C–O bond distances of $[2(OH_2)(3,5\text{-}Bu_2q)]^{2+}$ are much longer than an average quinone complex (1.22 Å) [76] and are slightly shorter than an average semiquinone complex (1.30 Å) [76]. This complex was originally assigned as a Ru^{III}(sq) complex based on the

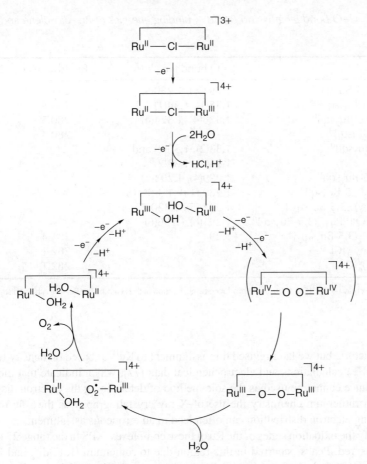

Scheme 5.5 *Possible mechanism for water oxidation catalyzed by [1Cl(bpy)₂]³⁺ in water. Reproduced with permission from [69]. Copyright © 2012 WILEY-VCH Verlag GmbH & Co. KGaA, Weinheim*

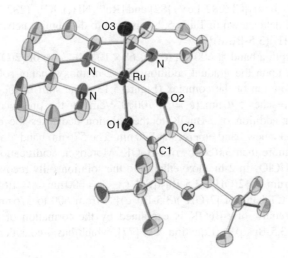

Figure 5.12 *Crystal structure of [2(OH₂)(3,5-Bu₂q)]²⁺*

Table 5.2   *C–O bond lengths and Ru $3d_{5/2}$ binding energies of Ru–dioxolene and other Ru complexes*

Complex	C–O bond length (Å)	Ru $3d_{5/2}$ (eV)	Reference
$[2(OH_2)(3,5\text{-}Bu_2q)]^{2+}$	1.293(5), 1.280(5)	281.5	[75]
$[2(O^{\bullet-})(3,5\text{-}Bu_2sq)]^{0}$ [a]	1.35(1), 1.34(1)	280.4	[75]
$[2(OAc)(3,5\text{-}Bu_2sq)]^{0}$	1.328(4), 1.324(4)	280.5	[75]
$[2(OAc)(4\text{-}Clsq)]^{0}$	–	280.4	[75]
$[2Cl(3,5\text{-}Bu_2sq)]^{0}$	1.33(2), 1.30(2) and 1.26(2), 1.33(2)[b]	–	[78]
$[2(CO)(3,5\text{-}Bu_2sq)]^{0}$	1.299(4), 1.293(4)	–	[65]
$[Ru^{II}(bpy)_2(3,5\text{-}Bu_2sq)]^{+}$	1.289 (14), 1.327 (15)	–	[62]
$[Ru^{II}(t\text{-}Bupy)_2(3,5\text{-}Bu_2sq)_2]$	1.322 (5), 1.320 (5)	–	[62]
$[Ru^{II}(Cl)(CO)(PPh_3)_2(3,5\text{-}Bu_2sq)]^{0}$	1.291(6), 1.296(6)	–	[79]
$[Ru^{III}(Clpy)_2(3,5\text{-}Bu_2sq)_2]^{+}$	1.29	281.4	[80, 81]
$[Ru^{III}Cl_2(bpy)_2]Cl$	–	282.1	[82]
$[Ru^{III}(NH_3)_6]Cl_3$	–	282.1	[83]

[a] This complex may be $[2(OH)(3,5\text{-}Bu_2sq)]^{0}$, produced by abstraction of the H atom. The Ru–(O$^{\bullet-}$) bond distance is rather long (2.043(7) Å)
[b] Two isomers

X-ray structure, but we later revised this assignment to Ru^{II}(q) based on density functional theory (DFT) calculations and electrochemical data [77], which indicate that the C–C or C–O distance cannot be used as the sole method of determining the electron distribution. In fact, in ruthenium chemistry the use of X-ray crystallography as the sole method of determining electron distribution can often lead to an erroneous assignment.

To clarify the oxidation states of the Ru of these complexes, XPS in the range 274–296 eV were measured. Peaks occurred in this region due to ruthenium (Ru $3d_{3/2}$ and Ru $3d_{5/2}$) and carbon (C 1s) electron transitions. Binding energies for the Ru $3d_{5/2}$ peaks of the three complexes are listed in Table 5.2. The complex $[2(OH_2)(3,5\text{-}Bu_2q)]^{2+}$ showed an Ru $3d_{5/2}$ peak at 281.5 eV: significantly larger than those of Ru^{II}(sq) complexes and smaller than those of $[Ru^{III}Cl_2(bpy)_2]Cl$ (282.1 eV) [82] and $[Ru^{III}(NH_3)_6]Cl_3$ (282.1 eV) [83]. On the basis of the XPS data shown in Table 5.2, it is hard to distinguish between Ru(III)sq and Ru(II)q for $[2(OH_2)(3,5\text{-}Bu_2q)]^{2+}$.

A strong absorption band at 600 nm ($\epsilon = 1.68 \times 10^4$ M^{-1}/cm) of $[2(OH_2)(3,5\text{-}Bu_2q)]^{2+}$ shifts to 576 nm upon the gradual addition of a 2-methoxyethanol solution of t-BuOK to the CH$_2$Cl$_2$ solution of the complex (Figure 5.13). The shift of the absorbance band around 600 nm reaches 576 nm ($\epsilon = 16\,700$ M^{-1}/cm) in the presence of 1.0 equiv. of t-BuOK. Further addition of t-BuOK to the solution decreases the absorption of the 576 nm band, and a new band emerges at 870 nm. The 576 nm band almost disappears in the presence of more than 3.0 equiv. of t-BuOK. Moreover, acidification by the addition of 3.0 equiv. of HClO$_4$ in 2-methoxyethanol to the solution fully recovers the electronic absorption spectrum of $[2(OH_2)(3,5\text{-}Bu_2q)]^{2+}$ ($\lambda_{max} = 600$ nm, $\epsilon = 1.68 \times 10^4$ M^{-1}/cm). The shift of the CT band of $[2(OH_2)(3,5\text{-}Bu_2q)]^{2+}$ from 600 to 576 nm in the presence of less than 1.0 equiv. of t-BuOK is explained by the formation of the deprotonated product $[2(OH)(3,5\text{-}Bu_2q)]^{+}$ (Equation 5.1) [77], which has a considerable contribution

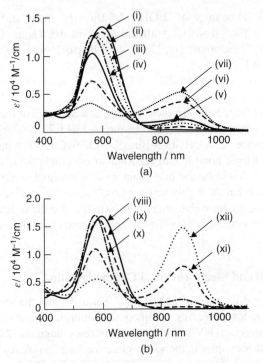

Figure 5.13 *(a) pH-dependent electronic absorption spectra of [2(OH₂)(3,5-Bu₂q)]²⁺: (i) pH 3.2; (ii) pH 4.5; (iii) pH 5.6; (iv) pH 7.1; (v) pH 10.1; (vi) pH 11.0; (vii) pH 12.0. (b) Electronic absorption spectra of [2(OH₂)(3,5-Bu₂q)]²⁺ in the presence of various amounts of t-BuOK in CH₂Cl₂: (viii) 0 equiv.; (ix) 0.5 equiv.; (x) 1.0 equiv.; (xi) 2.0 equiv.; (xii) 3.0 equiv. Reprinted with permission from [75]. Copyright © 2003, American Chemical Society*

from $[2(OH)(3,5-Bu_2sq)]^+$, as we describe in Section 5.4.2. The appearance of the 870 nm band in the treatment of $[2(OH_2)(3,5-Bu_2q)]^{2+}$ with more than 2.0 equiv. of t-BuOK is an indication of the occurrence of the one-electron reduction of the $Ru^{II}(q)$ core of $[2(OH_2)(3,5-Bu_2q)]^{2+}$ presumably affording $[Ru^{II}(tpy)(O^{\bullet-})(3,5-Bu_2sq)]^0$.

$$[Ru^{II}(tpy)(OH_2)(3,5-Bu_2q)]^{2+} \underset{+H^+}{\overset{-H^+}{\rightleftharpoons}} [Ru^{II}(tpy)(OH)(3,5-Bu_2q)]^+$$
$$[2(OH_2)(3,5-Bu_2q)]^{2+}$$

$$\underset{+H^+}{\overset{-H^+}{\rightleftharpoons}} [Ru^{II}(tpy)(O^{\bullet-})(3,5-Bu_2q)]^0$$

$$\overset{+H^{\bullet}}{\longrightarrow} [Ru^{II}(tpy)(OH)(3,5-Bu_2sq)]^0 \qquad (5.1)$$

The aqua complex $[2(OH_2)(3,5-Bu_2q)]^{2+}$ is soluble in H_2O and the pK_a value of the aqua ligand has been determined as 5.5 by means of pH titration and changes of the electronic absorption spectra in H_2O at 25 °C. We observed the occurrence of the second deprotonation of $[2(OH_2)(3,5-Bu_2q)]^{2+}$ at a pH higher than 10, but reddish-purple solids precipitated out of the aqueous solution in that pH region. Reddish-purple crystals of presumably $[2(O^{\bullet-})(3,5-Bu_2sq)]^0$ were obtained by slow evaporation of CH_3OH from a

CH$_3$OH/H$_2$O (1 : 2 v/v) solution of [**2**(OH$_2$)(3,5-Bu$_2$q)]$^{2+}$ under strong basic conditions. Treatment of [**2**(O$^{\bullet-}$)(3,5-Bu$_2$sq)]^0 with 2.0 equiv. of HClO$_4$ in CH$_2$Cl$_2$ regenerated [**2**(OH$_2$)(3,5-Bu$_2$q)]$^{2+}$ quantitatively. [**2**(O$^{\bullet-}$)(3,5-Bu$_2$sq)]^0 was isolated, and its crystal structure is shown in Figure 5.14.

The two C–O bond distances of [**2**(O$^{\bullet-}$)(3,5-Bu$_2$sq)]^0 are 1.35(1) and 1.34(1) Å, which are apparently longer than those of [**2**(OH$_2$)(3,5-Bu$_2$q)]$^{2+}$ (Figure 5.12) and equivalent to Ru^{II}(sq) complexes (Table 5.2). The most striking characteristic of the molecular structure of [**2**(O$^{\bullet-}$)(3,5-Bu$_2$q)]^0 is the Ru–O bond length of 2.043(7) Å, which is shorter than the Ru–OH$_2$ bond distance of [**2**(OH$_2$)(3,5-Bu$_2$q)]$^{2+}$ (2.099(3) Å) but much longer than that of any Ru–OH or Ru=O bond length reported so far (Table 5.3). The neutral complex [**2**(O$^{\bullet-}$)(3,5-Bu$_2$sq)]^0 might be the first example of a terminal metal–O complex with a single-bond character, but the highly reactive [**2**(O$^{\bullet-}$)(3,5-Bu$_2$sq)]^0 species might be plausible due to the strong acceptor character of the NIL ligand in the ruthenium oxo species; in fact, some conflicting evidence comes from electrochemical and theoretical investigations, as discussed later.

5.4.2   Theoretical and Electrochemical Characterization

More recently, the redox states of [**2**(OH$_2$)(3,5-Bu$_2$q)]$^{2+}$ and the species resulting from deprotonation and redox processes in aqueous solution have been investigated through experimental and theoretical UV/vis spectra and Pourbaix diagrams [77]. The formal oxidation states of the redox couples in the various intermediate complexes were systematically assigned using electronic structure theory. [**2**(OH$_2$)(3,5-Bu$_2$Q)]$^{2+}$ is definitely a closed-shell singlet (i.e. [Ru^{II}(OH$_2$)(Q)]$^{2+}$), based on the calculated energies in the second column of Table 5.4. While the broken-symmetry, antiferromagnetically coupled open-shell singlet

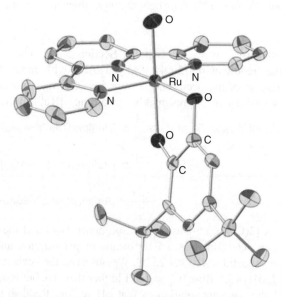

Figure 5.14   *Crystal structure of [2(O$^{\bullet-}$)(3,5-Bu$_2$sq)]^0*

Table 5.3 *Ru–O bond lengths of Ru–aqua, –hydroxo, and –oxo complexes. S and T indicate singlet and triplet species*

Complex	Observed Ru–O, Å	Calculated Ru–O, Å	Reference
$[2(OH_2)(3,5\text{-}Bu_2q)]^{2+}$	2.099(3)	S 2.148, T 2.146	[75, 77]
$[2(OH)(3,5\text{-}Bu_2q)]^+$	–	S 1.977, T 1.956	
$[2(O^{\bullet-})(3,5\text{-}Bu_2sq)]^0$	2.043(7)	S 1.802, T 1.824	[75, 77]
$[2(OH)(3,5\text{-}Bu_2sq)]^0$	–	2.018	[77]
$[Ru^{III}(OH)(Metacn)(acac)]^+$	1.971(9)		[84]
$[Ru^{IV}(O)(bpy)(DAMP)]^{2+}$	1.805(3)		[85]
$[Ru^{IV}(O)Cl(py)_4]^+$	1.862(8)		[86]
$[Ru^{IV}(O)(NCO)(TMC)]^+$	1.765(5)		[87]
$[Ru(O)(H^+TPA)(bpy)]^{3+}$	1.771(4)		[88]

Metacn = 1,4,7-trimethyl-1,4,7-triazacyclononane, DAMP = 2,6-bis((dimethylamino)methyl)pyridine, TMC = 1,4,8,11-tetramethyl-1,4,8,11-tetraazacyclotetradecane, TPA = tris(2-pyridylmethyl)amine

Table 5.4 *Calculated energies of the (sing, trip, and bs/afc) species formed by sequential deprotonation of $[2(OH_2)(3,5\text{-}Me_2q)]^{2+}$ at the B3LYP/LANL2DZ level of theory. The geometries of the singlet and triplet states were optimized and the bs/afc energy was obtained by a spin flip at the geometry of the triplet state. Reprinted with permission from [77]. Copyright © 2009, American Chemical Society*

Property	$[2(OH_2)(3,5\text{-}Me_2q)]^{2+}$	$[2(OH)(3,5\text{-}Me_2q)]^+$	$[2(O)(3,5\text{-}Me_2q)]^0$
E(sing)	−1608.187459	−1607.873216	−1607.407810
E(trip)	−1608.174151	−1607.876280	−1607.422998
E(bs/afc)	−1608.179669	−1607.877738	−1607.409370
E(trip)−E(sing)	8.35 kcal/mol	−1.92 kcal/mol	−9.53 kcal/mol
E(bs/afc)−E(sing)	4.89 kcal/mol	−2.84 kcal/mol	−0.98 kcal/mol

Note: Reproduced from [77] with permission of the copyright holder

$[Ru^{III}(OH)(SQ)]^+$ is slightly more stable (Table 5.4, third column), we believe the hydroxy species is probably (within the error of the calculation) $[Ru^{II}(OH)(Q)]^+$, with a considerable contribution from $[Ru^{III}(OH)(SQ)]^+$. The oxo species is the clearly the triplet oxyl radical, $[Ru^{II}(O^{\bullet-})(SQ)]^0$ (Table 5.4, fourth column), with a significant contribution from the intermediate-spin triplet $[Ru^{IV}(=O)(Cat)]^0$, as can be seen in Figure 5.15.

The bonding in these complexes can be described in terms of either single electronic configurations with pure versus "broken symmetry" spin states [77, 89, 90] or pure spin states involving multiple electronic configurations [51]. In the particular case of $[Ru(OH_2)(q)]^{2+}$ complexes, the $O(2p_\pi)$-orbitals, the Ru d_π-orbitals, and the $q(\pi^*)$-orbitals mix to form a set of doubly occupied π and unoccupied π^*-molecular orbitals. The multiconfigurational description of the bonding involves a wave function with electronic configurations corresponding to the principle (closed-shell) configuration, with additional configurations in which a pair of electrons in one of the set of bonding π-orbitals is promoted to the corresponding π^*-molecular orbital. The contribution of these additional configurations is

Figure 5.15 *Calculated structure of the doubly deprotonated species formed from* $[2(OH_2)(3,5\text{-}Me_2q)]^{2+}$. *There is excess alpha spin density (blue) on the Ru atom, the O, and the q ligands, confirming the assignment of the resonance between the* $^3[Ru^{II}(O^{\bullet-})(sq)]^0$ *and the intermediate-spin* $^3[Ru^{IV}(=O)(cat)]^0$ *species. See color plate*

determined variationally by the multiconfiguration self-consistent field (MCSCF) method. This restricted type of multiconfigurational approach, in which only pairs of electrons are shared between corresponding pairs of bonding and antibonding orbitals, is known as the generalized valence bond configuration interaction (GVB CI) method, which we will discuss again in Section 5.5 [51, 68]. The essential point is that the oxidation-state labels applied to these complexes with NILs are often not precise. For example, in a $Ru^{II}(Q)$ complex with strong mixing of a Ru d_π- and a $Q(\pi^*)$-orbital to form a doubly occupied π- and vacant π^*- pair of molecular orbitals, a pair of electrons in the π-orbital might be shared with the corresponding vacant π^*-orbital, thereby mixing $Ru^{II}(Q)$ and $Ru^{IV}(Cat)$ configurations (in fact, the principal configuration would also mix them). The result of this GVB CI might be labeled an antiferromagnetically coupled open-shell singlet $Ru^{III}(SQ)$ complex, even though there are no unpaired electrons.

TD-B3LYP calculated $[Ru^{II}(OH_2)(q)]^{2+}$, $[Ru^{II}(OH)(q)]^+$ (singlet), and $[Ru^{III}(OH)(sq)]^+$ (triplet) spectra are shown in Figure 5.16 [77]. While both calculated spectra of the singlet and triplet hydroxy complexes have absorption bands around 570 nm, it seems that the spectrum of the singlet species matches better with the observed spectrum in terms of the absorption intensity. The calculated triplet $[Ru^{III}(OH)(sq)]^+$ spectrum exhibits a very weak absorption around 700 nm. TD-B3LYP calculations for triplet $[Ru^{II}(O^{\bullet-})(sq)]^0$ indicate a near-infrared (NIR) absorption band. However, base titration experiments to measure absorption bands in the NIR region did not show any absorption, as predicted by TD-B3LYP. Our TD-B3LYP calculations of the electronic spectra of $Ru^{II}(O^{\bullet-})(sq)^0$ and $Ru^{II}(OH)(sq)^0$ indicate a much better match between the latter and the observed spectrum (positions and relative intensities of experimental peaks indicated by black bars), indicating that the $Ru^{II}(O^{\bullet-})(sq)^0$ species probably abstracts an H atom from somewhere (possibly trifluoroethanol, as shown in Figure 5.17) to form doublet $Ru^{II}(OH)(sq)^0$.

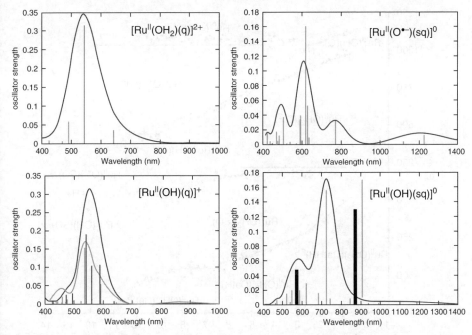

Figure 5.16 *Calculated spectra of [Ru(OH₂)(q)]²⁺, [Ru(OH)(q)]⁺, [Ru(O⁻⁻)(sq)]⁰, and [Ru(OH)(sq)]⁰. The vertical lines are the discrete transitions from a TD-B3LYP/LANL2DZ calculation and the smooth curves are the result of broadening with a Gaussian function. For the [Ru(OH)(q)]⁺ species, the spectra of singlet [Ruᴵᴵ(OH)(q)]⁺ (dark) and triplet [Ruᴵᴵᴵ(OH)(sq)]⁺ (light) species are shown. For [Ru(OH)(sq)]⁰, the black vertical bars indicate the (unnormalized) peaks of the experimental spectrum [91]. See color plate*

The electrochemical measurements for $[2(OH_2)(3,5\text{-}Bu_2q)]^{2+}$, carried out using only one direction (from the rest potential to positive or negative voltage) in a trifluoroethanol/H_2O mixture (v/v 0.02), show a pH-dependent behavior. An experimental Pourbaix diagram is shown in Figure 5.17, along with $E_{1/2}$ lines and oxidation-state and spin-state assignments either corroborated or predicted by the theoretical results [77]. If the $[Ru^{II}(O^{\bullet-})(sq)]^0$ species is stable on the CV timescale, we should see pH-independent oxidation of $[Ru^{II}(O^{\bullet-})(sq)]^0$ to $[Ru^{II}(O^{\bullet-})(q)]^+$ in the Pourbax diagram. Instead we observe a slope of $-59\,mV$ with increasing pH, indicating that the oxidation is coupled with a proton removal. If the reactive $[Ru^{II}(O^{\bullet-})(sq)]^0$ species abstracts an H atom from trifluoroethanol (added to increase the solubility of the neutral species) and forms $[Ru^{II}(OH)(sq)]^0$, the Pourbax diagram can be explained with various species, including the reduced species, as shown in Figure 5.17. We should mention that the pH-dependent points separating $^3[Ru^{II}(O^{\bullet-})(sq)]^0$ and $^2[Ru^{II}(O^{\bullet-})(q)]^+$ in the Pourbaix diagram can be explained by the 1e⁻ oxidation of $^2[Ru^{II}(OH)(sq)]^0$ to give $^1[Ru^{II}(OH)(q)]^+$, followed by the proton-coupled electron transfer (PCET) oxidation of $^1[Ru^{II}(OH)(q)]^+$ to give $^2[Ru^{II}(O^{\bullet-})(q)]^+$ (alternatively assigned as $^2[Ru^{III}(O^{\bullet-})(sq)]^+$ or $^2[Ru^{IV}(=O)(sq)]^+$). Also, the observed spectrum of this "doubly deptonated species" formed from base titration matches well with the calculated spectrum of the $^2[Ru^{II}(OH)(sq)]^0$ species. Furthermore, the calculated structure shows an

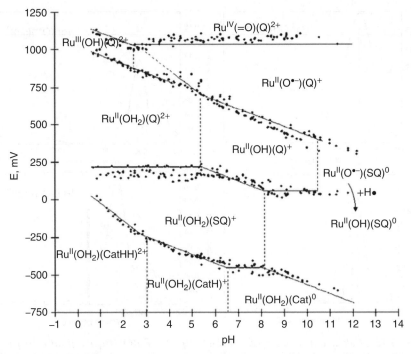

Figure 5.17   *Experimental Pourbaix diagram of [2(OH$_2$)(3,5-Bu$_2$q)]$^{2+}$, denoted as Ru^{II}(OH$_2$)(Q)$^{2+}$. $E_{1/2}$ is relative to the SCE. CatH and CatHH indicate that one and two oxygen atoms of Cat are protonated, respectively [77]*

extremely short Ru–(O$^{\bullet-}$) bond of 1.824 Å, as compared to that of the X-ray structure (2.043 ± 0.007 Å). This difference is beyond the error in the theoretical calculations.

Several questions now arise: What is the species shown in Figure 5.14? Is this a structure of [2(O$^{\bullet-}$)(3,5-Bu$_2$sq)]^0 or [2(OH)(3,5-Bu$_2$sq)]^0? As mentioned earlier, the observed Ru–(O$^{\bullet-}$) bond length (2.043 ± 0.007 Å) is considerably longer that the calculated distance of 1.824 Å. The hydrogen atom attached to O may be difficult to locate, especially if it is disordered. The crystal was obtained by slow evaporation of CH$_3$OH from a CH$_3$OH/H$_2$O (1 : 2 v/v) solution of [2(OH$_2$)(3,5-Bu$_2$q)]$^{2+}$ under strong basic conditions. Is the reactivity of [2(O$^{\bullet-}$)(3,5-Bu$_2$sq)]^0 different between trifluoroethanol and methanol? Does the EPR signal of the species provide a clue to its identification?

While the complex [2(OH$_2$)(3,5-Bu$_2$q)]$^{2+}$ did not show any EPR signals in CH$_2$Cl$_2$, an addition of more than 3.0 equiv. of *t*-BuOK to the solution resulted in the appearance of an isotropic broad signal (g= 2.030, ΔH_{msl} = 8.0 mT) without a hyperfine structure at 193 K. When a spin trapping experiment using 5,5-dimethl-1-pyrroline N-oxide (DMPO) was carried out for the detection and identification of the oxyl radical moiety of [Ru^{II}(O$^{\bullet-}$)(sq)]^0 ([2(O$^{\bullet-}$)(3,5-Bu$_2$sq)]^0), sharp 12-line signals centered at g = 2.006 were observed, indicating the formation of [Ru^{II}(O-DMPO)(3,5-Bu$_2$sq)(tpy)]^0 [75]. Using an ESI-MS, [Ru^{II}(^{16}O-DMPO)(3,5-Bu$_2$sq)(tpy)]^0 and [Ru^{II}(^{18}O-DMPO)(3,5-Bu$_2$sq)(tpy)]^0 were observed upon the addition of 3.0 equiv. of *t*-BuOK and 100 equiv. of DMPO into CH$_2$Cl$_2$ solution of

$[2(OH_2)(3,5-Bu_2q)]^{2+}$ containing a small amount of $H_2^{16}O$ and $H_2^{18}O$, respectively [75]. Furthermore, the EPR spectrum of a mixture of $[2(OH_2)(3,5-Bu_2q)]^{2+}$ and 3.0 equiv. of *t*-BuOK at 3.9 K exhibited an isotropic broad signal with a hyperfine structure in the Δm_s = 1 region and an isotropic signal at $\Delta m_s = 2$, indicating the triplet state of a biradical compound [75]. These EPR and mass spectrometer results appear to confirm the formation of biradical species in CH_2Cl_2 with more than 3 equiv. of *t*-BuOK. Unfortunately, these experiments were carried out using CH_2Cl_2 without any CH_3OH or CF_3CH_2OH. Therefore, the true nature of this species in CF_3CH_2OH is still debatable.

The mononuclear Ru–aqua complex $[Ru^{II}(OH_2)(3,5-Bu_2q)(tpy)]^{2+}$ (i.e. $[2(OH_2)(3,5-Bu_2q)]^{2+}$) reversibly dissociates protons upon base titration. $[Ru^{II}(OH_2)(tpy)(bpy)]^{2+}$ also dissociates protons in the presence of an applied potential near the rest potential [70, 75]. In the case of $[Ru^{II}(OH_2)(tpy)(bpy)]^{2+}$, an oxidation is coupled with the proton removal to form $[Ru^{III}(OH)(tpy)(bpy)]^{2+}$ and $[Ru^{IV}(O)(tpy)(bpy)]^{2+}$ [70], but in the case of $[2(OH_2)(3,5-Bu_2q)]^{2+}$ only proton removal takes place, meaning the total charge changes from 2+ to + to 0. When $[Ru^{IV}(O)(tpy)(bpy)]^{2+}$ is further oxidized to form the Ru(V) oxo species, a nucleophilic attack by water on the oxo O takes place to form an O–O bond [92, 93]. Ru complexes with an NIL not only provide Ru–oxyl radical species without becoming highly valent, such as Ru(IV) or Ru(V), due to the NIL acting as an electron reservoir, but also stabilize oxyl and possibly hydroxyl radicals of the complexes. So far, we have not found evidence that $[2(OH_2)(3,5-Bu_2q)]^{2+}$ acts as a catalyst for water oxidation, however dinuclear Ru–aqua complexes with dioxolene ligands are quite advantageous for the formation of an O–O bond via the intramolecular coupling of two Ru–oxyl radical species.

5.5 Mechanistic Investigation of Water Oxidation by Dinuclear Ru Complexes with NILs: Characterization of Key Intermediates

Electronic absorption spectra of the oxidized and reduced forms of the complex, $[1(OH)_2(3,6-Bu_2q)_2]^{2+}$ and $[1(OH)_2(3,6-Bu_2sq)_2]^0$, display strong bands at 576 and 870 nm, assignable to the MLCT band of the Ru–dioxolene (Figures 5.18 and 5.19). We assigned the formal electronic state of $[1(OH)_2(3,6-Bu_2q)_2]^{2+}$ and $[1(OH)_2(3,6-Bu_2sq)_2]^0$ as $[Ru^{II}(q), Ru^{II}(q)]^{2+}$ and $[Ru^{II}(sq), Ru^{II}(sq)]^0$ respectively [66] on the basis of the assignment of absorption spectra of Ru–dioxolene complexes reported previously [58, 59, 81], with the caveat that contributions from $Ru^{III}(sq)$ and $Ru^{IV}(cat)$ go towards the actual electronic states of these complexes, as was the case for the mononuclear complex $[2(OH)(3,5-Bu_2q)]^+$.

The complexes $[1(OH)_2(3,5-Cl_2sq)_2]^0$ and $[1(OH)_2(4-NO_2sq)_2]^0$, having dioxolene ligands substituted with electron-withdrawing groups, 3,5-Cl_2 and 4-NO_2, show MLCT bands at 884 and 828 nm, respectively (Figure 5.19). Therefore, the electronic states of $[1(OH)_2(3,5-Cl_2sq)_2]^0$ and $[[1(OH)_2(4-NO_2sq)_2]^0$ are also assigned as $[Ru^{II}(sq), Ru^{II}(sq)]^0$. A decrease in the absorption coefficients of the bands at around 800–900 nm with an increase in electron-withdrawing ability of dioxolene substituents implies a gradual decrease of orbital mixing between the $d\pi(Ru)$ and $\pi^*(sq)$ of the complexes [64]. Introduction of electron-withdrawing groups into dioxolene therefore shifts the resonance equilibrium from the $Ru^{II}(sq)$ framework to the $Ru^{III}(cat)$ one, and the contribution of the latter increases in the order $[1(OH)_2(3,6-Bu_2sq)_2]^0 < [1(OH)_2(3,5-Cl_2sq)_2]^0 < [1(OH)_2(4-NO_2sq)_2]^0$ [68].

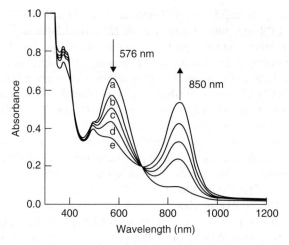

Figure 5.18 *Electronic absorption spectra of [1(OH)₂(3,6-Bu₂q)₂](SbF₆)₂ in MeOH and in the presence of various amounts of t-BuOK: (i) 0 equiv.; (ii) 0.5 equiv.; (iii) 1.0 equiv.; (iv) 1.5 equiv.; (v) 2.0 equiv. Reproduced with permission from [66], The Royal Society of Chemistry*

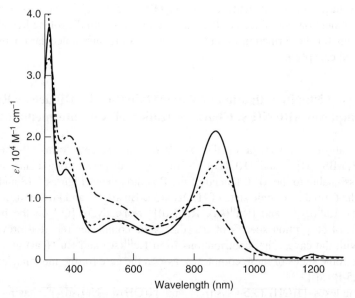

Figure 5.19 *UV/vis/NIR spectra of [1(OH)₂(3,6-Bu₂sq)₂]⁰ (–), [1(OH)₂(3,5-Cl₂sq)₂]⁰ (− − −), and [1(OH)₂(4-NO₂sq)₂]⁰ (−·−·−) in CH₂Cl₂ [68]. Reproduced with permission from [68], The Royal Society of Chemistry*

Addition of *t*-BuOK to the solution of $[1(OH)_2(3,6-Bu_2q)_2]^{2+}$ resulted in the appearance of a new band at 850 nm, which gradually increased with an increase in the amount of *t*-BuOK at the expense of the band at 576 nm (Figure 5.18). The 576 nm band completely disappeared in the presence of 2.0 equiv. of *t*-BuOK, and further addition of *t*-BuOK did not cause any spectral changes. Moreover, acidification of the resultant purple solution by the addition of 2.0 equiv. of $HClO_4$ resulted in the complete restoration of the 576 nm. band and disappearance of the 850 nm band. The 850 nm band is not completely consistent with but is close to the MLCT band of the reduced form, $[1(OH)_2(3,6-Bu_2sq)_2]^0$. Therefore, the 850 nm band is reasonably assigned to the MLCT bands of the $Ru^{II}(sq)$ moieties. Such a reversible change of the electronic absorption spectra is attributable to the acid–base equilibria between the mononuclear aqua complex $[2(OH_2)(3,5-Bu_2q)]^{2+}$, the hydroxo complex $[2(OH)(3,5-Bu_2q)]^+$, and the corresponding oxyl radical complex, $[2(O^{\cdot-})(3,5-Bu_2sq)]^0$ (Equation 5.1). Thus, deprotonation of the hydroxo groups is coupled with the redox reaction of the quinones. We had initially expected that the O–O bond formation would prove to occur in the deprotonated form of the dinuclear complex $[1(OH)_2(3,6-Bu_2sq)_2]^0$ due to production of the Ru–oxyl radical complex $[2(O^{\cdot-})(3,5-Bu_2sq)]^0$ from the Ru–aqua complex $[2(OH_2)(3,5-Bu_2sq)]^{2+}$. However, we did not succeed in observing the O–O bond vibration band using resonance Raman spectral measurement (vide infra).

On the other hand, the 870 nm band of the reduced form, $[1(OH)_2(3,6-Bu_2sq)_2]^0$, did not change at all upon addition of 2 equiv. of *t*-BuOLi in CH_2Cl_2. The addition of *t*-BuOLi to a CH_2Cl_2 solution of $[1(OH)_2(3,5-Cl_2sq)_2]^0$ caused a decrease in the absorbance of the 882 nm band, and two isosbestic points appeared at 738 and 443 nm in a range of <1.0 equiv. of *t*-BuOLi (Figure 5.20).

The absorbance of the 882 nm band further decreased upon the addition of >1.0 equiv. of *t*-BuOLi, and new isosbestic points emerged at 705 and 485 nm. The spectral changes stopped after 2.0 equiv. of *t*-BuOLi had been added. Acidification of the resultant solution

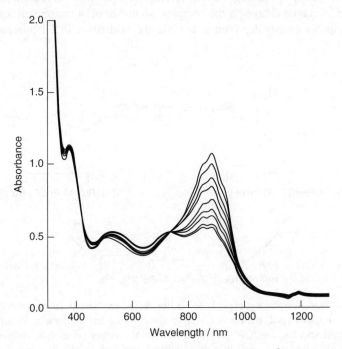

Figure 5.20 *Electronic absorption spectra of $[1(OH)_2(3,5-Cl_2sq)_2]^0$ in MeOH with the presence of various amounts of t-BuOLi: 0.00, 0.25, 0.50, 0.75, 1.00, 1.25, 1.50, 1.75, and 2.00 equiv [68]. Reproduced with permission from [68], The Royal Society of Chemistry*

by addition of 2.0 equiv. of $HClO_4$ resulted in complete recovery of the 882 nm band, indicating reversible dissociation of the two-hydroxyl protons of $[1(OH)_2(3,5-Cl_2sq)_2]^0$. The 4-nitro complex, $[1(OH)_2(4-NO_2sq)_2]^0$, also displayed two-stage changes of the electronic absorption spectra in a range from 0 to 1 and from 1 to 2 equiv. of added t-BuOLi, as well as complete regeneration of the electronic spectrum of $[1(OH)_2(4-NO_2sq)_2]^0$ after neutralization by $HClO_4$. The remarkable difference between the electronic spectra of $[1(OH)_2(3,6-Bu_2sq)_2]^0$, $[1(OH)_2(3,5-Cl_2sq)_2]^0$, and $[1(OH)_2(4-NO_2sq)_2]^0$ in the cyclic base–acid treatments shows that the two hydroxo groups of $[1(OH)_2(3,6-Bu_2sq)_2]^0$ are nondissociable, while $[1(OH)_2(3,5-Cl_2sq)_2]^0$ and $[1(OH)_2(4-NO_2sq)_2]^0$ reversibly liberate two hydroxo protons in the presence of strong base. The dissociation of a hydroxo proton from $Ru^{II}(OH)$ and $Ru^{III}(OH)$ complexes inevitably brings about an accumulation of negative charges on the oxygen atom. The fully occupied t_{2g}-orbitals of Ru^{II} are not able to accept these π-character electrons [94]. Proton dissociation from $Ru^{II}(OH)$ complexes therefore hardly takes place unless the negative charges generated on the oxygen atom are effectively transferred to another ligand. On the other hand, the incompletely occupied t_{2g}-orbitals of Ru^{III} can afford to receive an electron and will serve to prevent an accumulation of too much negative charge on the oxygen atom of the resultant Ru–oxo bond.

A proposed mechanism for the water oxidation catalyzed by $[1(OH)_2(3,6-Bu_2q)_2]^{2+}$ modified on ITO in water is depicted in Scheme 5.6. Even though the protons of the hydroxo groups of the $[1(OH)_2(3,6-Bu_2q)_2]^{2+}$ (and certainly of the $[1(OH)_2(3,6-Bu_2q)_2]^0$) complex have such high pK_as that it requires a very strong base to remove them, deprotonation can be coupled to the oxidation of the complex. Our calculations indicate that while the pK_a of $[1(OH)_2(3,6-Bu_2q)_2]^{2+}$ is large, that of $[1(OH)_2(3,6-Bu_2q)_2]^{3+}$ is considerably smaller, owing to the increased charge on the complex, so that at pH 4 the one-electron-oxidized species will spontaneously deprotonate to make the oxidation a PCET process. The same

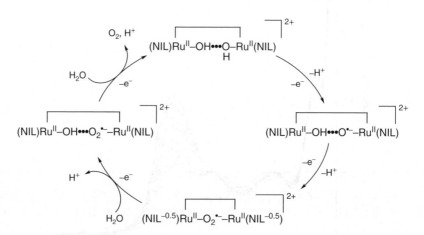

Scheme 5.6 *Proposed catalytic cycle for water oxidation by the Tanaka catalyst in aqueous solution at pH 4. NIL denotes 3,6-Bu$_2$q Because of the molecular symmetry of the μ-O–O-bridged species, we have indicated fractional charges on its NIL. Reproduced from [51] with permission of the copyright holder. Reproduced with permission from [51]. Copyright © Elsevier, 2010*

is true even for the second oxidation. The $[1(OH)(O)(3,6-Bu_2q)_2]^{3+}$ pK_a is lower than that of $[1(OH)(O)(3,6-Bu_2q)_2]^{2+}$, so that it spontaneously ejects a proton in a PCET process. In fact, there is persuasive evidence [68, 89] that the first two oxidation steps correspond to a two-electron, two-proton process.

Although $[1(O^{\bullet-})_2(3,6-Bu_2sq)_2]^{2+}$ must have an enormous Coulomb repulsion between the two negatively charged oxyl radicals, electron transfer from an anionic radical to quinone, affording $[qRu^{II}-(O^{\bullet-})\cdots(O)-Ru^{II}sq]^{2+}$ (for clarity, the btpyan ligand is not shown), will effectively relax much of the electrostatic strain and will facilitate O–O bond formation through the coupling reaction between anionic and neutral oxyl radicals. Once the O–O bond is formed in $[qRu^{II}-(\mu-O_2^{\bullet-})-Ru^{II}sq]^{2+}$ (or $[qRu^{II}-(\mu-O_2^{2-})-Ru^{II}q]^{2+}$ as a resonance isomer), the oxidation of the complex will induce a nucleophilic attack of H_2O on an Ru atom and cleave one of the Ru–O_2–Ru bonds. Further oxidation of the complex causes a second nucleophilic attack of H_2O on Ru, which leads to O_2 evolution and the regeneration of $[1(OH)_2(3,6-Bu_2q)_2]^{2+}$. In accordance with this, the catalytic currents for O_2 evolution begin to flow at potentials slightly more positive than the Ru^{II}-localized oxidation wave at $E_{pa} = +1.39$ V in the CV of $[1(OH)_2(3,6-Bu_2q)_2]^{2+}$ in CF_3CH_2OH/ether (Figure 5.4). The higher oxidation states of the complexes, such as Ru^{IV} and Ru^{V}, may not participate in the oxidation of O_2 catalyzed by $[1(OH)_2(3,6-Bu_2q)_2]^{2+}$. While there are differences in the details of previous theoretical studies of the overall water oxidation mechanism of the Tanaka catalyst, there is general agreement on the mechanism shown in Scheme 5.6 [51, 68, 90].

The presence of non-innocent quinone ligands affords $[qRu^{II}-(\mu-O_2^{\bullet-})-Ru^{II}sq]^{2+}$ upon two-electron oxidation, accompanied by the loss of two protons of $[1(OH)_2(3,6-Bu_2q)_2]^{2+}$. The Coulomb repulsion between the two oxyl radical ions may be reduced through the mediation of resonance structures such as $[\{Ru^{IV}-(O^{\bullet-})\}_2(cat)_2]^{2+}$ or $[(Ru^{IV}=O)_2(sq)_2]^{2+}$, but there must be a balance between reducing this repulsion, which keeps the two oxyl radicals spatially separated, and maintaining the radical character on the oxo groups, which is required for the coupling reaction that forms O_2. Once the coupling reaction has occurred, forming a bridged peroxo or superoxo species $([(q)Ru^{II}-O_2^{2-}-Ru^{II}(q)]^{2+}$ or $[(q)(Ru^{II}-O_2^{\bullet-}-Ru^{II}(sq)]^{2+})$, two additional PCET oxidation steps, in which the removal of the electron is accompanied by the nucleaphilic attach of a water molecule on first one and then the other metal center, are required to release the O_2 molecule. The presence of the non-innocent quinone ligands helps to delocalize the negative charge on the hydroxo ligands, and especially on the oxyl radial ion ligands, as the first PCET oxidation steps occur, avoiding high oxidation states on the metal centers.

This same effect of delocalizing the negative charge over the three-component framework of aqua-ligand moiety, metal center, and quinone ligand makes it difficult to assign precise oxidation states to the metal center and other components. So-called "broken symmetry" single-configuration calculations, usually employing DFT, are one approach to providing insight into the electronic structures of such complicated systems. They yield quite accurate relative energetics but do not produce spin eigenfunctions in the cases of interest. On the other hand, GVB CI and, more generally, MCSCF approaches yield an excellent qualitative description of the electronic structure and bonding, but since they do not incorporate dynamic correlation effects they are not as quantitatively accurate.

Complete active space self-consistent field (CASSCF) calculations involving 12 electrons and 8 valence orbitals (i.e. CAS(12,8) calculations) in all possible configurations carried out on the 2+ singlet state of the dihydroxo complex with 3,6-Me$_2$q ligands have revealed that of the 336 possible singlet configurations, only 4 are important. DFT (and other single-configuration methods) assume that there is only one important configuration. Figure 5.21 shows the final active space orbitals in order of increasing orbital energy. These eight orbitals are numbered according to their orbital energies in the molecule. The starting orbitals in each case were obtained from a (single-configuration) Hartree–Fock calculation. The eight active space orbitals were taken to be the six highest doubly occupied orbitals plus the two lowest unoccupied orbitals. There were only two active pairs of electrons responsible for the four important configurations. These active pairs were in fact GVB pairs that were "shared" or "split" between a bonding molecular orbital and the corresponding antibonding molecular orbital. It is this kind of valence configuration interaction that allows for a molecular orbital description of bond dissociation to radical products. It

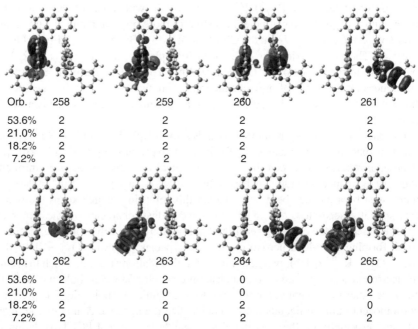

Orb.	258	259	260	261
53.6%	2	2	2	2
21.0%	2	2	2	2
18.2%	2	2	2	0
7.2%	2	2	2	0

Orb.	262	263	264	265
53.6%	2	2	0	0
21.0%	2	0	0	2
18.2%	2	2	2	0
7.2%	2	0	2	2

Figure 5.21 *Active space orbitals and important configurations of [Ru$_2$(OH)$_2$(3,6-Me$_2$q)$_2$ (btpyan)]$^{2+}$ for a CAS(12,8) calculation. Beneath each of the active space orbitals, its orbital number and its occupation numbers in each of the four important configurations are listed. The percentage of each configuration contributing to the total wave function is shown at the left of each line. The "principal configuration" with the first six active space orbitals doubly occupied constitutes only 53.6% of the total CASSCF wave function. The configuration that promotes the pair of electrons in orbital 263 to orbital 265 (left side) constitutes 21.0% of the total wave function and indicates that the pair of electrons in the highest occupied molecular orbital (HOMO) is one of the two GVB pairs. The configuration that promotes the pair of electrons in orbital 261 to orbital 264 (right side) constitutes 18.2% of the CASSCF wave function and indentifies the second GVB pair of electrons. In the final configuration, which constitutes only 7.2% of the CASSCF wave function, both GVB pairs are promoted [68]. Reproduced with permission from [68], The Royal Society of Chemistry. See color plate*

should be emphasized that there were no open-shell singlet contributions to the MCSCF wave function, although such configurations were included in the MCSCF active space (which is more general than the GVB CI space).

The monomeric class B systems of the generic formula $[Ru(tpy)(OH_2)(NIL)]^{2+}$ and their oxidized products exhibit non-integer oxidation states for the metal and NIL, with a particular Ru oxidation state displaying a predominant contribution. Due to the extensive mixing of the Ru and NIL orbitals in these complexes, the Ru centers and NIL are best described by non-integer oxidation states in certain systems. In these systems, the electron distribution is delocalized not only over the Ru−NIL fragments but also over the H_2O-derived ligand *trans* to the NIL. In particular, the utility of the class B $[Ru(tpy)(OH)(NIL)]^{+}$ motif has been demonstrated in the $[Ru_2(btpyan)(OH)_2(NIL)_2]^{2+}$ (i.e. $[\mathbf{1}(OH)_2(3,6-Bu_2q)_2]^{2+}$) systems. The NIL = $3,6-Bu_2q$ system catalyzes the oxidation of water with high turnover number when immobilized on an ITO electrode, while the NIL = $3,5-Cl_2q$ and $4-NO_2q$ systems are not active for water oxidation. This ability to spread electrons over the framework enables the Ru−NIL's fragments to act as electron reservoirs and help stabilize reactive radical intermediates, all while avoiding a high-valent metal center in the catalytic reactions. Compared to $[\mathbf{1}(OH)_2(3,6-Bu_2q)_2]^{2+}$, the bpy analog $[\mathbf{1}(OH)_2(bpy)_2]^{2+}$ is much less efficient for water oxidation, but it allows the observation of O−O bond formation via resonance Raman spectroscopy.

References

[1] Moonshiram, D., Jurss, J. W., Concepcion, J. J., Zakharova, T., Alperovich, I., Meyer, T. J., Pushkar, Y. (2012) J. Am. Chem. Soc., 134: 4625–4636.

[2] Jurss, J. W., Concepcion, J. J., Butler, J. M., Omberg, K. M., Baraldo, L. M., Thompson, D. G., Lebeau, E. L., Hornstein, B., Schoonover, J. R., Jude, H., Thompson, J. D., Dattelbaum, D. M., Rocha, R. C., Templeton, J. L., Meyer, T. J. (2012) Inorg. Chem., 51: 1345–1358.

[3] Concepcion, J. J., Jurss, J. W., Templeton, J. L., Meyer, T. J. (2008) Proc. Natl. Acad. Sci. U. S. A.: 1–4.

[4] Liu, F., Concepcion, J. J., Jurss, J. W., Cardolaccia, T., Templeton, J. L., Meyer, T. J. (2008) Inorg. Chem., 47: 1727–1752.

[5] Liu, F., Cardolaccia, T., Hornstein, B. J., Schoonover, J. R., Meyer, T. J. (2007) J. Am. Chem. Soc., 129: 2446–2447.

[6] Lebeau, E. L., Adeyemi, S. A., Meyer, T. J. (1998) Inorg. Chem., 37: 6476–6484.

[7] Chronister, C. W., Binstead, R. A., Ni, J., Meyer, T. J. (1997) Inorg. Chem., 36: 3814–3815.

[8] Schoonover, J. R., Ni, J., Roecker, L., White, P. S., Meyer, T. J. (1996) Inorg. Chem., 35: 5885–5892.

[9] Geselowitz, D., Meyer, T. J. (1990) Inorg. Chem., 29: 3894–3896.

[10] Raven, S. J., Meyer, T. J. (1988) Inorg. Chem., 27: 4478–4483.

[11] Gilbert, J. A., Geselowitz, D., Meyer, T. J. (1986) J. Am. Chem. Soc., 108: 1493–1501.

[12] Gilbert, J. A., Eggleston, D. S., Murphy, W. R. Jr, Geselowitz, D. A., Gersten, S. W., Hodgson, D. J., Meyer, T. J. (1985) J. Am. Chem. Soc., 107: 3855–3864.

[13] Limburg, J., Vrettos, J. S., Liable–Sands, L. M., Rheingold, A. L., Crabtree, R. H., Brudvig, G. W. (1999) Science, 283: 1524–1527.

[14] Naruta, Y., Sasayama, M.-A., Sadaki, T. (1994) Angew. Chem. Int. Ed., 106: 1964–1965.

[15] Seidler–Egdal, R. K., Nielsen, A., Bond, A. D., Bjerrum, M. J., McKenzie, C. J. (2011) Dalton Trans., 40: 3849–3858.

[16] Zong, R., Thummel, R. P. (2005) J. Am. Chem. Soc., 127: 12 802–12 803.

[17] Mola, J., Dinoi, C., Sala, X., Rodriguez, M., Romero, I., Parella, T., Fontrodona, X., Llobet, A. (2011) Dalton Trans., 40: 3640–3646.

[18] Francas, L., Sala, X., Escudero–Adan, E., Benet–Buchholz, J., Escriche, L., Llobet, A. (2011) Inorg. Chem., 50: 2771–2781.

[19] Sens, C., Romero, I., Rodriguez, M., Llobet, A., Parella, T., Benet–Buchholz, J. (2004) J. Am. Chem. Soc., 126: 7798–7799.

[20] Xu, Y., Fischer, A., Duan, L., Tong, L., Gabrielsson, E., Aakermark, B., Sun, L. (2010) Angew. Chem. Int. Ed., 49: 8934–8937.

[21] Li, L., Duan, L., Xu, Y., Gorlov, M., Hagfeldt, A., Sun, L. (2010) Chem. Commun., 46: 7307–7309.

[22] Xu, Y., Duan, L., Tong, L., Aakermark, B., Sun, L. (2010) Chem. Commun., 46: 6506–6508.

[23] Xu, Y., Aakermark, T., Gyollai, V., Zou, D., Eriksson, L., Duan, L., Zhang, R., Aakermark, B., Sun, L. (2009) Inorg. Chem., 48: 2717–2719.

[24] Yagi, M., Ogino, I., Miura, A., Kurimura, Y., Kaneko, M. (1995) Chem. Lett.: 863–864.

[25] Sartorel, A., Carraro, M., Scorrano, G., De Zorzi, R., Geremia, S., McDaniel, N. D., Bernhard, S., Bonchio, M. (2008) J. Am. Chem. Soc., 130: 5006–5007.

[26] Geletii, Y. V., Besson, C., Hou, Y., Yin, Q., Musaev, D. G., Quinonero, D., Cao, R., Hardcastle, K. I., Proust, A., Kogerler, P., Hill, C. L. (2009) J. Am. Chem. Soc., 131: 17 360–17 370.

[27] Cao, R., Ma, H., Geletii, Y. V., Hardcastle, K. I., Hill, C. L. (2009) Inorg. Chem., 48: 5596–5598.

[28] Dzik, W. I., Calvo, S. E., Reek, J. N. H., Lutz, M., Ciriano, M. A., Tejel, C., Hetterscheid, D. G. H., de Bruin, B. (2011) Organometallics, 30: 372–374.

[29] Concepcion, J. J., Jurss, J. W., Templeton, J. L., Meyer, T. J. (2008) J. Am. Chem. Soc., 130: 16 462–16 463.

[30] Chen, Z., Concepcion, J. J., Meyer, T. J. (2011) Dalton Trans., 40: 3789–3792.

[31] Concepcion, J. J., Tsai, M.-K., Muckerman, J. T., Meyer, T. J. (2010) J. Am. Chem. Soc., 132: 1545–1557.

[32] Zhang, G., Zong, R., Tseng, H.-W., Thummel, R. P. (2008) Inorg. Chem., 47: 990–998.

[33] Duan, L., Fischer, A., Xu, Y., Sun, L. (2009) J. Am. Chem. Soc., 131: 10 397–10 399.

[34] Duan, L., Bozoglian, F., Mandal, S., Stewart, B., Privalov, T., Llobet, A., Sun, L. (2012) Nat. Chem., 4: 418–423.

[35] Yoshida, M., Masaoka, S., Sakai, K. (2009) Chem. Lett., 38: 702–703.

[36] Masaoka, S., Sakai, K. (2009) Chem. Lett., 38: 182–183.

[37] Hetterscheid, D. G. H., Reek, J. N. H. (2011) Chem. Commun., 47: 2712–2714.

[38] Brewster, T. P., Blakemore, J. D., Schley, N. D., Incarvito, C. D., Hazari, N., Brudvig, G. W., Crabtree, R. H. (2011) Organometallics, 30: 965–973.

[39] Blakemore, J. D., Schley, N. D., Balcells, D., Hull, J. F., Olack, G. W., Incarvito, C. D., Eisenstein, O., Brudvig, G. W., Crabtree, R. H. (2010) J. Am. Chem. Soc., 132, 16 017–16 029.

[40] Hull, J. F., Balcells, D., Blakemore, J. D., Incarvito, C. D., Eisenstein, O., Brudvig, G. W., Crabtree, R. H. (2009) J. Am. Chem. Soc., 131: 8730–8731.

[41] DePasquale, J., Nieto, I., Reuther, L. E., Herbst-Gervasoni, C. J., Paul, J. J., Mochalin, V., Zeller, M., Thomas, C. M., Addison, A. W., and Papish, E. T. (2014) Inorg. Chem., submitted.

[42] Fillol, J. L., Codola, Z., Garcia–Bosch, I., Gomez, L., Pla, J. J., Costas, M. (2011) Nat. Chem., 3: 807–813.

[43] Dogutan, D. K., McGuire, R., Nocera, D. G. (2011) J. Am. Chem. Soc., 133: 9178–9180.

[44] Lewis, N. S., Nocera, D. G. (2007) Proc. Natl. Acad. Sci. U. S. A., 1: 1.

[45] Romain, S., Vigara, L., Llobet, A. (2009) Acc. Chem. Res., 42: 1944–1953.

[46] Lin, X., Hu, X., Concepcion, J. J., Chen, Z., Liu, S., Meyer, T. J., Yang, W. (2012) Proc. Natl. Acad. Sci. U. S. A., 1–4.

[47] Chen, Z., Concepcion, J. J., Hu, X., Yang, W., Hoertz, P. G., Meyer, T. J. (2010) Proc. Natl. Acad. Sci. U. S. A., 107: 7225–7229.

[48] Romain, S., Bozoglian, F., Sala, X., Llobet, A. (2009) J. Am. Chem. Soc., 131: 2768–2769.

[49] Maji, S., Vigara, L., Cottone, F., Bozoglian, F., Benet–Buchholz, J., Llobet, A. (2012) Angew. Chem. Int. Ed., 51: 5967–5970.

[50] Polyansky, D. E., Muckerman, J. T., Rochford, J., Zong, R., Thummel, R. P., Fujita, E. (2011) J. Am. Chem. Soc., 133: 14 649–14 665.

[51] Boyer, J. L., Rochford, J., Tsai, M.-K., Muckerman, J. T., Fujita, E. (2010) Coord. Chem. Rev., 254: 309–330.

[52] Ward, M. D., McCleverty, J. A. (2002) J. Chem. Soc. Dalton Trans., 275–288.

[53] Lever, A. B. P. (2010) Coord. Chem. Rev., 254: 1397–1405.

[54] Lever, A. B. P., Gorelsky, S. I. (2004) Struct. Bond., 107: 77–114.

[55] Gorelsky, S. I., Lever, A. B. P., Ebadi, M. (2002) Coord. Chem. Rev., 230: 97–105.

[56] Lever, A. B. P., Gorelsky, S. I. (2000) Coord. Chem. Rev., 208: 153–167.

[57] Gorelsky, S. I., Dodsworth, E. S., Lever, A. B. P., Vlcek, A. A. (1998) Coord. Chem. Rev., 174: 469–494.

[58] Stufkens, D. J., Snoeck, T. L., Lever, A. B. P. (1988) Inorg. Chem., 27, 953–956.

[59] Haga, M., Dodsworth, E. S., Lever, A. B. P. (1986) Inorg. Chem., 25, 447–453.

[60] Pierpont, C. G. (2001) Coord. Chem. Rev., 219–221: 415–433.

[61] Pierpont, C. G., Lange, C. W. (1994) Prog. Inorg. Chem., 41: 331–442.

[62] Boone, S. R., Pierpont, C. G. (1987) Inorg. Chem., 26: 1769–1773.

[63] Pierpont, C. G., Buchanan, R. M. (1981) Coord. Chem. Rev., 38: 45–87.

[64] Wada, T., Yamanaka, M., Fujihara, T., Miyazato, Y., Tanaka, K. (2006) Inorg. Chem., 45: 8887–8894.

[65] Wada, T., Fujihara, T., Tomori, M., Ooyama, D., Tanaka, K. (2004) Bull. Chem. Soc. Jpn, 77: 741–749.

[66] Wada, T., Tsuge, K., Tanaka, K. (2001) Inorg. Chem., 40: 329–337.

[67] Wada, T., Tsuge, K., Tanaka, K. (2000) Angew. Chem. Int. Ed., 39, 1479–1482.

[68] Wada, T., Muckerman, J. T., Fujita, E., Tanaka, K. (2011) Dalton Trans., 40: 2225–2233.

[69] Wada, T., Ohtsu, H., Tanaka, K. (2012) Chem.-Eur. J., 18: 2374–2381.

[70] Takeuchi, K. J., Thompson, M. S., Pipes, D. W., Meyer, T. J. (1984) Inorg. Chem., 23: 1845–1851.

[71] Suzuki, M., Ishiguro, T., Kozuka, M., Nakamoto, K. (1981) Inorg. Chem., 20: 1993–1996.

[72] Kim, K., Lippard, S. J. (1996) J. Am. Chem. Soc., 118, 4914–4915.

[73] Zhang, X., Furutachi, H., Fujinami, S., Nagatomo, S., Maeda, Y., Watanabe, Y., Kitagawa, T., Suzuki, M. (2005) J. Am. Chem. Soc., 127: 826–827.

[74] Khan, M. M. T., Ramachandraiah, G. (1982) Inorg. Chem., 21: 2109–2111.

[75] Kobayashi, K., Ohtsu, H., Wada, T., Kato, T., Tanaka, K. (2003) J. Am. Chem. Soc., 125: 6729–6739.

[76] Bhattacharya, S., Gupta, P., Basuli, F., Pierpont, C. G. (2002) Inorg. Chem., 41: 5810–5816.

[77] Tsai, M.-K., Rochford, J., Polyansky, D. E., Wada, T., Tanaka, K., Fujita, E., Muckerman, J. T. (2009) Inorg. Chem., 48: 4372–4383.

[78] Kurihara, M., Daniele, S., Tsuge, K., Sugimoto, H., Tanaka, K. (1998) Bull. Chem. Soc. Jpn, 71: 867–875.

[79] Bhattacharyya, I., Shivakumar, M., Chakravorty, A. (2002) Polyhedron, 21: 2761–2765.

[80] Boone, S. R., Pierpont, C. G. (1990) Polyhedron, 9: 2267–2272.

[81] Auburn, P. R., Dodsworth, E. S., Haga, M., Liu, W., Nevin, W. A., Lever, A. B. P. (1991) Inorg. Chem., 30: 3502–3512.

[82] Weaver, T. R., Meyer, T. J., Adeyemi, S. A., Brown, G. M., Eckberg, R. P., Hatfield, W. E., Johnson, E. C., Murray, R. W., Untereker, D. (1975) J. Am. Chem. Soc., 97: 3039–3048.

[83] Shepherd, R. E., Proctor, A., Henderson, W. W., Myser, T. K. (1987) Inorg. Chem., 26: 2440–2444.

[84] Schneider, R., Weyhermueller, T., Wieghardt, K., Nuber, B. (1993) Inorg. Chem., 32: 4925–4934.

[85] Welch, T. W., Ciftan, S. A., White, P. S., Thorp, H. H. (1997) Inorg. Chem., 36: 4812–4821.

[86] Yukawa, Y., Aoyagi, K., Kurihara, M., Shirai, K., Shimizu, K., Mukaida, M., Takeuchi, T., Kakihana, H. (1985) Chem. Lett.: 283–286.

[87] Che, C. M., Lai, T. F., Wong, K. Y. (1987) Inorg. Chem., 26: 2289–2299.

[88] Kojima, T., Nakayama, K., Ikemura, K.-I., Ogura, T., Fukuzumi, S.-I. (2011) J. Am. Chem. Soc., 133: 11 692–11 700.

[89] Ghosh, S., Baik, M.-H. (2011) Inorg. Chem., 50: 5946–5957.

[90] Ghosh, S., Baik, M.–H. (2012) Angew. Chem. Int. Ed., 51: 1221–1224.
[91] Muckerman, J. T., Polyansky, D. E., Wada, T., Tanaka, K., Fujita, E. (2008) Inorg. Chem., 47: 1787–1802.
[92] Wasylenko, D. J., Ganesamoorthy, C., Henderson, M. A., Koivisto, B. D., Osthoff, H. D., Berlinguette, C. P. (2010) J. Am. Chem. Soc., 132: 16094–16106.
[93] Wasylenko, D. J., Ganesamoorthy, C., Henderson, M. A., Berlinguette, C. P. (2011) Inorg. Chem., 50: 3662–3672.
[94] Meyer, T. J., Huynh, M. H. V. (2003) Inorg. Chem., 42: 8140–8160.

6

Recent Advances in the Field of Iridium-Catalyzed Molecular Water Oxidation

James A. Woods[1], Stefan Bernhard[1], and Martin Albrecht[2]

[1]*Department of Chemistry, Carnegie Mellon University, Pittsburgh, PA, USA*
[2]*School of Chemistry & Chemical Biology, University College Dublin, Dublin, Ireland*

6.1   Introduction

Solar energy, efficiently converted into a reservoir of chemical potential, offers a means towards radically expanding the scale and accessibility of energy production. The lynchpin of this scheme is the reversal of an exergonic reaction. Developing a process that reverts water, an omnipresent product of combustion reactions, back to its initial state as dioxygen represents a significant step towards the successful implementation of a fuel-producing electrolysis or photocatalysis process. Several decades of research have been invested in the development of materials and techniques designed to improve the overall net efficiency of electrolysis. The main goal of these efforts has been the development of materials with low oxidation overpotentials and high faradaic efficiency. Several materials with these criteria have been identified. Nickel, either in its elemental state or as mixed oxides ($NiCo_2O_4$), RuO_2, and IrO_2 have been demonstrated to possess very small oxidation overpotentials with excellent faradaic efficiencies.

While most commercially available electrolyzers employ nickel species, due to their cost and availability, it is important to remember that the less abundant transition metals have equally favorable electrocatalytic properties and the potential for extremely interesting chemistry. Early reviews of anodic oxygen evolution by RuO_2 or IrO_2 rationalized this behavior with correlations to O_2 exchange energies and heat-of-formation changes upon

Molecular Water Oxidation Catalysis: A Key Topic for New Sustainable Energy Conversion Schemes,
First Edition. Edited by Antoni Llobet.
© 2014 John Wiley & Sons, Ltd. Published 2014 by John Wiley & Sons, Ltd.

increasing the metal redox state in oxides [4, 5]. Second- and third-row transition metals, due to the breadth of their acceptable oxidation states, would then be natural candidates over first-row metals for the oxidation of water, for two primary reasons: the increased number of redox states accessible to platinum-group metals allows a diversity of reaction intermediates and the relative stability of high oxidation states facilitates oxidative transformations.

Since only a thin slice of atoms are active participants in chemical transformations at an electrode surface, ruthenium and iridium are not cost-efficient as bulk electrode materials. However, reframing the metals as reactive coordination complexes in solution [6] drastically improves atom economy and allows a nascent field to build on the well-founded successes of molecular-orbital and ligand-field theories. Following the pioneering work on dinuclear ruthenium catalysts by Meyer [7, 8] and Grätzel [9, 10], hundreds of innovative articles focused on the structure–activity relationships in these oxidation catalysts. Since Ir(III) and Ru(II) are isoelectronic, substituting metals provides a logical and effective way of discovering new and interesting chemistry.

The field of molecular water oxidation by homogeneous iridium catalysts has developed significantly over the last few years. This review is divided between a chronological depiction of work in the field and an in-depth analysis based on these primary sources. In composing the chronological section, careful attention was allocated to the *conditions* of each experimental work, as reported reaction conditions were found to vary considerably. Specifically, catalyst, oxidant, and supporting ion concentrations were reported across several orders of magnitude, which, when coupled with an apparently complex reaction mechanism, prevents trivial comparisons of activity and longevity between experiments.

6.2   Bernhard 2008 [11]

The first instance of mononuclear iridium complexes capable of oxidizing water to dioxygen was seen in 2008 by the family of complexes [Ir(5-R^1, 4′-R^2, 2-phenylpyridine)$_2$ $(OH_2)_2$]OTf. The catalytic cycle was driven with Ce(IV) as a chemical oxidant and the work explored the role of ligand modification on the catalysts' activities and lifetimes (Figure 6.1). Initial turnover frequencies (TOFs) varied from less than 4 to 16h^{-1} through modification of ligand substituents, while catalyst lifetimes had turnover numbers (TONs) as high as 2760. Electrochemical analysis uncovered a distinct structure–activity relationship, demonstrating the strong involvement of the supporting ligand in the overall water oxidation mechanism. The highly tunable ligand field splitting observed in these 5d metal complexes proved to be an advantage over their Ru(II) cousins. This redox tuning was greatly aided by the diverse ligand environments possible with the Ir(III/IV/V) center. Consequently, the structural and electronic diversity of these catalysts influenced the catalytic activity greatly. Additionally, free sites on the metal complex were critical, as demonstrated by the absence of water oxidation from bipyridine-substituted analogs and mixed-solvent systems in which the added solvent is strongly ligating, as in acetonitrile or dimethyl sulfoxide.

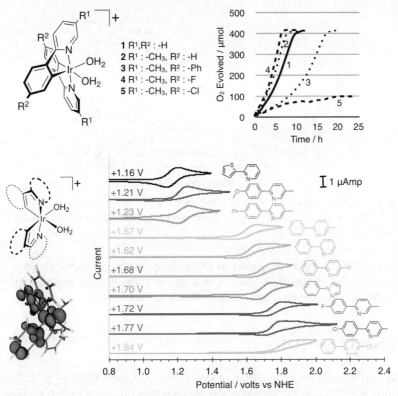

Figure 6.1 *Ligand substitution on the [Ir(ppy)$_2$(OH$_2$)$_2$]OTf framework produces a distinct structure–activity relationship with the rate of oxygen production. Bottom: Experiment and calculations show that substitution affecting the ligand-centered highest occupied molecular orbital (HOMO) plays an important and predictable role in tuning the electrochemical behavior. See color plate.*

6.3 Crabtree 2009 [12]

Inferring that strongly donating ligands are required to stabilize the putative Ir(V) oxo species, Crabtree was able to accelerate the rate of water oxidation an order of magnitude over the initial [Ir(ppy)$_2$(OH$_2$)$_2$]OTf complexes by substituting a Cp* ligand for a phenylpyridine to give (Cp*)Ir(ppy)Cl (Figure 6.2). A continued proof of concept was afforded by a decreased rate of oxygen formation on exchanging phenylpyridine for the less-donating phenylpyrimidine, (Cp*)Ir(ppm)Cl. Chloride and triflate analogs of (Cp*)Ir(ppy)X were found to have identical rates of reaction [13], suggesting fast counter-ion exchange for solvent water, with no effect on the rate-limiting step. On the basis of a first-order reaction rate and density functional theory (DFT) calculations, the authors posited [(Cp)Ir(O)(ppy)]$^+$ as an intermediate leading to O–O bond formation from a solvent H$_2$O. The ppy-containing catalyst was fully recoverable, displayed reaction rates significantly in excess of the heterogeneous IrO$_2$, did not have an induction period, and did not form visible deposits or particulates. The authors excluded the formation

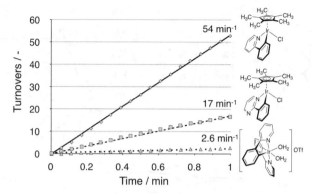

Figure 6.2 *Initial rates for catalysts (Cp*)Ir(ppy)Cl, (Cp*)Ir(pym)Cl, and [Ir(ppy)$_2$(OH$_2$)$_2$]OTf as measured by a Clark electrode using 38 nmol (5.43 μM) of catalyst and 0.55 mmol (78 mM) of CAN in H$_2$O (7 mL) at 25 °C.*

of IrO$_2$ or other heterogeneous catalysts on the basis of the combined evidence. Cyclic voltammetry (CV) of the complexes in acetonitrile showed three and two irreversible oxidation waves respectively, suggesting oxidation to OCl$^-$ or Cl$_2$.

6.4   Crabtree 2010 [13]

Crabtree further explored complexes incorporating (Cp*)Ir(N-C)X, [(Cp*)Ir(N-N)X]X, and [(Cp)Ir(N-N)X]X as catalyst precursors for homogeneous water oxidation (Figure 6.3). The N–N chelate ligands in particular offered an improvement as the cationic complexes were fully soluble, in contrast to the previously studied (Cp*)Ir(ppy)X, which was only soluble after oxidant addition. Clear first-order kinetics were observed for (Cp*)Ir(ppy)Cl, [(Cp*)Ir(bpy)Cl]Cl, and [(Cp)Ir(bpy)I]NO$_3$, with initial TOFs of 10, 14, and 9 min^{-1}, respectively. Rapid solvolysis is believed to occur due to the limited effects of varying counter ions. Rationale for the Cp-containing complex stemmed from the presence of potentially oxidizable methyl groups on the Cp* ligand. Overall, the Cp complexes showed decreased initial TOF compared to Cp* complexes at 5 μM; however, the oxygen evolution traces of (Cp*)Ir(ppy)Cl showed a decrease in the rate over the span of 1 minute, while [(Cp)Ir(bpy)I]NO$_3$ showed no such decrease over the span of 10 minutes. At low catalyst loadings, all three classes of iridium complex exhibited an indistinguishable linear relationship between concentration and initial TOF. At higher catalyst loadings of

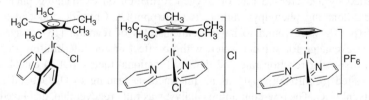

Figure 6.3 *(Cp*)Ir(ppy)Cl, [(Cp*)Ir(bpy)Cl]Cl, and [(Cp)Ir(bpy)I]PF$_6$ pre-catalysts for homogenous water oxidation investigated by Crabtree and coworkers.*

7–30 μM, [(Cp)Ir(bpy)I]NO$_3$ tended towards zero-order behavior, suggesting a catalyst self-passivation deactivation pathway.

Electrochemical analysis of [(Cp*)Ir(bpy)Cl]Cl revealed the expected 63 mV/pH unit shift in oxidation potential when conducting pH-dependent CVs. Further, repeated scans did not show the decrease in intensity expected of a stoichiometric oxidation of chloride. The oxygen evolution rate dependence on Ce(IV) was reported without specifying the concentration of catalyst, but the results suggested a change in the rate-determining step based on Ce(IV) concentration. A study of the H/D kinetic isotope effect revealed low concentrations of cerium (8 mM), leading to an inverse kinetic isotope effect (KIE) of 0.65, while at higher concentrations (243 mM) a KIE of 1.2 was observed. The authors contrasted this with the behavior of IrO$_2$ nanoparticles, which have a KIE of 1.6 at 8 mM cerium and, at 243 mM, do not catalytically produce oxygen beyond a brief burst. A plausible series of intermediates leading to O–O bond formation was explored via DFT calculations involving hydrogen bonding between a substrate and a solvent water molecule. In addition to the aforementioned (Cp*)Ir(N − C)X, [(Cp*)Ir(N-N)X]X, and [(Cp)Ir(N-N)X]X complexes, Crabtree examined several other catalyst precursors. [(Cp*)Ir(tmeda)Cl]Cl (tmeda, tetramethylethylenediamine), [(Cp*)Ir(OH$_2$)$_3$]SO$_4$, and the dimer [{(Cp*)Ir}$_2$(μ-OH)$_3$]OH exhibit non-first-order reaction rates with similarities in kinetic studies, suggesting that the tmeda is quickly lost to form the [(Cp*)Ir(OH$_2$)$_3$]$^{n+}$ complex, which goes on to form a monomer–dimer equilibrium, affecting the rate of catalysis.

6.5 Macchioni 2010 [14]

Macchioni reported the synthesis of a complex using 2-benzoyl pyridine, (Cp*)Ir(bzpy) NO$_3$, in an effort to improve water solubility. The authors continued with a comparison of activity between (Cp*)Ir(ppy)Cl and [(Cp*)Ir(OH$_2$)$_3$](NO$_3$)$_2$, exploring the potential for loss of the chelating ligand in both complexes, which may be a primary degradation pathway (Figure 6.4). Turnover frequencies of 8.5, 4.7, and 15.7 min^{-1} were observed by spectrophotometric monitoring of cerium at 340 nm and first-order rate constants were calculated of 0.14, 0.08, and 0.26 s^{-1}, respectively.

6.6 Albrecht/Bernhard 2010 [15]

The next development came when Albrecht and coworkers explored the potential for abnormal N-heterocyclic carbene chelating ligands to stabilize the large changes in metal

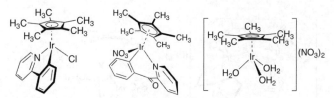

Figure 6.4 *(Cp*)Ir(ppy)Cl, (Cp*)Ir(bzpy)NO$_3$, and [(Cp*)Ir(OH$_2$)$_3$](NO$_3$)$_2$ investigated by Macchioni, yielding evidence of Cp* degradation.*

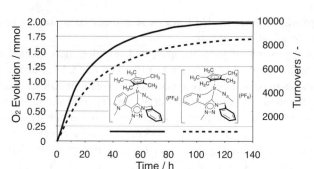

Figure 6.5   *(Cp*)Ir triazole carbene complexes investigated by Bernhard and coworkers.*

oxidation state required for mononuclear water oxidation. Two new Cp^* iridium carbene-containing complexes were synthesized (Figure 6.5), differing in the attachment of the chelating methyl pyridine. Initial TOFs were similar to $[Ir(ppy)_2(OH_2)_2]OTf$ but greatly accelerated following an induction period. TOFs at 1000 seconds were 5.2 and 2.5 min^{-1} and maximum TONs were 10 000 and 8350 for the ring-chelated and N-methyl-chelated complexes, respectively. Several oxidation waves were observed between +0.7 and +1.0 V, ascribed to non-innocent ligand (NIL) behavior, which served as one potential reason for the extreme catalyst lifetimes. The TONs reported in this paper correspond to 1.2 L of O_2 per mg of iridium, a significant step towards the goal of producing a water oxidation catalyst capable of positively impacting the global demand for energy.

6.7   Hetterscheid/Reek 2011 [16, 17]

Borrowing inspiration from the binuclear ruthenium complexes pioneered by Meyer in the 1980s, Hetterscheid and Reek developed an asymmetric binuclear iridium complex bridged by a bispyridyl imine (Figure 6.6, left). The bispyridyl imine moiety added an interesting resonance contribution, with the imine π-coordination leading to a 2 e^- oxidation state ambiguity in the associated iridium. The reaction was found to be first-order in pre-catalyst and in cerium, with a combined second-order rate constant of 4.7 1/(M · s) and a TOF of 57 min^{-1}. The authors did not speculate over the mechanism or the potential involvement of the second iridium, beyond making a comparison to a similar, mononuclear catalyst with 2-pyridinal-1-ethylimine as a ligand and an order of magnitude less activity.

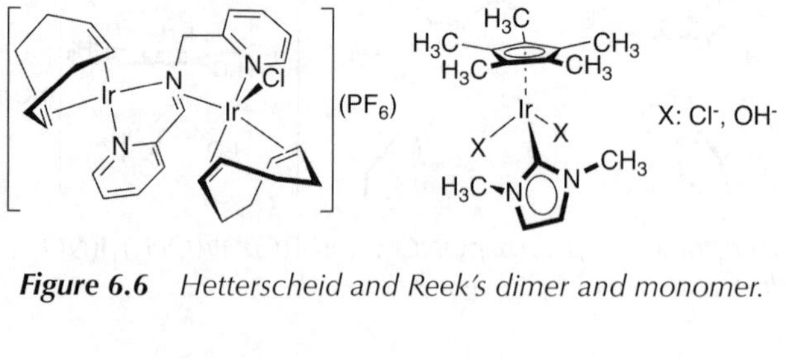

Figure 6.6   *Hetterscheid and Reek's dimer and monomer.*

In a separate publication, they also reported the synthesis and characterization of (Cp^*) Ir(NHC-Me$_2$)Cl$_2$ and (Cp^*)Ir(NHC-Me$_2$)(OH)$_2$(NHC = 1, 3-dimethylimidazolium-2-ylidene) as capable water-oxidation catalysts when driven by Ce(IV) (Figure 6.6, right). Interestingly, the dihydroxy complex has a TOF of 90 min^{-1} : approximately three times faster than the dichloro complex, with an order of magnitude greater lifetime at > 2000 TON. The dihydroxy complex was also found to rapidly disproportion hydrogen peroxide, suggesting potential mechanisms incorporating a peroxo intermediate. Importantly, preactivation was not required, suggesting that the dihydroxy or di-aquo complex is the active catalytic species. Increasingly, reversible CVs with faster scan rates suggest that a quick chemical reaction occurs just after oxidation, with voltammograms broadening significantly in the presence of small amounts of water. No electrode deposition occurred, unlike with $[(Cp^*)Ir(OH_2)_3]SO_4$ and $[\{(Cp^*)Ir\}_2(\mu\text{-OH})_3]OH$, as observed by Crabtree and others. Electrospray mass spectrometry of catalytically active solutions indicated the presence of several species that were distinctly different from the parent molecule, potentially corresponding to various stages of a proposed catalytic cycle involving Ir-OOH or Ir-OO-Ce moieties akin to similar iron and ruthenium complexes [1–3].

6.8 Crabtree 2011 [18]

Continued work by Crabtree saw the synthesis and characterization of a competent pre-catalyst of (Cp^*)Ir(NHC)Cl $(NHC = \kappa^2 C^2, C^{2'}$-1, 3-diphenylimidazol-2-ylidene). Both cerium(IV) ammonium nitrate and sodium periodate were evaluated as chemical oxidants. When driven by cerium, Clark electrode measurements indicated a slight consumption of oxygen followed by the expected oxygen production. The authors attribute this to the oxidation of Cl$^-$ to OCl$^-$ or another oxyanion under the harsh conditions of aqueous Ce(IV). In contrast, oxidation by periodate at pH 5 showed no consumption prior to activity. Rate experiments gave TOFs of 8 and 12–16 min^{-1}, with reaction orders of 1.85 and 0.67, when driven by cerium and periodate, respectively. Compared to previous studies of $[(Cp^*)Ir(OH_2)_3]SO_4$, this suggests extended exposure to the harsh conditions of cerium may lead to loss of the NHC moiety. Despite the milder conditions afforded by periodate, higher oxidant loadings led to deactivation, albeit with well-behaved oxygen traces and shorter lag phases than with cerium. The reaction order of 0.67 suggested the presence of complicated kinetics, due in part to periodate's potential to serve as a one- or two-electron oxidant.

A comparison of CVs of (Cp^*)Ir(NHC)Cl with (Cp^*)Ir(ppy)Cl in acetonitrile revealed the carbene's stabilizing effect: the CV of the NHC-containing complex showed that neither a lower Ir(III/IV) couple nor a reversible inner-sphere chloride oxidation wave was present in the ppy-containing complex. Changing the solvent to a 50/50 water/acetonitrile mix clearly indicated that water had replaced the inner-sphere chloride. Neither the NHC- nor the ppy-containing catalyst degraded after repeated cycling; nor did the electrode activity differ from baseline activity when placed in a blank solution, unlike in Crabtree's work involving the anodic deposition of $[(Cp^*)Ir(OH_2)_3]SO_4$. The stabilizing effect of the NHC was used to good effect, enabling observation of an Ir(IV) species via electron paramagnetic resonance (EPR) spectroscopy. $[Ru(bpy)_3]^{3+}$ was used as an oxidant in a 1 : 1 acetonitrile/toluene glass to give the first observed rhombic spectra from a (Cp^*)Ir species with

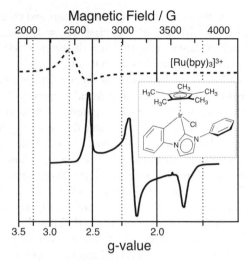

Figure 6.7 *Solid line: Electron paramagnetic resonance (EPR) spectra of $(Cp^*)Ir(\kappa^2 C^2,$ $C^{2'} - 1, 3 - diphenylimidazol - 2 - ylidene)Cl$ oxidized by 0.75 equiv. of $[Ru^{III}(bpy)_3]^{3+}$. Dashed line: $[Ru^{III}(bpy)_3]^{3+}$ as a control.*

$g_x = 2.53, g_y = 2.17$, and $g_z = 1.85$ (Figure 6.7). The Ir(IV) NHC species was found to be stable on a minute timescale under an inert atmosphere but became EPR silent after 30 minutes. Similar attempts using room-temperature ^{1}H nuclear magnetic resonance (NMR) and paramagnetic ^{1}H obtained at $-35\,°C$ were unsuccessful at observing resolved spectra.

6.9 Crabtree 2011 [19]

Crabtree later compared the behavior of the previously studied $[(Cp^*)Ir(H_2O)_3]^{2+}$ with a new water-soluble $(Cp^*)Ir$ catalyst chelated with 2-(2'-pyridyl)-2-propanol, $(Cp^*)Ir(pyr-CMe_2O)X$ (X = Cl$^-$, OTf$^-$). An electrochemical quartz nanobalance, rotating ring-disk electrode, and Clark-type oxygen electrode experiments were used to probe the electrodeposition of metal oxides from molecular precursors. The deposition of the $[(Cp^*)Ir(H_2O)_3]SO_4$ species was observed in real time on the electrode surface, in contrast to $(Cp^*)Ir(pyr-CMe_2O)X$, which produced no change in mass for linear potential sweeps between 0.2 and 1.5 V at 2.5–3.0 mM in complex (Figure 6.8). A single voltage-dependent ring current was observed in rotating ring disk electrode (RRDE) experiments using this electrochemically stable complex. The ring current was ascribed to dioxygen reduction and not peroxide reduction on the basis of pH independence; this was further confirmed by measurements with a Clark-type electrode.

6.10 Lin 2011 [20]

An interesting adaptation was explored by Lin and coworkers. Three bis-carboxy phenylpyridine (bcppy) analogs of previously studied homogeneous catalysts, $(Cp^*)Ir(bcppy)Cl$, $[(Cp^*)Ir(bcbpy)Cl]Cl$, and $[Ir(bcppy)_2(H_2O)_2]Cl$, were doped inside

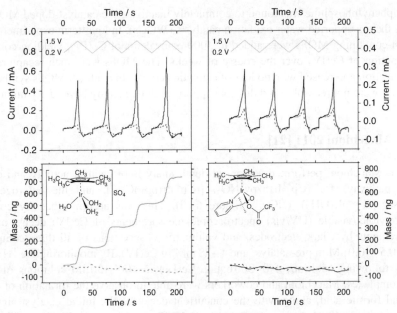

Figure 6.8 *Comparison of current and mass responses of solutions containing complexes [(Cp*)Ir(OH₂)₃]SO₄ (left panels) or (Cp*)Ir-(pyr-CMe₂O)OTf (right panels) as a function of time. Upper panels: Gray solid lines represent applied potential; black solid lines represent current response of [(Cp*)Ir(OH₂)₃]SO₄ (left) and (Cp*)Ir – (pyr – CMe₂O)OTf (right) recorded by the potentiostat; dashed lines are background current. Lower panels: Solid lines represent mass response in each experiment; dashed lines represent the corresponding catalyst-free backgrounds. Conditions: 0.1 M KNO₃ in air-saturated deionized water; 3 mM [(Cp*)Ir(OH₂)₃]SO₄; ca. 2.5 mM (Cp*)Ir-(pyr-CMe₂O)OTf. Scan rate: 50 mV/s.*

highly stable porous metal–organic frameworks (MOFs; Figure 6.9). Using Ce(IV) as a chemical oxidant, TOFs of 4.8, 1.9, and 0.4 h⁻¹ were observed in the MOF-supported catalysts, as compared to 37, 15.7, and 5.6 h⁻¹ in the unsupported catalysts. The decrease in activity was explained by size exclusion; the cerium nitrate anions are ~ 11.3 Å while the MOF channels are ~ 6.7 Å, as determined by nitrogen isotherms. Surface poisoning

Figure 6.9 *Left: Structural model of UiO-67 framework doped with (Cp*)Ir(ppy)Cl analogs. Right: SEM micrograph showing intergrown nanocrystals of the doped MOF. See color plate.*

with triphenylphosphine was found to completely inactivate the catalyst-doped MOFs, as well as the unsupported water oxidation catalysts (WOCs). In long-term experiments, the catalyst-containing MOFs were added to 4000 equivalents of Ce(IV), yielding complete consumption of Ce(IV) over the course of weeks. The MOFs were then recaptured via centrifugation and reused with no loss of catalytic activity. In contrast, MOFs synthesized with IrO_2 were not reuseable and did not support catalytic activity beyond 30 minutes.

6.11   Macchioni 2011 [21]

In 2011, Macchioni performed a more in-depth analysis of catalyst activity and degradation pathways for $(Cp^*)Ir(bzpy)NO_3$, $[(Cp^*)Ir(bpy)Cl]Cl$, and the heterogeneous precursor $[(Cp^*)Ir(OH_2)_3](NO_3)_2$ (Figure 6.10). This was accomplished by comparing ultraviolet/visible (UV/vis) spectroscopic measurements of Ce(IV) decay, oxygen measurements by Clark electrodes, and volumetric observations of all three complexes across $0.5–10.0\,\mu M$ in pre-catalyst and $1–28\,mM$ in Ce(IV). By monitoring the ^1H NMR spectra following addition of $0–300$ equiv. Ce(IV) to $100–130\,\mu M$ solutions of these three complexes, Macchioni and coworkers were able to observe the formation of acetic acid and formic acid, along with the concurrent decay of the initial catalyst signal in all three complexes. A series of 1D- and 2D-NMR experiments following addition of 6, 10, and 12 molar equivalents of Ce(IV) to $30\,mM$ solutions of $(Cp^*)Ir(bzpy)NO_3$ and $[(Cp^*)Ir(OH_2)_3](NO_3)_2$ at pH 1 by HNO_3 in $1:3$ CD_3CN/D_2O and neat D_2O, respectively, identified two different intermediates to decomposition products of the two catalysts.

Ce(IV) consumption was found to be first-order for $[(Cp^*)Ir(bpy)Cl]Cl$ and zero-order for $(Cp^*)Ir(bzpy)NO_3$ and $[(Cp^*)Ir(OH_2)_3](NO_3)_2$ for $1\,mM$ Ce(IV) across $0.5–10.0\,\mu M$ catalyst, while catalyst reaction orders were 1.0, 1.0, and 1.5, respectively. Sequential addition of cerium aliquots to catalyst solutions found that both $(Cp^*)Ir(bzpy)NO_3$ and $[(Cp^*)Ir(bpy)Cl]Cl$ decreased in activity by 20% between additions, regardless of delay between additions, while $[(Cp^*)Ir(OH_2)_3](NO_3)_2$ decayed only when there was a delay between additions. A second body of work by Macchioni and coworkers described the oxidative degradation of $[(Cp^*)Ir(OH_2)_3](NO_3)_2$ and $(Cp^*)Ir(bzpy)NO_3$ by hydrogen peroxide in acidic media and characterized several degradation products similar to the previously identified species. NMR analysis required much higher catalyst concentrations

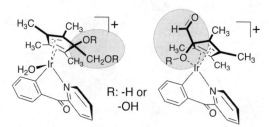

Figure 6.10   *Schematic depiction of two $(Cp^*)Ir(bzpy)NO_3$ intermediates proposed by Macchioni and coworkers in their investigation of Cp^* degradation.*

of 20–30 mM, oxidized with successive aliquots of 9.8 M H_2O_2. ^{1}H NMR spectra were recorded over 20 minute intervals and clearly indicated the susceptibility of quarternary carbon atoms of the Cp^* moiety to oxidative attack under these conditions.

6.12 Grotjahn 2011 [22]

Later in 2011, Grotjahn and coworkers illustrated a few issues incipient to the use of iridium and cerium in the study of catalyzed water oxidation. They reported observation of iridium–cerium nanoparticles by scanning transmission electron microscopy (STEM) and energy-dispersive X-ray spectroscopy (EDX) following evaporation of a 1.35 mM acetonitrile/water solution of $(Cp^*)Ir(ppy)Cl$ treated with 78 mM Ce(IV). Attempts to separate the cerium and iridium materials via dialysis and ultracentrifugation were not described beyond the fact they were unsuccessful.

The authors next turned to powder X-ray diffraction studies of lyophilized aliquots withdrawn at regular intervals from a 0.46 mM solution of $(Cp^*)Ir(ppy)Cl$ in 85 : 15 v/v acetonitrile/water following treatment with Ce(IV) at 8 mM active concentration (18 : 1 Ce(IV) to catalyst). These studies did not clearly indicate the presence of iridium oxide with low treatments of Ce(IV), but treatment with 78 mM Ce(IV) (176 : 1 Ce(IV) to catalyst) showed the presence of quite small bands associated with iridium oxide. A large portion of this work attempted to relate the UV/vis spectra of the investigated compounds and the development of absorbance peaks centered between 550 and 650 nm with similar bands observed in IrO_2 and IrO_x nanoparticles; however, they conducted no *in situ* analysis that positively or negatively indicated the presence of nanoparticles. As with Macchioni, they reported the presence of formic acid and acetic acid following the addition of 15 equivalents of Ce(IV), which may indicate the loss of the Cp^* moiety but does not conclusively indicate nanoparticle formation. Unfortunately, the role of cosolvent acetonitrile on the reaction mechanism was not explored or assessed.

6.13 Fukuzumi 2011 [23]

Rounding out 2011, Fukuzumi and coworkers conducted an excellent structure–activity relationship study of substituted $[(Cp^*)Ir(4, 4'-R_2-bpy)(OH_2)]SO_4$ complexes (R = OH, OMe, Me, and COOH). After adding Ir-containing complexes at 5 µM active concentration to a 10 mM solution of Ce(IV), the authors observed accelerating rates of catalytic activity for R = OH but not for R = OMe, Me, or COOH. ^{1}H NMR experiments on a 5 mM solution of the 4, 4'-(OH)$_2$-bpy complex revealed a decrease in signal intensity from the bpy ligand without signal broadening that was proportional to the quantity of cerium added (0–10 mol equiv. Ce(IV)). Dynamic light scattering (DLS) measurements on a 50 µM solution of 4, 4'-(OH)$_2$-bpy complex following addition of 10 mM CAN indicated the formation of particles with an average size of 348 nm, distributed over 180–1000 nm. Increasing the complex concentration to either 250 or 500 µM caused an increase of the average particle size to 600 nm, distributed over 300–1100 nm. Experiments were performed employing simultaneous solution UV/vis and gas chromatographic (GC) headspace analysis, which allowed the researchers to observe carbon dioxide production and cerium consumption for three of the catalysts. The catalyst with $4,4'$-(COOH)$_2$-bpy was not evaluated, due to poor aqueous solubility. All three catalysts produced CO_2 to varying degrees, with $4,4'$-(OH)$_2$-bpy producing significantly more than by $4,4'$-(OMe)$_2$-bpy or $4,4'$-Me$_2$-bpy. The onset of CO_2 production was instantaneous for $4,4'$-(OH)$_2$-bpy, with a correlating decrease in Ce(IV) concentration. Thermogravimetric/differential thermal analyzer (TG/DTA) and X-ray photoelectron spectra (XPS) analysis of the nanoparticles obtained from oxidation of a 2 mM solution of the $4,4'$-(OH)$_2$-bpy complex indicated a structure comprising Ir(OH)$_3$ and carbonaceous residues.

6.14   Lin 2012 [24]

Following investigations of MOF-supported molecular catalysts, Lin and coworkers investigated the properties of molecular catalysts covalently bonded to carbon electrodes as diazonium-grafted analogs of [(Cp*)Ir(5-NH$_2$-bpy)Cl]Cl, [(Cp*)Ir(4-NH$_2$-bpy)Cl]Cl, and (Cp*)Ir(5-NH$_2$-ppy)Cl (Figure 6.11). The orientation of the grafting site appears to significantly affect surface coverage at 2.18, 0.55, and 0.26 molecules/nm^2, respectively. This was ascribed to solubility issues and ambiguity in the reduction potential necessary to initiate grafting. Initial TOFs at pH 5 in acetate buffer were found to be 1.67, 0.59, and 3.31 s^{-1}, respectively, as measured by CV. Electrolysis using a [(Cp*)Ir(5-NH$_2$-bpy)Cl]Cl-grafted electrode at 1.6 V yielded sustained TOFs of 0.113–0.178 s^{-1} over 1 hour, in good agreement with cerium-driven TOFs reported in 2010 [13]. The grafted complex completes 400–650 TON over 1 hour, as compared to 150 total TON after 7.5 hours for a 25 μM solution of [(Cp*)Ir(5-NH$_2$-bpy)Cl]Cl with 50 mM CAN at pH 1 in 0.1 N HNO$_3$. Electrode-derived TOFs were difficult to determine beyond 1 hour due to surface degradation in both the background and the grafted carbon electrodes. Phosphorescent oxygen sensor measurements of the solutions surrounding the background and grafted carbon electrodes confirmed the absence and presence of oxygen, respectively. Additionally, the binding energy of the Ir 4f$_{7/2}$ and 4f$_{5/2}$ energy levels in the grafted complex did not change over 900 minutes of continuous electrolysis, beyond decreasing in intensity as a result of surface degradation. The limiting factor in carbon-grafted electrolysis appears to be carbon oxidation and subsequent loss of grafted catalyst.

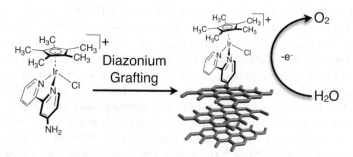

Figure 6.11   *Diazonium-grafted (Cp*)Ir(ppy)Cl analog for electrode-driven water oxidation.*

6.15   Crabtree 2012 [25–27]

In 2011 [25] and 2012 [26], Crabtree gathered evidence suggesting that sodium periodate serves as a suitable primary oxidant for catalysts with low kinetic barriers by examining its efficacy versus a variety of complexes. Specifically, $[(Cp^*)Ir(bpy)(OH_2)]SO_4$, $[(Cp^*)Ir(OH_2)_3]SO_4$, and $(Cp^*)Ir(pyr-CMe_2O)Cl$ were found to be effective as precatalysts for oxygen production using periodate. The rate of oxygen evolution was found to be first-order in periodate from 2.5 to 20 mM at 0.5 mM catalyst, which is consistent with a rate-determining step dictated by either oxidation of the catalyst or reaction of free periodate with a periodate–catalyst oligomer. First-order dependence would manifest at low concentrations of periodate with the former mechanism and at high concentrations with the latter; however, the authors described with some worry how limited reactivity and solubility combined with the ineffectiveness of isotope labeling prevented an unambiguous depiction of the rate-determining step. Due to fast oxygen exchange between water and periodate, it is not possible to exclude periodate as a source of oxygen. The use of periodate is additionally complicated by its role as either a one- or a two-electron oxidant and by the potential for the resulting iodate to serve as a ligand or as a weakly oxidizing species. Despite these complications, the authors showed periodate has the potential to provide significant insight into pH-dependent catalyst behavior and to enable a broader window for UV/vis measurements of catalytic intermediates. Results from periodate studies should therefore be analyzed in conjunction with experiments employing other oxidizing conditions.

The complex $[(Cp^*)Ir(bpy)(OH_2)]SO_4$ was shown to be significantly more resistant to periodate than cerium, as carbon dioxide is observed during reactions employing the latter but not the former. DLS measurements and ethanol-reduction tests indicated the absence of iridium dioxide nanoparticles, unlike reactions performed with $IrCl_3$ and with $[(Cp^*)Ir(OH_2)_3]SO_4$. NMR spectra of recovered $[(Cp^*)Ir(bpy)(OH_2)]SO_4$ following addition of 100 equiv. periodate indicated partial oxidation had occurred, with three distinct Cp^* methyl peaks and broad, shifted bpy peaks. Further, infrared (IR) spectra indicated the formation of a carbonyl resonance, which the authors ascribed to a carboxylic acid in the absence of an aldehyde signal from the NMR. While the reaction of $[(Cp^*)Ir(bpy)(OH_2)]SO_4$ exhibited an induction period, reactions with $(Cp^*)Ir(pyr-CMe_2O)Cl$ did not have a measurable one and proceeded 2 orders of magnitude faster (TOFs of 2.200 and $0.042 s^{-1}$, respectively). The authors attributed the rate increase to the non-innocent involvement of the deprotonated alkoxy group in close proximity.

6.16   Albrecht/Bernhard 2012 [28]

In 2012, Albrecht employed a pH-dependent reversibly cyclometalated $(Cp^*)Ir$ complex to oxidize water by both a hematite-driven photoelectrochemical cell and a long-lived Ce(IV)-driven dark reaction. A 5 μM aqueous solution of $(Cp^*)Ir(trz)Cl_2$

(trz = 1, 3-dimethyl-4-phenyl-1, 2, 3-triazolylidene) and 0.83 M Ce(IV) was found to produce 1.2 mmol oxygen over 250 hours, corresponding to a TON of ca. 22 800 based on the original catalyst loading. Under these conditions, the rate of oxygen evolution was constant at $0.14 h^{-1}$ for roughly 100 hours, producing close to 0.7 mmol O_2. Aqueous catalyst solutions change from a straw color to light blue following oxidation with 10–100 equiv. CAN. This blue decays back to the solution's original straw with a second-order rate dependence.

Adaptation of this cerium-driven reaction to a photocatalytic one was afforded by coating fluorine-doped tin oxide with a thin layer of hematite, which served as a working electrode and photoabsorber. Activity was measured by illuminating the electrode with an AM 1.5 spectrum while it was immersed in a solution of catalyst (using 3.5 wt% NaCl as an electrolyte) and comparing the photocurrents of photoelectrode–catalyst systems with photoelectrode-only systems. These experiments yielded pH- and catalyst concentration-dependent responses, which suggested that at high pH values the catalyst would be susceptible to degradation and that at catalyst concentrations above $100 \mu M$ photocurrents would be suppressed in a manner consistent with electrodeposition. The highest light-induced activity was found at between 5 and $30 \mu M$ catalyst and with pH values below 7. Extended photoelectrochemcial measurements, beginning at pH 0, showed an increase up to pH 5 and a corresponding decrease in photocurrent, which is regained on reacidification with HCl.

6.17   Crabtree 2012 [29]

Conclusively ascribing the predominant reaction responsible for water oxidation to a molecular species is a nearly intractable problem, clouded by the fitness of various iridium oxide species for water oxidation. Despite this, further work by Crabtree effectively illustrated the import role reaction conditions play in the formation of nanoparticles relevant to the oxidation of water by iridium precursors. It also conclusively showed that absorbance at 580 nm is more likely the result of various Ir(IV) species than of nanoparticle formation through the use of *in situ* time-resolved DLS. DLS measurements of $[\{(Cp^*)Ir\}_2(\mu\text{-OH})_3]OH$, $[(Cp^*)Ir(bpy)(OH)]BF_4$, $(Cp^*)Ir(ppy)OH$, and $(Cp^*)Ir(pyr\text{-}CMe_2O)OH$ following oxidation with sodium periodate in water underline these differences. In contrast to the other mentioned catalysts, $[\{(Cp^*)Ir\}_2(\mu\text{-OH})_3]OH$ readily forms nanoparticles effected by buffers, ions, and large quantities of oxidant. At 2.5 mM, each catalyst was oxidized with 100–200 equivalents of sodium periodate and developed UV/vis absorption peaks centered at 580–610 nm but only $[\{(Cp^*)Ir\}_2(\mu\text{-OH})_3]OH$ produced a recoverable precipitate.

The blue-black precipitate recovered from solutions of $[\{(Cp^*)Ir\}_2(\mu\text{-OH})_3]OH$ was initially amorphous but converted to tetragonal IrO_2 following calcination. Importantly, both the supernatant and the recovered nanoparticles were shown to be competent oxidation catalysts. A key finding in this work was the reappearance of a deep-blue color and the absence of scattering upon addition of concentrated HCl to a colorless suspension of the recovered nanoparticles. This is in contrast to anhydrous IrO_2, Crabtree's blue layer, IrO_x nanoparticles, and bulk IrO_2 catalysts, all of which do not dissolve in concentrated HCl and do not produce a color change. The blue solutions so obtained could be reduced with ethanol or sulfite to particle-free bright-yellow solutions. No particles were observed when only 50 equiv. periodate was added to neat aqueous [{(Cp*)Ir}$_2$(μ-OH)$_3$]OH solutions. The addition of either NaNO$_3$ or NaCl was found to increase both the rate and extent of particle formation. Furthermore, catalyst concentration played a significant role in the kinetics of particle formation. Light scattering intensity was nonexistent over the course of 18 hours following addition of 100 equiv. NaIO$_4$ to 0.6 mM neat aqueous [{(Cp*)Ir}$_2$(μ-OH)$_3$]OH solutions. Doubling the catalyst concentration to 1.2 mM caused the scattering intensity to increase sigmoidally following a brief induction period. Doubling it again drastically increased scattering intensity and completely eliminated the induction period.

6.18 Beller 2012 [30]

Beller and coworkers examined the activity of several Ir–phenylazole complexes as chloro-bridged dimers and Cp* or di-ppy containing monomers and compared that activity with those of Ir(acac)$_3$, IrCl$_3$·xH$_2$O, and IrO$_2$. Through the use of X-ray absorption spectroscopy (XAS) techniques, X-ray absorption near-edge structure (XANES), and extended X-ray absorption fine structure (EXAFS), the authors were able to estimate the ratio of Ir(IV) to Ir(III) in both the pre-catalysts and *in situ* oxidation reactions. On the basis of the initial catalytic activity observed with the various catalysts examined in conjunction with investigations involving nanopore-supported IrO$_2$, Beller outlined a general kinetic model with interconversion between five species, three of which are active water oxidation catalysts (Figure 6.12). An initial induction period separates the molecularly defined pre-catalyst from the active homogeneous catalyst, which then converts either to nanoparticles in the 1–3 nm range or to an inactive iridium species. Increased nanoparticle loadings cause agglomeration to bulky, active heterogeneous species that can also decay to an inactive iridium species. Expanding on this concept, Beller and coworkers conjectured that relative rates of activity could be used to identify the predominant active species as a reaction mixture progressed. More quantitatively, however, they said that *in situ* reaction particle sizing coupled with observations of degradation products should be able to be used to identify the speciation of reactive intermediates. The general kinetic scheme described in this work,

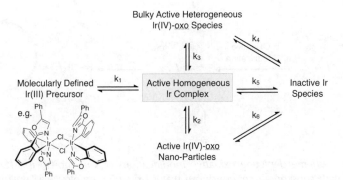

Figure 6.12 *Proposed general kinetic scheme describing the transformation of molecularly defined Ir(III) precursors into active catalytic species with interconversion between inactive Ir species.*

when linked with the DLS results presented by Crabtree, offers a potentially unifying concept that could explain the reports of varied activity and the array of statements proscribing homogeneous/heterogeneous mechanisms.

6.19   Lin 2012 [31]

Late in 2012, Lin and coworkers incorporated analogs of [(Cp*)Ir(bpy)Cl]Cl and (Cp*)Ir(ppy)Cl as structural components in MOFs and investigated their kinetics in cerium-driven water oxidation reactions (Figure 6.13). This approach offers two benefits over a completely homogeneous one: intermolecular degradation pathways are minimized due to the rigid linking groups and the heterogeneous MOFs are easily separated from solutions for post-reaction characterization. Time-dependent measurements of cerium consumption and detected oxygen were in good agreement for the bpy-containing catalyst and did not vary significantly between runs where the catalyst was recovered and reused. Negligible quantities of iridium were detected in the supernatant with 3 mM Ce(IV), but supernatant leaching significantly increased at 10 mM Ce(IV) and above. In contrast, the ppy-containing catalyst was found to be much more prone to decay, with 6% of iridium leaching into the supernatant following treatment by 3 mM Ce(IV). The initial rate of cerium consumption was an order of magnitude greater than measured oxygen production, suggesting significant oxidative transformations occurred. These transformations increased TOFs for the second and third run by a factor of five over the first.

Samples of the bpy-containing MOF were analyzed following prolonged treatment with cerium and indicated loss of the Cp* fragment with replacement by either a formate or acetate ligand. Furthermore, the absence of an induction period seems to suggest similar levels of catalytic activity at both the Cp* -containing sites and the oxidatively degraded sites. These results motivated the synthesis and investigation of K[Ir(bpy)Cl$_4$], which was

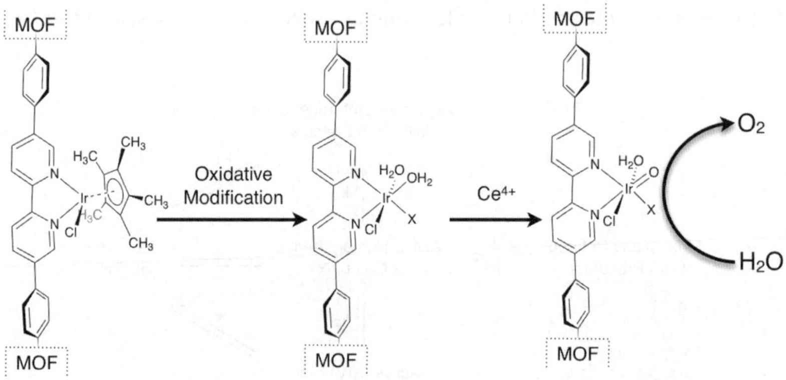

Figure 6.13 *Scheme depicting oxidative modifications of Cp* moiety in MOF-supported pre-catalyst with subsequent cerium-driven water oxidation, employing a putative Ir(V) oxo species.*

found *not* to be active. In contrast, the mixture of species present following chloride abstraction was highly active, with an initial TOF of 2.5 min^{-1} at 25 μM catalyst and 10 mM Ce(IV) at pH 1. Reactions longer than an hour yielded a decrease in activity and the formation of dark-blue solids, suggesting that intermolecular deactivation pathways serve a limiting role on catalyst lifetimes. Further instances of catalyst deactivation are afforded by cerium concentrations higher than 3 mM, which degrade catalyst-containing MOFs, but not an analogous UiO [29] MOF lacking catalytic sites. In any event, MOF-supported catalysis appears to be a valuable investigative technique for molecular species.

6.20 Lloblet and Macchioni 2012 [33]

Llobet and Macchioni synthesized several highly active Cp* Ir complexes based on a pyridinecarboxylate framework. The activity of (Cp*)Ir $κ^2$-N, O 2-pyridinecarboxylic acid and analogs functionalized with uncoordinated carboxyl groups at position 4 or 6 was compared to the previously studied (Cp*)Ir(ppy)Cl and (Cp)Ir(bzpy)NO$_3$. Cyclic voltammetry (CV) and differential pulse voltammetry (DPV) experiments revealed several interesting redox properties suggesting the ligand is an active participant in the catalytic scheme. The reaction order of all three new compounds was found to be very close to one, with catalyst concentrations ranging from 0.5 to 5.0 μM at 1 mM Ce(IV). Long-term turnover frequencies derived from these kinetic experiments ranged from 2.6 to 5.0 min^{-1}. In contrast, initial TOF values were significantly higher than long-term activity and differed strongly between (Cp*)Ir($κ^2$-N, O 2-carboxypyridine)NO$_3$, (Cp*)Ir(bzpy)NO$_3$, and (Cp*)Ir(ppy)Cl, at 70, 31, and 15 min^{-1}, respectively (Figure 6.14). Lloblet and Macchioni noted significant effects due to experimental conditions and suggested a means of countering the observed variance. While different experimental conditions produced appreciable differences in catalyst activity, the ratio of catalyst activities was extremely consistent from reaction to reaction.

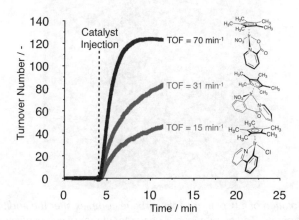

Figure 6.14 *Evolved oxygen from 20 mM solutions of CAN (pH 1 by HNO$_3$) containing (Cp*)Ir(k^2-N, O 2-carboxypyridine)NO$_3$, (Cp*)Ir(bzpy)NO$_3$, and (Cp)Ir(ppy)Cl at 1 μM.*

6.21   Analysis

After examining the evidence that has been confidently obtained thus far, there are several key points to be made in regards to the field of homogeneous iridium-based water oxidation. First, conditions matter. Certain experimental conditions directly affect the formation of nanoparticles in an otherwise homogeneous system [23, 29, 30]. Catalyst and oxidant concentrations in particular have well-documented effects on reaction homogeneity. Reaction mechanisms are more likely than not affected by the choice of oxidant and likely differ again from electrode-driven reactions [24]. Furthermore, ionic strength and species, pH, and the presence of cosolvents all have the potential to affect equilibrium concentrations of these primary species or their collective intermediates. The solution conditions necessitated by different oxidants may anorthogonally affect reaction conditions.

There appears to be a moderately variable upper threshold for catalyst and oxidant concentrations before homogeneous activity breaks down [34]. Resistance to concentration-induced agglomeration appears to largely be related to the specific catalyst and oxidant species [29]. As evidenced by Fukuzumi's work [23] with $4,4'$ hydroxyl-substituted bipyridines, easily oxidizable catalyst substructures exacerbate catalyst agglomeration at catalyst loadings of $50\,\mu M$ with $10\,mM$ CAN. With the related $[\{(Cp)^*Ir\}_2(\mu\text{-}OH)_3]OH$, Crabtree was able to selectively initiate particle formation by varying catalyst concentrations between 0.6 and $2.5\,mM$ with $100\,equiv$. $NaIO_4$.[26] In contrast, Crabtree observed no particle formation for $2.5\,mM$ solutions of $[(Cp^*)Ir(bpy)(OH)]BF_4$, $(Cp^*)Ir(ppy)(OH)$, or $(Cp^*)Ir(pyr\text{-}CMe_2O)(OH)$ with $250\,mM$ $NaIO_4$. Concentration-dependent particle formation is not a novel concept and is exploited with great effect in the field of nanocluster synthesis [35].

One interpretation of the current body of evidence suggests that the several recently identified Cp^*-containing pre-catalysts lose the capping Cp^* to form $[(L)Ir(OH_2)_3]^{n+}$ or

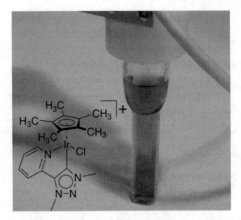

Figure 6.15   *A solution of a (Cp*)Ir triazole carbene complex after the addition of 160 equiv. Ceric ammonium nitrate. The homogeneous character of the oxygen-evolving catalyst solution was established by DLS. See color plate.*

similar hydroxyl analogs. At this point, however, the weight of this interpretation is unclear (Figure 6.15). Work by Crabtree has shown several water oxidation catalysts to also be capable of stereoselectively oxidizing C–H bonds under conditions similar to water oxidation, indicating retention of the Cp* moiety [36]. Subsequent mechanistic and computational studies support the notion of a homogeneous active catalyst derived from the initial pre-catalyst [37]. Llobet and Macchioni's recent water oxidation work with various supporting ligands that yield different levels of kinetic activity suggests a molecular active species, although DLS measurements are needed to confirm these results [33]. Additionally, their suggestion of evaluating *ratios* of TOF values of new compounds with a known standard under identical conditions is a significant step forward, provided the known standard possesses similar active-site speciation or is at least homotopic [38] under the chosen conditions. Despite recent activity in the field of homogeneous iridium-catalyzed water oxidation, conclusively identifying catalytically active species has remained problematic. This challenge may become less problematic once conditions where homogeneous activity dominates have clearly been established.

The discussion seems to have reached a similar point to that reached by palladium-catalyzed cross-coupling some 10 years ago. It is not prudent to categorically affirm a homogeneous or heterogeneous mode of action for iridium-catalyzed water oxidation. If the general kinetic scheme outlined by Beller proves to be an effective model, there could be as many as five primary iridium-containing species present in a reaction mixture with numerous intermediates and several intersecting catalytic cycles. *In operando*, rather than post-reaction, analysis is necessary to confidently obtain the necessary data. The challenges ahead involve accurately identifying active molecular species capable of water oxidation, understanding the chemical transformations occurring in the catalytic cycle, and applying that understanding to the intelligent design of ligands capable of supporting the homogeneously active catalyst. Additionally, the chemical potential currently provided by cerium and other chemical oxidants must be replaced with photochemical analogs that are compatible with a fuel-producing half-reaction.

References

[1] Codolá, Z., Garcia-Bosch, I., Acuña-Parés, F., Prat, I., Luis, J. M., Costas, M., Lloret-Fillol, J. (2013) Chem. Eur. J., 19: 8042–8047.

[2] Yoshida, M., Masaoka, S., Abe, J., Sakai, K. (2010) Chemistry – An Asian Journal, 5: 2369–2378.

[3] Kimoto, A., Yamauchi, K., Yoshida, M., Masaoka, S., Sakai, K. (2012) Chem. Commun., 48: 239–241.

[4] Trasatti, S. (1984) Electrochim. Acta, 29: 1503–1512.

[5] Trasatti, S. J. (1980) Electroanal. Chem., 111: 125–131.

[6] McDaniel, N. D., Bernhard, S. (2010) Dalton Trans., 39: 10 021–10 030.

[7] Moyer, B. A., Meyer, T. J. (1978) J. Am. Chem. Soc., 100: 3601–3603.

[8] Gilbert, J. A., Eggleston, D. S., Murphy, W. R., Geselowitz, D. A., Gersten, S. W., Hodgson, D. J., Meyer, T. J. (1985) J. Am. Chem. Soc., 107: 3855–3864.

[9] O'Regan, B., Grätzel, M. (1991) Nature, 353: 737–740.

[10] Hagfeldt, A., Grätzel, M. (2000) Acc. Chem. Res., 33: 269–277.

[11] McDaniel, N. D., Coughlin, F. J., Tinker, L. L., Bernhard, S. (2008) J. Am. Chem. Soc., 130: 210–217.

[12] Hull, J. F., Balcells, D., Blakemore, J. D., Incarvito, C. D., Eisenstein, O., Brudvig, G. W., Crabtree, R. H. (2009) J. Am. Chem. Soc., 131: 8730–8731.

[13] Blakemore, J. D., Schley, N. D., Balcells, D., Hull, J. F., Olack, G. W., Incarvito, C. D., Eisenstein, O., Brudvig, G. W., Crabtree, R. H. (2010) J. Am. Chem. Soc., 132: 16017–16029.

[14] Savini, A., Bellachioma, G., Ciancaleoni, G., Zuccaccia, C., Zuccaccia, D., Macchioni, A. (2010) Chem. Commun., 46: 9218–9219.

[15] Lalrempuia, R., McDaniel, N. D., Müller-Bunz, H., Bernhard, S., Albrecht, M. (2010) Angew. Chem. Int. Ed., 49: 9765–9768.

[16] Dzik, W. I., Calvo, S. E., Reek, J. N. H., Lutz, M., Ciriano, M. A., Tejel, C., Hetterscheid, D. G. H., Bruin, B. (2011) de Organometallics, 30: 372–374.

[17] Hetterscheid, D. G. H., Reek, J. N. H. (2011) Chem. Commun., 47: 2712–2714.

[18] Brewster, T. P., Blakemore, J. D., Schley, N. D., Incarvito, C. D., Hazari, N., Brudvig, G. W., Crabtree, R. H. (2011) Organometallics, 30: 965–973.

[19] Schley, N. D., Blakemore, J. D., Subbaiyan, N. K., Incarvito, C. D., D'Souza, F., Crabtree, R. H., Brudvig, G. W. (2011) J. Am. Chem. Soc., 133: 10473–10481.

[20] Wang, C., Xie, Z., de Krafft, K. E., Lin, W. (2011) J. Am. Chem. Soc., 133: 13445–13454.

[21] Savini, A., Belanzoni, P., Bellachioma, G., Zuccaccia, C., Zuccaccia, D., Macchioni, A. (2011) Green Chem., 13: 3360–3374.

[22] Grotjahn, D. B., Brown, D. B., Martin, J. K., Marelius, D. C., Abadjian, M.-C., Tran, H. N., Kalyuzhny, G., Vecchio, K. S., Specht, Z. G., Cortes-Llamas, S. A., Miranda-Soto, V., Niekerk, C. van, Moore, C. E., Rheingold, A. L. (2011) J. Am. Chem. Soc., 133: 19024–19027.

[23] Hong, D., Murakami, M., Yamada, Y., Fukuzumi, S. (2012) Energy Environ. Sci., 5: 5708–5716.

[24] deKrafft, K. E., Wang, C., Xie, Z., Su, X., Hinds, B. J., Lin, W. (2012) ACS Appl. Mater. Interfaces, 4: 608–613.

[25] Parent, A. R., Blakemore, J. D., Brudvig, G. W., Crabtree, R. H. (2011) Chem. Commun., 47: 11745–11747.

[26] Parent, A. R., Brewster, T. P., Wolf, W. De, Crabtree, R. H., Brudvig, G. W. (2012) Inorg. Chem., 51: 6147–6152.

[27] Parent, A. R., Crabtree, R. H., Brudvig, G. W. (2012) Chem. Soc. Rev, 42: 2247–2252.

[28] Petronilho, A., Rahman, M., Woods, J. A., Al-Sayyed, H., Müller-Bunz, H., Don MacElroy, J. M., Bernhard, S., Albrecht, M. (2012) Dalton Trans., 41: 13074–13080.

[29] Hintermair, U., Hashmi, S. M., Elimelech, M., Crabtree, R. H. (2012) J. Am. Chem. Soc., 134: 9785–9795.

[30] Junge, H., Marquet, N., Kammer, A., Denurra, S., Bauer, M., Wohlrab, S., Gärtner, F., Pohl, M.-M., Spannenberg, A., Gladiali, S., Beller, M. (2012) Chem. Eur. J., 18: 12749–12758.

[31] Wang, C., Wang, J.-L., Lin, W. (2012) J. Am. Chem. Soc., 134: 19895–19908.

[32] Cavka, J. H., Jakobsen, S., Olsbye, U., Guillou, N., Lamberti, C., Bordiga, S., Lillerud, K. P. (2008) J. Am. Chem. Soc., 130: 13 850–13 851.

[33] Bucci, A., Savini, A., Rocchigiani, L., Zuccaccia, C., Rizzato, S., Albinati, A., Llobet, A., Macchioni, A. (2012) Organometallics, 31: 8071–8074.

[34] Zuccaccia, C., Bellachioma, G., Bolaño, S., Rocchigiani, L., Savini, A., Macchioni, A. (2012) Inorg. Chem.: 1462–1468.

[35] Jin, R. (2010) Nanoscale, 2: 343–362.

[36] Zhou, M., Schley, N. D., Crabtree, R. H. (2010) J. Am. Chem. Soc., 132: 12 550–12 551.

[37] Zhou, M., Balcells, D., Parent, A. R., Crabtree, R. H., Eisenstein, O. (2012) ACS Catalysis, 2: 208–218.

[38] Crabtree, R. H. (2012) Chem. Rev., 112: 1536–1554.

7

Complexes of First Row d-Block Metals: Manganese

Philipp Kurz
Institute for Inorganic and Analytical Chemistry, Albert-Ludwigs-University of Freiburg, Freiburg, Germany

7.1 Background

Over 75 years ago, Hopkins and Pirson reported the results of fundamental experiments in photosynthesis research: growing the unicellular green algae *Chlorella* under manganese-depleted conditions, they could show that the presence of manganese(II) ions was essential for both culture growth [1] (but not survival) and the algae's ability to evolve oxygen [2]. As other cell parameters (e.g. the chlorophyll concentration) were not greatly affected by the manganese concentration, Pirson correctly assumed that the manganese in the plant might somehow directly be involved in O_2 formation [2]. This fact was then confirmed in much more elegant experiments by Kessler in 1955 [3]. Firmly establishing that the catalyst for the biological light-driven oxidation of water (Equation 7.1) was indeed a manganese site took even longer and depended on the development of modern analytical tools such as electron paramagnetic resonance (EPR) and X-ray absorption spectroscopy (XAS) in the 1970s and 1980s. Even then it remained unclear where and how this manganese-based water oxidation reaction might precisely occur. For example, a review by the esteemed photosynthesis researcher Kenneth Sauer as late as 1980 stated that "to date it has proved impossible to detect directly in the photosynthetic membranes the Mn-containing entity that is responsible for mediating O_2 evolution" [4].

$$2H_2O \longrightarrow O_2 + 4H^+ + 4e^- \qquad (1)$$

In the 30 years that followed, our knowledge of water-oxidation by photosystem II (PSII) has come a long way, especially as protein crystallography has become available as an additional technique to EPR and XAS. As discussed in Chapter 1 of this book and in numerous reviews, many key features of the oxygen-evolving complex (OEC; Figure 7.1), the manganese-containing active site for water oxidation in PSII, are today known and

Molecular Water Oxidation Catalysis: A Key Topic for New Sustainable Energy Conversion Schemes,
First Edition. Edited by Antoni Llobet.
© 2014 John Wiley & Sons, Ltd. Published 2014 by John Wiley & Sons, Ltd.

agreed upon [5–14]. The reader is referred to the extensive special literature on this topic for details, but nevertheless some points of special importance for this chapter should be summarized here:

- The OEC contains four manganese and one calcium ion, interconnected by bridging water-derived ligands (WDLs; i.e. H_2O, OH^-, and O^{2-}).
- In addition to these μ-WDLs, the metal centers are almost exclusively coordinated to terminal WDLs and carboxylate moieties of the protein chain. As a result, for all four manganese centers a five- or six-fold all-oxygen coordination sphere results.
- The OEC can exist in five redox states (the S-states: S_0 to S_4), which differ in their total oxidation state by one electron from one to the next.
- Even as some disagreement remains concerning oxidation state details, most S-state transitions are best described as equivalent to single-electron oxidations of manganese(III) centers ($Mn^{III} \rightarrow Mn^{IV}$).
- In addition to being the redox-active part of the OEC, the manganese ions are also the binding sites for water, the substrate molecules for the reaction. As a water molecule coordinated to an Mn^{III} or Mn^{IV} ion is much more acidic than a water molecule in aqueous solution, the manganese centers also directly facilitate the removal of protons, which is a key process in water oxidation.
- Many suggestions for the last, O–O bond-forming step of water oxidation catalysis by the OEC involve at least one μ-WDL.

As can be seen from this list above (which is far from complete), many mechanistic details on biological water oxidation are known today due to the fact that intensive research on this

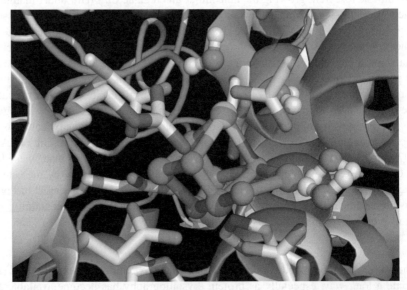

Figure 7.1 *The oxygen-evolving complex (OEC) and its immediate surroundings as found by the X-ray crystallography analysis of the PSII enzyme ensemble of Thermosynechococcus vulcanus. Color code: Ca in green, Mn in purple, O in red, N in blue, C in light gray, H in white. The figure is drawn using the atom coordinates given in [12]. See color plate*

process, which is one of the most fundamental reactions of the biosphere, has been going on for at least the last 100 years.

Consequently, one might dare to say that no catalytic water oxidation process is as well understood as the one catalyzed by the OEC within PSII. In addition to this high degree of mechanistic knowledge about its mode of action, the OEC is also a good catalyst from a purely technical perspective: PSII can reach turnover frequencies of 100 O_2 per second (but only maximal turnover numbers of 100 000), the overpotential for water oxidation is moderate (200–300 mV), and last but not least the catalytic site contains only the abundant, affordable metals calcium and manganese [8, 11, 14]. Therefore, manganese should probably be *the* metal of choice for the preparation of synthetic, molecular water oxidation catalysts (WOCs). However, as will be argued in this chapter, a convincing example of *homogeneous* water oxidation catalysis by a manganese coordination compound is still lacking today, and an attempt will be made to explain this fact.

But first it seems important to focus on the redox chemistry of manganese in water, because H_2O is both the solvent and the substrate of water oxidation catalysis. Additionally, and similarly to the OEC, water-derived entities are also present as ligands to manganese in all prominent examples of potential molecular WOCs discussed in Sections 7.3 and 7.4.

7.2   Oxidation States of Manganese in an Aqueous Environment

The element manganese is a textbook example of the fact that d-block metals, unlike the metals of the s-, p-, and f-blocks, can show a large variation in possible metal oxidation states depending on reaction partners and conditions. For manganese, compounds of all oxidation states between Mn^0 and Mn^{VII} are known [15, 16]. With the exceptions of Mn^I and Mn^V, all positive oxidation states of manganese are observed at ambient temperature and pressure for the simple reaction system of manganese/water, the reaction medium of importance in the context of water oxidation.

Pourbaix diagrams offer an overview of the dominant metal-containing species in an aqueous environment as a function of pH and redox potential. It has to be noted that there is a certain degree of uncertainty concerning the data in such diagrams, as it is especially difficult to assign correct redox potentials to insoluble species like oxides and hydroxides. Thus, slightly differing versions of the Pourbaix diagram for manganese exist [17, 18]. The species with by far the largest stability range within the window of pH and E values shown is the hydrated Mn^{2+} ion, present in solution as the hexa-aqua complex $[Mn(OH_2)_6]^{2+}$. The protons of this species are not very acidic, so an initial deprotonation in order to obtain the hydroxido species $[Mn(OH)(OH_2)_5]^+$ occurs only above pH $\approx$ 10.5. When the redox potential is increased, manganese oxidation occurs. At extremely acidic pHs and very high potentials, the oxidation directly results in the formation of the highest accessible oxidation state, Mn^{VII}. However, for pH > 2, the first oxidation products of Mn^{2+} in water are insoluble manganese oxides or hydroxides of various, mixed oxidation states between Mn^{II} and Mn^{IV}. Although all of these phases exist as stable, crystalline materials, it must be noted that in reality the chemistry of manganese oxides and hydroxides of oxidation states +II, +III, and +IV is extremely complex and very much dependant on subtle differences in reaction conditions, which is known from decades of geochemical and mineralogical research [19, 20].

7.3   Dinuclear Manganese Complexes: Syntheses and Structures

The basic manganese chemistry in aqueous solution, which was outlined in the previous section, has been well known for a long time. It must also have been clear to researchers at an early stage that any manganese compound able to mimic key features of the OEC's reactivity must be able to undergo multiple oxidation transitions of the type $Mn^{II} \rightarrow Mn^{III}$ and/or $Mn^{III} \rightarrow Mn^{IV}$. As the Pourbaix diagram indicates that in the presence of water such oxidations tend to end in the thermodynamic sink of Mn^{III} or Mn^{IV} (hydr)oxides, it was obvious that additional, non-water-derived ligands would have to be coordinated to the manganese centers in order to obtain complexes stable against (hydr)oxide formation. Additionally, monodentate ligands like water coordinated to an Mn^{II} center are known to be rather weakly bound and quick to exchange [21]. So it was apparently clear at an early stage that chelate ligands had to be used in order to prepare reasonably stable manganese coordination compounds for reactivity studies in solution and in the presence of water.

Most likely with at least some of these considerations in mind, a number of research teams (among them some of the most prominent groups of the then rather young field of bioinorganic model chemistry) started in the 1980s to synthesize and study Mn complexes. Most of the well-defined and thoroughly studied systems contain two manganese centers of oxidation states +II, +III, or +IV connected by WDLs, but a multitude of variations exist. In a review article from 2004 (the last of its kind), Armstrong and coworkers tried to list all known Mn_2 complexes of this compound family [22]; a total count of about 100(!) was reached, and by now there are likely dozens more. It is thus impossible to review this chemistry completely here. Rather, an attempt will be made to present some general features of such Mn_2 coordination compounds.

The ionic radii of Mn^{3+} and Mn^{4+} ions in an octahedral environment are given as about 72 and 67 pm, respectively [16]. In combination with their rather high positive charge, the manganese centers can thus be categorized as hard Lewis acids, explaining their affinity for hard donors like the O^{2-} ion.

This strong bonding of WDLs to manganese in higher oxidation states is important for water oxidation but also very useful for the synthesis of Mn_2 complexes. Scheme 7.1 shows a very much generalized synthetic route to μ-oxido dimanganese coordination compounds. Of course, a large number of synthetic approaches exist, but many follow the depicted reaction sequence [22–26]: (i) Mn^{2+} or Mn^{3+} ions are either dissolved or generated in a polar solvent (H_2O, MeOH, MeCN), which also contains weakly coordinating anions X (most often halides or acetates); (ii) chelating ligands are added to the mixture and coordinates to the manganese centers (due to the low Mn oxidation state, ligand substitution is often fast); (iii) finally, the manganese centers are often oxidized in the presence of water to yield $Mn_2^{\geq III,III}$ units, mostly containing at least one μ-WLD linking the manganese ions.

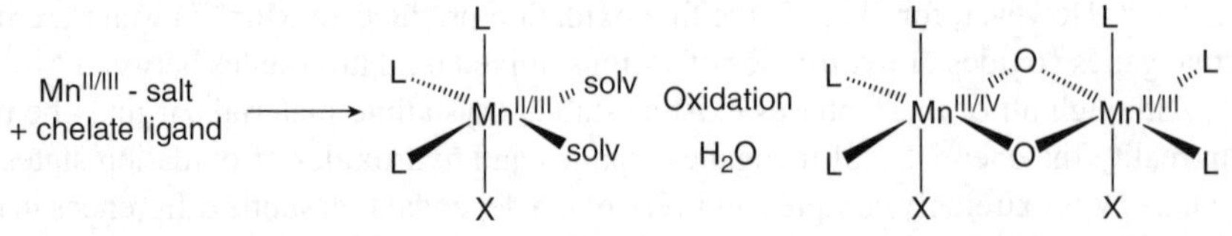

Scheme 7.1   *A common synthetic route to μ-oxido dimanganese coordination compounds*

Due to the slow ligand exchange (especially for Mn^{IV}) and the strong $Mn^{III,IV}$ oxygen bonds, the composition of the coordination compound is now somewhat fixed. In particular, the $[Mn_2(\mu\text{-}O)_2]$ unit shown in Scheme 7.1 is very stable and thus a very common motif used in bioinorganic manganese model chemistry [22]. If in addition the chelate ligands are sufficiently inert against oxidation (N and O donors have proved to be particularly good choices), the thus prepared compounds are well suited to studies concerning redox properties and oxygen-evolution catalysis, which will be the topics of Sections 7.4 and 7.5.

Scheme 7.2 shows four prominent examples of Mn_2 complexes prepared following this or a related preparation strategy. Without going into too many details, the structures and spectroscopic properties of compounds such as **1–4** [27–30] (and many other related Mn_2 species [22–25]) demonstrate a number of important points for this field of manganese coordination chemistry:

- Through the right choice of co-ligands and conditions, examples of Mn_2 complexes of all redox states between $Mn_2{}^{II,II}$ and $Mn_2{}^{IV,IV}$ can be stabilized against (hydr)oxide formation, so that an investigation of their properties in water-containing solution is often possible.
- Detailed analyses of the spectroscopic properties of such compounds (using infrared (IR), Raman, ultraviolet/visible (UV/vis), magnetic circular dichroism (MCD), EPR, and XAS methods, among others) is of great importance in obtaining reference data for comparisons with data on the OEC (and also other multinuclear manganese metalloenzymes like catalases or arginases).
- For cases where the structural features of such complexes can be elucidated using X-ray diffraction or other techniques, the determined distances for Mn–Mn (2.5–3.5 Å for $Mn^{>II}$) and $Mn^{>II}$–O bonds (~ 2 Å) are similar to the values found for the OEC, as well as to those observed in manganese(III/IV) oxides. The complexes are thus valuable structural OEC-models, especially for the characterization of the bonding situation between manganese and WDLs.
- Finally (and of special importance to the topic of this chapter), due to their molecular character, it is possible to investigate redox, acid–base, and catalytic properties for compounds like the ones shown in Scheme 7.2 in detail, which will be the topic of the following two sections.

7.4   Redox and Acid–Base Chemistry of Mn_2-μ-WDL Systems

The synthesis of compounds like **1–4** made it possible to determine redox potentials for manganese-centered oxidations and reductions in solution. Because of its large potential range of electrochemical stability, studies in acetonitrile solution were often the first choice, but later electroanalytical measurements of Mn_2 complexes in the presence of water, or even in 100% H_2O, were carried out. In most cases, a number of redox events are observed for such dinuclear complexes, but assignment of the individual waves detected by, for example, cyclic voltammetry (CV) is often only possible if the electroanalysis is combined with additional techniques. In particular, the combinations of electrolyses with UV/vis and EPR spectroscopy have proven to be very useful in this context.

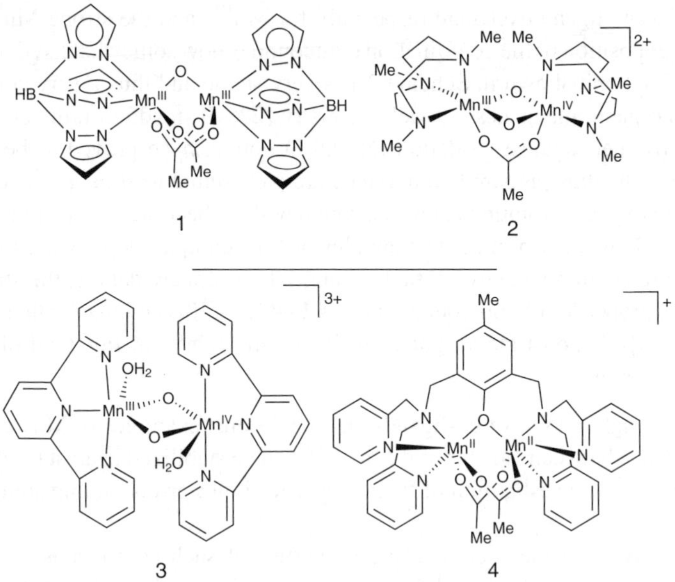

Scheme 7.2   *Examples of Mn$_2$ complexes*

As an example of one of the very first electrochemical measurements on a Mn$_2$ complex, Figure 7.2 shows the original plot of the CV reported for the Mn$_2^{III,III}$ species **1** in 1987 [29]. Besides the ferrocene reference signal, an irreversible reduction at about −0.5 V and a partially reversible oxidation at ∼ +1.2 V were observed. (Note: An attempt has been made to give all redox potentials in this chapter versus standard hydrogen potentials. The author is aware of the fact that comparisons of electrochemical data, especially between different solvent systems, are far from trivial [31].) While the reduction of **1** yielded unidentified Mn^{II}-containing species, the oxidation event could be clearly assigned to involve the transformation Mn$_2^{III,III}$ → Mn$_2^{III,IV}$ by using a combination of bulk electrolysis and EPR spectroscopy. Additionally, chemical oxidation experiments showed that when the oxidation was carried out in the presence of water, ligand exchange at the manganese centers occurred, and one additional μ-O ligand was incorporated to yield the very stable [Mn$_2(\mu$-O)$_2$] unit from the [Mn$_2(\mu$-O)(μ-AcO)$_2$] precursor [29]. Very similar events were also detected for reductions and oxidations of complex **2**, for which even the Mn$_2^{IV,IV}$ state could be reached [30, 32]. In this way these very early examples of molecular OEC mimics already showed reaction steps that were only much later identified as key events for the S-state transitions of the OEC: manganese-centered oxidations, ligand exchange, and deprotonation of water after coordination to manganese.

On the other hand, the events observed in the voltammogram of Figure 7.2 are in quite some contrast to the manganese redox chemistry found for manganese in water earlier. First, it can be clearly seen that the coordination of chelating ligands prevents oxide formation. Even in 100% aqueous solution, it is sometimes possible to oxidize Mn$_2$ complexes to an all-manganese(IV) state (e.g. this is well documented for complex **3**) without forming any oxide or hydroxide precipitates [33]. On the other hand, the manganese centers in complexes such as **1–4** are now coordinated to much better electron donors than the WDLs

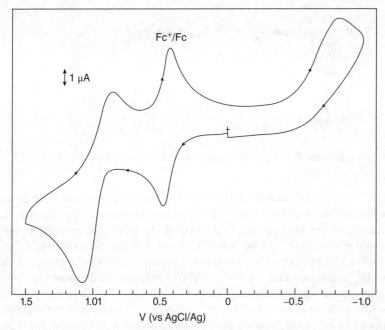

Figure 7.2 *CV of **1** in dry MeCN containing 0.1 M tetrabutylammonium perchlorate (NBu_4 ClO_4) as electrolyte. Reprinted with permission from ref. [29]. Copyright (1987) American Chemical Society*

in the system Mn/H_2O. As a consequence, the oxidation events (especially those involving oxidations of Mn^{II}) are often shifted to much lower potentials than the $\sim +0.7$ V found for the oxidation of $[Mn(OH_2)_6]^{2+}$ at pH 7 in water. This seems to be a general trend: a recent, detailed review on the redox properties of a large number of Mn_2 compounds lists only a few examples where oxidation potentials for processes involving the Mn^{II} state are significantly higher than +0.5 V [33]. In contrast, potentials well above +1 V are often necessary for the oxidizations of Mn^{III} centers to Mn^{IV}.

More detailed studies on the constitution of the Mn_2 complexes in various oxidation states revealed that manganese oxidation in many cases had profound consequences for the nature of the bridging ligands. Two general trends are observed:

- With increasing manganese oxidation states, the preference of the manganese centers to coordinate WDLs increases (the Mn ions become more oxophilic). Thus, non-water-derived ligands like acetates and halides coordinated to Mn^{II}-containing compounds are often exchanged for WDLs when the Mn^{IV} state is reached. The case of **1** is a typical example.
- With increasing manganese oxidation states, water or hydroxide coordinated to manganese becomes more acidic. As a result, the dominant WDL for Mn^{IV}-containing complexes is the oxido ligand (O^{2-}), especially in its bridging form between two manganese centers (μ-O^{2-}).

Both trends have an important consequence: the overall charge of the Mn_2 complex is often not increased upon oxidations beyond the Mn^{III} state, because exchanges such as

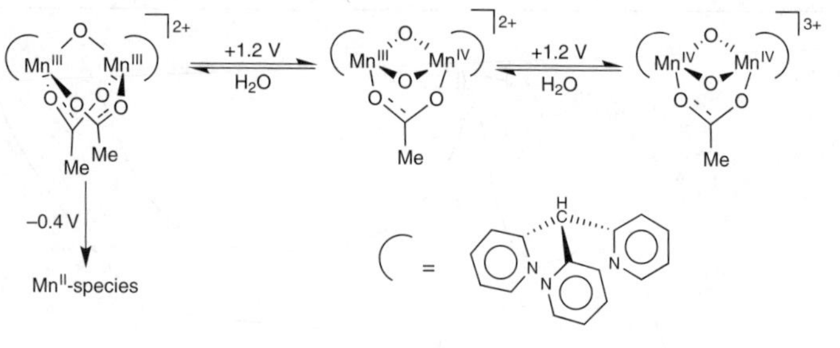

Scheme 7.3 *Typical reaction sequence for an Mn₂ complex*

$AcO^-/Hal^- \rightarrow OH^-/O^{2-}$ and $H_2O/OH^- \rightarrow O^{2-}$ often result in an accumulation of additional negative charges in the ligand sphere, which compensates for the increased total positive charge of the Mn centers. As the overall charge of the complex strongly influences its oxidation potential, the net result is a leveling of the redox potentials. Because of the factors described before, this is often much more pronounced for processes involving $Mn^{III} \rightarrow Mn^{IV}$ oxidations than for $Mn^{II} \rightarrow Mn^{III}$ processes. To illustrate this, Scheme 7.3 shows a very typical reaction sequence for an Mn₂ complex [34]. Due to a ligand exchange $AcO^- \rightarrow O^{2-}$ upon oxidation, the $Mn_2^{III,III}$ and $Mn_2^{III,IV}$ forms of the complex have the same charge and their oxidations occur at the same potential. In contrast, the reduction of the Mn^{III} centers requires reduction potentials well below 0 V.

It is important to note here that for dinuclear manganese complexes, charge compensation and the resulting redox leveling are often the result of a combination of ligand exchange and deprotonation of H_2O or OH^-. The processes thus often have features resembling "classical" proton-coupled electron transfer (PCET), but are often complicated by additional ligand-exchange steps. The important overall result of the reactions described in this section is the ability of Mn₂ complexes to accumulate multiple oxidation equivalents (some at oxidation potentials well above +1 V). These might be used for catalytic oxidation reactions (such as water oxidation), which are the topic of the next section.

7.5   Mn₂ Systems: Oxygen Evolution (but not Water Oxidation) Catalysis

Interestingly, even as most early reports on the synthesis and properties of Mn₂ complexes mentioned their model character for the OEC as a central motivation for their study, investigations concerning the actual abilities of the compounds to catalyze water oxidation were not reported for quite some time. A reason for this might be that such experiments are not as straightforward to design as the simplicity of Equation 7.1 might suggest.

In a catalytic experiment on water oxidation, the potential manganese catalyst should be exposed to a strong oxidant (E > +1 V) in the presence of water. The evolving oxygen, as the product of Equation 7.1, is then detected as a measure of catalytic performance. Unfortunately, the possible choices for oxidation agents are to date not ideal [35]. Equations 7.2–7.4 summarize the most investigated reactions on manganese-catalyzed O_2 formation.

$$2 \, H_2O_2 \xrightarrow{Mn_{cat.}} O_2 + 2 \, H_2O \tag{2}$$

$$2 \, HOOSO_3^- \xrightarrow{Mn_{cat.}} O_2 + 2 \, H^+ + 2 \, SO_2^{2-} \tag{3}$$

$$4 \, Ox^+ + 2 \, H_2O \xrightarrow{Mn_{cat.}} O_2 + 4 \, H^+ + 4 \, Ox \tag{4}$$

$$(Ox^+ = Ce^{IV}, [Ru^{III}(bpy)_3]^{3+}, h^+ \text{ @ anode})$$

The very first experiments of this kind investigated O_2 formation as a result of reactions of manganese complexes with hydrogen peroxide [23]. This is formally a disproportionation reaction, generating two O^{-II} (black in the equations) and two O^0 (light grey) oxygen equivalents from four O^{-I} (dark grey) oxygens. *In vivo*, Mn-catalase enzymes are able to catalyze this reaction. As these metalloproteins also contain an Mn_2 moiety as their active sites, these reactions are thus excellent model processes for catalases and over time Mn_2 complexes with remarkable turnover frequencies for H_2O_2 disproportionation have been identified [36]. The process mechanistically demands that the metal site is able to coordinate oxygen-donor ligands and switches back and forth between redox states differing by two redox equivalents. Looking at Section 7.4, one can see why an Mn_2 complex is well suited to carrying out this catalytic process. On the other hand, even as O_2 is formed, this reaction has key mechanistic differences when compared to water oxidation: only two (and not four) redox equivalents need to be stored, the substrate already contains an O–O bond, and H_2O_2 is both a reasonable oxidation and reduction agent. Isotope labeling studies involving ^{18}O-labelled H_2O_2 or H_2O clearly showed that the evolving O_2 contains only oxygen atoms derived from hydrogen peroxide and not from water [37], thus demonstrating that water oxidation steps are not involved.

Very influential work starting in 1999 introduced reactions with hydrogen peroxosulphate (oxone) as new test systems for catalytic oxygen formation by manganese complexes in water [28]. It was found that the Mn_2 complex **3** (Scheme 7.2) very efficiently catalyzes O_2 evolution in reactions with HSO_5^-, which initiated a large number of studies concerning this catalytic process [24, 33]. Indeed, it might be no exaggeration to claim that **3** to date is the most thoroughly investigated Mn_2 complex. From these numerous investigations it could be shown that during the course of the reaction the complex cycles through $Mn_2^{III,III}$ and $Mn_2^{IV,IV}$ forms while retaining its $[Mn_2(\mu\text{-}O)_2]$ core. In higher oxidation states, dimerizations can occur to yield Mn_4 entities [38]. The analysis of isotope labeling experiments indicates that most likely only one of the oxygen atoms of the O_2 product originated from the inorganic peroxide [39]. In consequence, these catalytic reactions by a molecular manganese species may include one of the central steps of water oxidation catalysis, the formation of an O–O bond, starting from water as the substrate [28]. Even as they do not represent a "true" true water oxidation reaction according to Equation 7.1, the homogeneous catalytic oxygen-evolving reactions of **3** (and other Mn_2 complexes) with HSO_5^- provide very important insights concerning the chemistry of high-valent manganese species in an aqueous environment and especially prove that it might be possible to catalyze O–O bond formation using manganese active sites.

When compared to biological water oxidation, reactions according to Equations 7.2 and 7.3 feature two significant differences: firstly, the oxidation agent is a two-electron oxidant, whereas water oxidation catalysis directly coupled to light reactions requires the handling of one oxidation equivalent at a time; and secondly, the oxidation agents already contain activated oxygen atoms in the form of peroxides, while in proper water oxidation the rather redox-inert substrate H_2O is the only oxygen and electron source. To investigate a compound's ability to act as a "true" WOC, three test reactions are currently used (Equation 7.4) [26, 35, 40]: (i) chemical oxidations using solutions of Ce^{IV}; (ii) chemical oxidations using (mainly photochemically generated) $[Ru^{III}(bpy)_3]^{3+}$; and (iii) electrochemical experiments in which a catalytically non-active anode serves as oxidation agent for the potential catalyst. These systems have the advantage that (at least for (i) and (ii)) they surely only involve single-electron oxidation steps and activated oxygen species are absent. Consequently, if oxygen is produced in such reactions, they are generally considered to represent "real" water oxidation catalysis. To support this assumption, isotope labeling experiments have for such cases shown that the O_2 originates fully from water as the only substrate in these reactions [39, 41].

Unfortunately, and despite a number of claims, there seems so far to be no example of a manganese complex which beyond reasonable doubt is a homogenous catalyst for a process of the type summarized in Equation 7.4. There have been reports of sub-stoichiometric reactions, very small amounts of fully labeled O_2 product, and systems where a synthetic Mn_2 complex is irreversibly degraded to an undefined but active compound, but none of these examples so far fully satisfies the requirements of true homogenous water oxidation catalysis, which is the topic of this book. The scientific community active in this field seems to have realized that the approach of using Mn_2 complexes as potential catalysts, is most likely no route to success: following a short burst of publications in the last decade during which a number homogeneous Mn catalysts were claimed [39, 42–44], the area has become quiet of late.

Why, if rich redox chemistry, high oxophilicity, and the ability to form O–O bonds all characterize Mn_2 complexes, are they not able to catalyze water oxidation? And why, if one looks at other chapters of this book, are molecular Ru_2 systems and the manganese-containing OEC superb catalysts for water oxidation? Section 7.6 will try to offer a simplified, most likely incomplete, but hopefully still somewhat useful explanation.

7.6 Mn_2 Complexes/the OEC/Ru_2 Catalysts: A Comparison

An attempt to compare the events following single-electron oxidations in an aqueous environment for Mn_2 complexes, the OEC, and Ru_2 catalysts is made in Scheme 7.4. Of course, for each path there is not just one established mechanism and details differ when one looks at different studies for each case. However, by choosing accepted possible reaction routes described for the OEC, [45] a Mn_2 -compound [34] and the famous "blue ruthenium dimer", [46] the author hopes to have selected fairly representative examples of each case.

Looking at this scheme, one first has to note a number of similarities between the different cases, which have already been described in other reviews [8, 47]:

- The units can be oxidized multiple times in single-electron steps.
- The units have high affinities for oxygen ligands and thus bind water readily.
- As electrons are removed, charge compensation occurs via ligand exchange and/or deprotonation reactions for coordinated H_2O/OH^-.

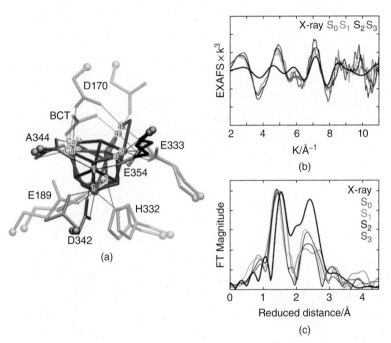

Figure 1.1 *Structural models of the OEC of PSII. (a) Superposition of the OEC in the XRD models of PSII at 3.5Å (red) and 1.9Å (blue) resolution. (b–c) Comparison between experimental isotropic EXAFS spectra of S_0 (green), S_1 (light blue), S_2 (dark gray), S_3 (brown), and calculated EXAFS spectra of the high-resolution XRD model (blue), including the k3-weighted EXAFS spectra (b) and the corresponding Fourier transform (FT) magnitudes (c)*

Molecular Water Oxidation Catalysis: A Key Topic for New Sustainable Energy Conversion Schemes,
First Edition. Edited by Antoni Llobet.
© 2014 John Wiley & Sons, Ltd. Published 2014 by John Wiley & Sons, Ltd.

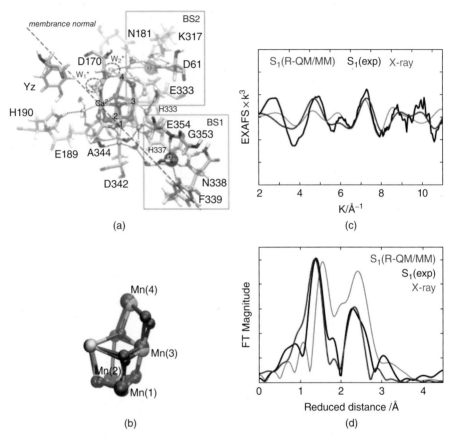

Figure 1.2 *DFT–QM/MM model of the dark-adapted S_1 state of the OEC of PSII. (a) Ligation scheme of the OEC oxomanganese cluster, including substrate water molecules (W_1^* and W_2^*), six carboxylate ligands (D170, E189, E333, D342, A344, and E354), one imidazole ligand (H332), surrounding amino acid residues (Y161 (YZ), D61, and K317) and the chloride binding sites (BS1 and BS2). (b) Superposition of the oxomanganese core of the high-resolution XRD (magenta) and the DFT–QM/MM (colored) models. (c–d) Comparison between experimental (black) isotropic EXAFS spectra of S_1 and calculated spectra of the refined R-QM/MM model (red) and the high-resolution XRD model (magenta)*

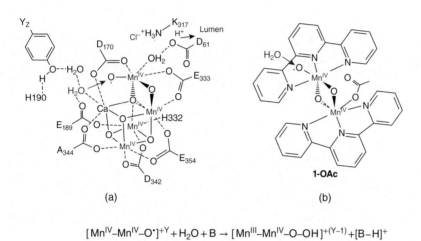

(a)   (b)

$$[Mn^{IV}\text{–}Mn^{IV}\text{–}O^\bullet]^{+Y} + H_2O + B \rightarrow [Mn^{III}\text{–}Mn^{IV}\text{–}O\text{–}OH]^{+(Y-1)} + [B\text{–}H]^+$$

(c)

Figure 1.3 *Schematic representation of (a) the Mn(IV) – O• oxyl radical in the oxidized forms of the OEC of PSII and (b) complex **1-OAc**, showing an analogous nucleophilic attack mechanism to that of a substrate water molecule (red). (c) O–O bond formation mechanism, involving production of the hydroperoxo Mn(IV)–Mn(III) species and proton release to a generic basic center (B)*

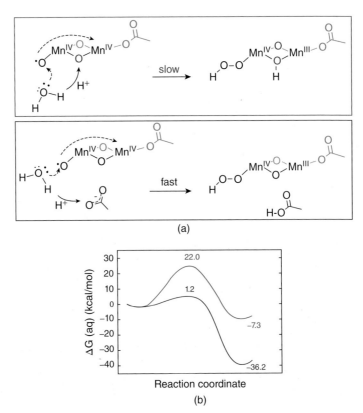

(a)

(b)

Figure 1.4 (a) O–O bond formation step during O_2 evolution catalyzed by **1-OAc** in the absence (red box) and presence (blue box) of an acetate proton acceptor. (b) Comparison of the free energy profiles for the O–O bond formation mechanism depicted in (a)

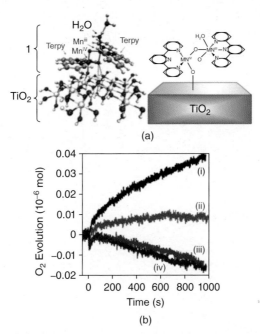

(a)

(b)

Figure 1.5 (a) DFT cluster model of a **1**–TiO$_2$(NP) covalent assembly, with a water ligand exchanged by the nanoparticles (NPs) and a terminal water exposed to the solvent. (b) O$_2$ evolution during water oxidation catalyzed by the **1**–TiO$_2$(NP) assembly, using Ce^{4+} as a single-electron oxidant. Complex **1** was loaded on three different TiO$_2$ (50 mg) samples: (i) P25, (ii) D450, and (iii) D70, as well as (iv) a control test using bare P25 NPs as the catalyst

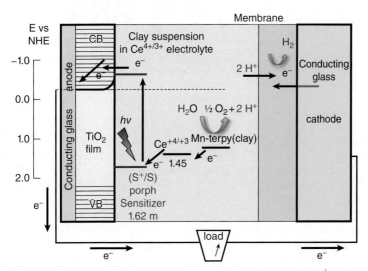

Figure 1.6 Schematic diagram of a photocatalytic cell used for direct solar water oxidation, based on the Mn–terpy catalyst in a clay suspension coupled to the sensitized TiO$_2$ electrode surface by a redox couple Ce$^{4+/3+}$ mediator

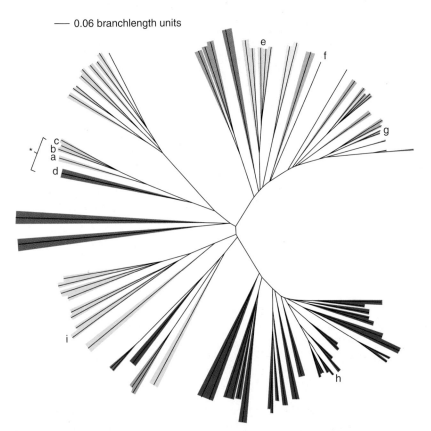

Figure 2.1 *Phylogenetic tree of Cld sequences from diverse hosts. The phylum/kingdom affiliation of each species is indicated by color: Proteobacteria, yellow; Firmicutes, orange; Nitrospirae, red; Actinobacteria, blue; Archaea, light blue; Deinococcus-Thermus, gray; Chloroflexi, green; Planctomycetes, dark purple; Verrucomicrobia, light purple; Acidobacteria, pink. The Halobacteriaceae, pictured near the bottom of the tree, form their own group distinct from the other archaea. Species known to carry out chlorite detoxification are indicated with a bracket/asterisk. Representative crystallographically characterized Clds are labeled as follows: (a) D. aromatica, (b) D. agitata, (c) Ideonella dechloratans, (d) N. defluvii, (e) Halobacterium sp. NRC-1, (f) T. thermophilus HB8, (g) G. stearothermophilus, (h) M. tuberculosis, and (i) T. acidophilum. A three-iteration PSI-BLAST search was performed using DaCld as the bait sequence. The top 500 result sequences were aligned by ClustalX and a phylogenetic tree was constructed. Representative sequences from each phylum were chosen for the display. Settings used for the tree building were random number = 111 and bootstrap maximum = 1000. The iTOL (Interactive Tree of Life) program was used for branch coloring and figure generation (http://itol.embl.de/, last accessed 13 December 2013). Reprinted with permission from [14]. Copyright © Elsevier, 2011*

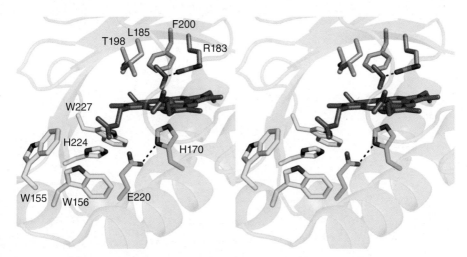

Figure 2.2 Active site of nitrite-bound DaCld. The proximal His–Glu pair is shown in green, the hydrophobic Thr/Leu/Phe triad in salmon, the Trp/His network in yellow, and the distal Arg in orange. Adapted with permission from [17] and generated using PyMOL. Reprinted from [17] with kind permission from Springer Science + Business Media

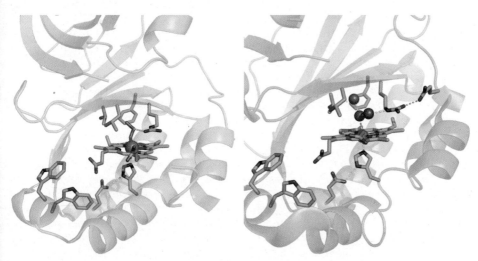

Figure 2.3 Active sites of the nitrite-bound D. aromatica Cld (left) [17] and water-bound N. winogradskyi Cld (right) [16], illustrating the conformational flexibility of the distal arginine. In the image on the left, this residue is hydrogen bonded to the axial nitrito ligand, and in that on the right, to water and an asparagine. These images represent the proposed "in" and "out" configurations of the arginine, respectively. Reproduced with permission from [36]. Copyright © 2012, American Chemical Society

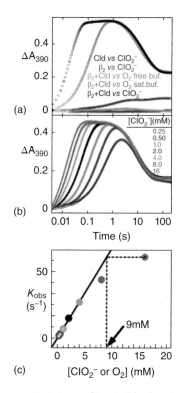

Figure 2.4 *A transient enzyme – O_2 intermediate (with absorbance at 390 nm) was prepared using DaCld and ClO_2^- to generate the O_2. (a) Absorbance versus time traces were measured following rapid mixing at 5°C of two solutions in a stopped-flow spectrometer. Syringe 1 contained an anaerobic/reduced solution of 0.2 mM of the Mn- and Fe-dependent ribonucleotide reductase β_2 subunit from Chlamydia tracomatis, 0.6 mM Mn^{II}, 0.2 mM Fe^{II}, and 0.01 mM DaCld (molarity given per heme-containing subunit). This was mixed with an equal volume of either 20 mM ClO_2^- (black trace), O_2 – saturated 100 mM HEPES buffer (pH 7.6) (green trace), or O_2 – free buffer (orange trace). The steep upward curve represents formation of the Mn^{IV}/Fe^{IV} intermediate and the subsequent downward curve shows its decomposition; the plateau region in between therefore represents the intermediate's lifetime, which is extended when DaCld/ClO_2^- rather than O_2 – saturated buffer is used as the O_2 generator, because higher O_2 concentrations are attainable. Traces from control reactions, from which DaCld and β_2 were omitted, are colored red and blue, respectively. (b) Delineation of the O_2 concentration dependence of the β_2 activation reaction by variation of ClO_2^- concentration. Reactions were conducted as described for the black trace in panel (a), but with the concentration of the ClO_2^- reactant solution varied to give the final ClO_2^- concentrations noted in the figure. Traces were analyzed by nonlinear regression using the equation for two exponential phases to extract observed first-order rate constants for formation of the Mn^{IV}/Fe^{IV} intermediate (k_{obs}). (c) Plot of these observed first-order rate constants versus ClO_2^- or O_2 concentration. The points at ≤ 4 mM ClO_2^- were fit by the equation for a line. Extrapolation of the k_{obs} for the reaction with 16 mM ClO_2^- to the linear-fit line (dashes) in this case gave an effective O_2 concentration of 9 mM (arrow). Reproduced with permission from [34]. Copyright © 2008, American Chemical Society*

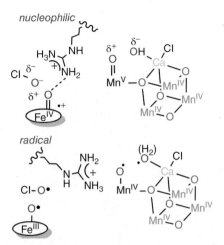

Scheme 2.2 *Comparison of proposed nucleophilic and radical mechanisms for the O–O bond joining step in DaCld (left) and PSII (right). The distal arginine residue is drawn in its "in" conformation in the top panel and its "out" conformation in the bottom. The positively charged residue is close to the heme iron and might play a role in stabilizing or positioning the various proposed intermediates and leaving groups*

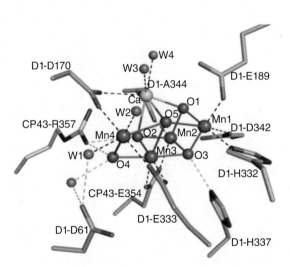

Figure 3.1 *X-ray structure of the OEC-PSII at 1.9Å resolution. Reprinted by permission from Macmillan Publishers Ltd: Nature, [6], copyright © 2011*

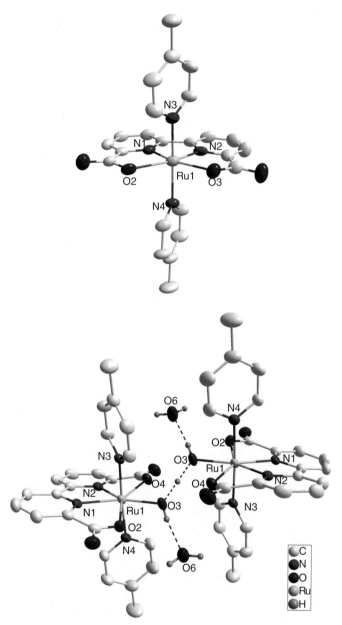

Figure 4.6 Crystal structures of **11** (upper) and **12** (lower) with thermal ellipsoids at 50% probability. Hydrogen atoms are omitted for clarity, except for the H–O type. Adapted with permission from [14]. Copyright © 2009, American Chemical Society

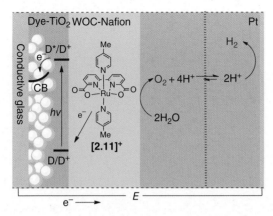

Figure 4.21 *Illustration of the working principle of a photoelectrochemical cell towards artificial water splitting. Reproduced with permission from [28], The Royal Society of Chemistry*

Figure 5.15 *Calculated structure of the doubly deprotonated species formed from* $[2(OH_2)(3,5\text{-}Me_2q)]^{2+}$. *There is excess alpha spin density (blue) on the Ru atom, the O, and the q ligands, confirming the assignment of the resonance between the* $^3[Ru^{II}(O^{\bullet-})(sq)]^0$ *and the intermediate-spin* $^3[Ru^{IV}(=O)(cat)]^0$ *species*

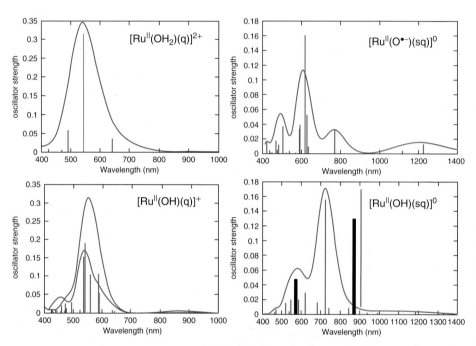

Figure 5.16 Calculated spectra of $[Ru(OH_2)(q)]^{2+}$, $[Ru(OH)(q)]^+$, $[Ru(O^{\bullet-})(sq)]^0$, and $[Ru(OH)(sq)]^0$. The vertical lines (blue) are the discrete transitions from a TD-B3LYP/LANL2DZ calculation and the smooth curves (red) are the result of broadening with a Gaussian function. For the $[Ru(OH)(q)]^+$ species, the spectra of singlet $[Ru^{II}(OH)(q)]^+$ (red) and triplet $[Ru^{III}(OH)(sq)]^+$ (blue) species are shown. For $[Ru(OH)(sq)]^0$, the black vertical bars indicate the (unnormalized) peaks of the experimental spectrum [91]

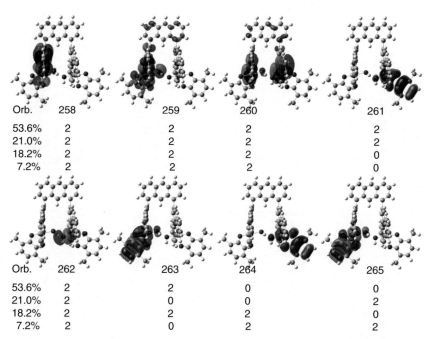

Orb.	258	259	260	261
53.6%	2	2	2	2
21.0%	2	2	2	2
18.2%	2	2	2	0
7.2%	2	2	2	0

Orb.	262	263	264	265
53.6%	2	2	0	0
21.0%	2	0	0	2
18.2%	2	2	2	0
7.2%	2	0	2	2

Figure 5.21 *Active space orbitals and important configurations of $[Ru_2(OH)_2(3,6\text{-}Me_2q)_2$ $(btpyan)]^{2+}$ for a CAS(12,8) calculation. Beneath each of the active space orbitals, its orbital number and its occupation numbers in each of the four important configurations are listed. The percentage of each configuration contributing to the total wave function is shown at the left of each line. The "principal configuration" with the first six active space orbitals doubly occupied constitutes only 53.6% of the total CASSCF wave function. The configuration that promotes the pair of electrons in orbital 263 to orbital 265 (left side) constitutes 21.0% of the total wave function and indicates that the pair of electrons in the highest occupied molecular orbital (HOMO) is one of the two GVB pairs. The configuration that promotes the pair of electrons in orbital 261 to orbital 264 (right side) constitutes 18.2% of the CASSCF wave function and indentifies the second GVB pair of electrons. In the final configuration, which constitutes only 7.2% of the CASSCF wave function, both GVB pairs are promoted [68]. Reproduced with permission from [68], The Royal Society of Chemistry*

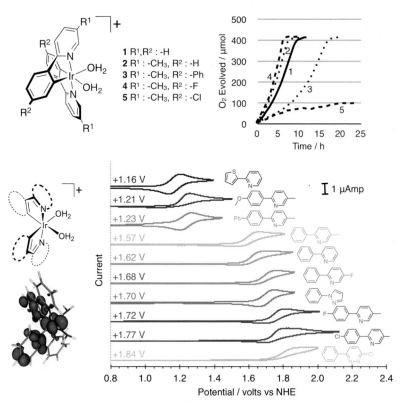

Figure 6.1 *Ligand substitution on the [Ir(ppy)₂(OH₂)₂]OTf framework produces a distinct structure–activity relationship with the rate of oxygen production. Bottom: Experiment and calculations show that substitution affecting the ligand-centered highest occupied molecular orbital (HOMO) plays an important and predictable role in tuning the electrochemical behavior*

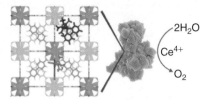

Figure 6.9 Left: Structural model of UiO-67 framework doped with (Cp*)Ir(ppy)Cl analogs. Right: SEM micrograph showing intergrown nanocrystals of the doped MOF

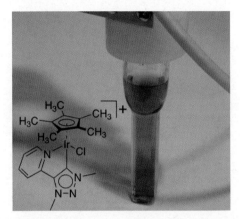

Figure 6.15 A solution of a (Cp*)Ir triazole carbene complex after the addition of 160 equiv. Ceric ammonium nitrate. The homogeneous character of the oxygen-evolving catalyst solution was established by DLS

Figure 7.1 The oxygen-evolving complex (OEC) and its immediate surroundings as found by the X-ray crystallography analysis of the PSII enzyme ensemble of Thermosynechococcus vulcanus. Color code: Ca in green, Mn in purple, O in red, N in blue, C in light gray, H in white. The figure is drawn using the atom coordinates given in [12]

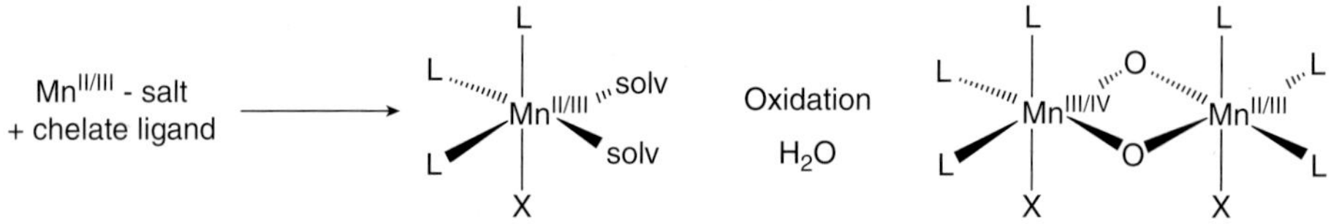

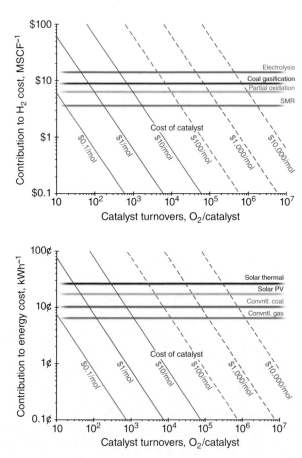

Scheme 7.1 *Synthetic route to μ-oxido dimanganese coordination compounds*

Figure 8.3 *The cost of artificially photosynthesized hydrogen is shown as a function of water oxidation catalyst robustness and price, in units of hydrogen cost (top) and corresponding power cost (bottom). Reference lines have been added to give context where available. The unit of hydrogen normalization, MSCF, is a thousand standard cubic feet. SMR refers to steam methane reforming, which accounts for > **90**% of all hydrogen produced today [33,34]*

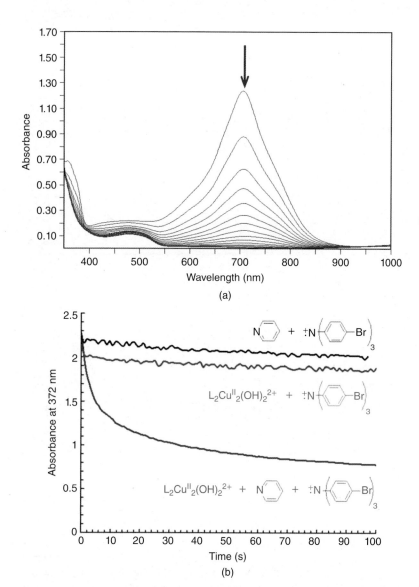

Figure 10.1 *(a) Optical spectra of the reaction of [(Me₃TACN)₂Cuᴵᴵ₂(OH)₂](OTf)₂ with [N(4 − Br − C₆H₄)₃]⁺• and pyridine in acetonitrile at 298 K, showing the decay of the aminium oxidant (L = Me₃TACN). (b) Changes in the absorbance at 372 nm for the first reaction (blue, bottom) and for control experiments omitting the copper (black, top) or the pyridine (red, middle). (c) (next page) Optical spectra of the reaction of [LCuᴵ(CH₃CN)]OTf with O₂ at −75°C in methylene chloride, showing the growth of the μ-oxo complex [L₂Cuᴵᴵᴵ₂(μ-O)₂]²⁺*

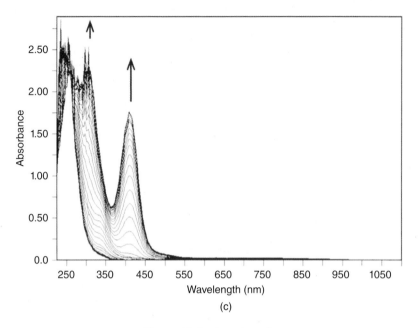

(c)

Figure 10.1 *(continued)*

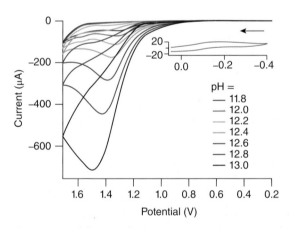

Figure 10.2 *CVs of a 0.5 mM solution of [(bpy)$_2$Cu$_2$(μ-OH)$_2$](OAc)$_2$ show increasing catalytic current as the solution becomes more alkaline. Arrow shows direction of potential sweep. Inset: The 1e$^-$ Cu$^{II/I}$ couple for the same solution, shown to scale. Conditions: 0.1 M electrolyte (NaOAc + NaOH), glassy carbon working electrode, platinum reference electrode, Ag/AgCl reference electrode (adjusted to normal hydrogen electrode, NHE). Reproduced with permission from [2]. Copyright © 2012, Rights Managed by Nature Publishing Group*

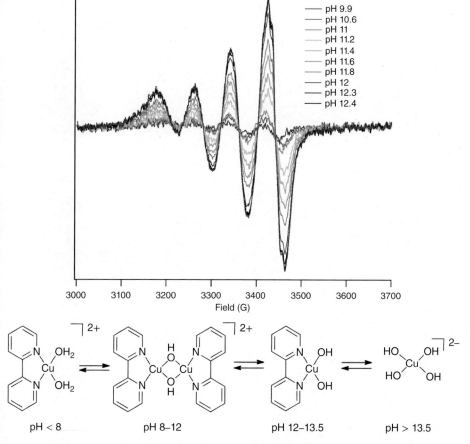

Figure 10.4 Top: As the pH increases, the EPR signal corresponding to the (bpy)Cu(OH)$_2$ monomer increases, at 1 mM Cu$^{(II)}$/bpy in 0.1 M electrolyte (NaOAc + NaOH). Bottom: Proposed speciation for the Cu$^{(II)}$/bpy system under these conditions

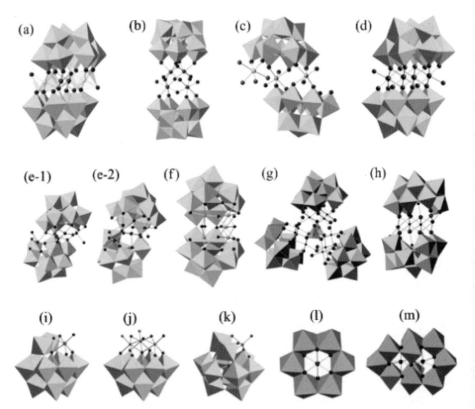

Figure 11.1 *X-ray crystal structure of polyoxometalates (POMs) in combined ball-and-stick and polyhedral representations corresponding to Table 11.1. Red, O; magenta, Ru; blue, Co; green, Ni; yellow, Ir; dark teal, Cl; Purple, Bi; orange tetrahedra, $PO_4/SiO_4/GeO_4$; white tetrahedra or polyhedra, ZnO_4/ZnO_6; light-blue polyhedra, WO_6; grey polyhedra, MoO_6*

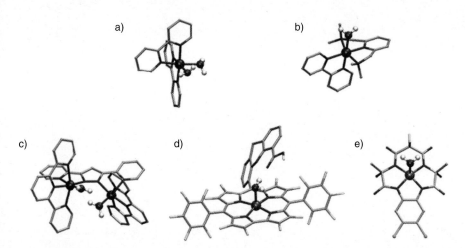

Figure 12.2 WOCs studied: (a) $cis - [Ru^{II}(bpy)_2(H_2O)_2]^{2+}$, (b) $[Ru^{II}(damp)(bpy)(H_2O)]^{2+}$, (c) $in,in - [(Ru^{II}(trpy)(H_2O))^2(\mu\text{-}bpp)]^{3+}$, (d) $CoH^{\beta F}\text{-}CO_2H$, and (e) $Fe^{III}\text{-}TAML$. Only hydrogen atoms in water and carboxylic acid groups are shown; ^tBu groups in the cobalt catalysts are not shown for clarity. Purple, ruthenium; ochre, cobalt; orange, iron; yellow, fluorine; green, chlorine; gray, carbon; white, hydrogen

Figure 12.8 Spin-density plots of $[H^{\beta F}CX\text{-}CO_2H\text{-}Co^{III}\text{-}H_2O](^3A)$ (left), $[H^{+\bullet\beta F}CX\text{-}CO_2H\text{-}Co^{III}\text{-}OH](^2A)$ (center), and $[H^{+\bullet\beta F}CX\text{-}CO_2H\text{-}Co^{III}\text{-}O^{\bullet -}](^3A)$ (right). Orange, cobalt; gray, carbon; blue, nitrogen; red, oxygen; teal, fluoride; white, hydrogen

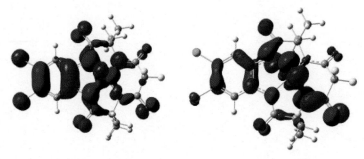

Figure 12.11 SOMOs for the $[TAML^{+\bullet}-Fe^V-O]$ triplet state from the DFT triplet state that are analogous to the CASSCF orbitals. White, hydrogen; gray, carbon; blue, nitrogen; red, oxygen; teal, fluorine; green, chlorine; mauve, iron

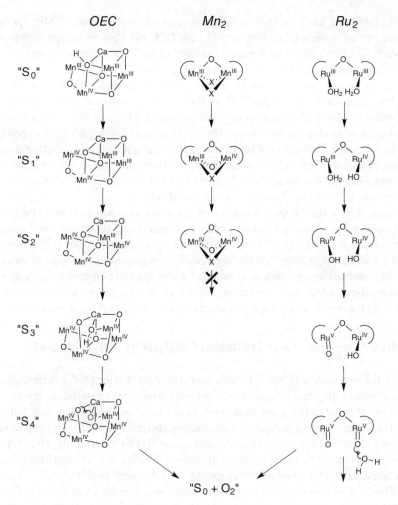

Scheme 7.4 *Comparison of events following single-electron oxidations in an aqueous environment for the OEC, Mn$_2$ complexes, and molecular Ru$_2$ water oxidation catalysts*

- In all cases, there is a strong electronic coupling between the metal centers, such that the units should be considered as single redox-active entities in each case.

However, and despite these common properties, the path towards water oxidation stops at an S$_2$-like intermediate for Mn$_2$ complexes, while the OEC and Ru$_2$ species are active catalysts. From the comparison shown, there seems to be one main reason for this: Mn$_2$ compounds are not able to store four oxidation equivalents of high enough oxidative power to drive water oxidation.

Thermodynamically, a redox potential of about +0.8 V is needed to oxidize water to O$_2$ at pH 7. In reality, a slight overpotential is always required, so that about +1.1 V is needed as an average oxidation potential for the four holes stored intermediately by the catalytic unit. Looking at the redox characteristics of Mn$_2$ complexes discussed before, only the processes involving $Mn^{III} \rightarrow Mn^{IV}$ oxidations meet this requirement. A Mn_2 species can only have two of these, while a Mn_4 unit like the OEC can have up to four (although only three $Mn^{III} \rightarrow Mn^{IV}$ oxidations for each turnover of the OEC seem to occur in vivo, see Scheme 7.4), and thus the Mn_2 compounds "get stuck at S_2."

As also shown in Scheme 7.4, for a dinuclear ruthenium complex this limitation does not exist. Here, it could be shown that $Ru^{II} \rightarrow Ru^{III}$ occurs at reasonably high potentials and that additionally the Ru^V oxidation state can be reached beyond Ru^{IV}. Thus a multitude of ways exist of storing the four necessary high-potential holes, and the Ru redox states drawn in Scheme 7.4 should only be seen as one of many possibilities. Furthermore, even examples of mononuclear ruthenium catalysts have been synthesised, where suggested catalytic cycles involve oxidation states from Ru^{II} all the way up to Ru^{VI}. Details about this chemistry can be found in Chapters 3, 4 and 5 of this book.

Of course, there might be more barriers to water oxidation catalysis beyond the described inability of the Mn_2 complexes to accumulate enough oxidation power. For example, the spin requirements for the O–O bond formation step might be crucial [48]. However, as Section 7.7 will show, besides the OEC there are by now numerous examples of manganese-catalyzed water oxidation being demonstrated using synthetic compounds. Strikingly, all of them involve catalytic units containing more than two manganese centers, which might indicate that the simple line of argumentation depicted in Scheme 7.4 is not entirely wrong.

7.7   Heterogeneous Water Oxidation Catalysis by $Mn_{>2}$ Systems

Taking a different look at Figure 7.1, one could also argue that the OEC can be seen as biology's solution to the immobilization of a heterogeneous oxide catalyst within an organic matrix [49]. Whether this was on their minds or not is today impossible to say, but as early as 1981 a research group presented a completely different approach to the synthesis of OEC model systems: the preparation of manganese(III,IV) oxides [50]. They found that the precipitates obtained by the comproportionation reaction of Mn^{2+} and MnO_4^- in water were active catalysts for water oxidation in catalysis runs using $[Ru^{III}(bpy)_3]^{3+}$ as oxidation agent. Thus, such easily obtained manganese-containing materials had the catalytic activity that all elaborately synthesized Mn_2 complexes prepared so far lack. Today, it is known that many manganese(III,IV) oxides are WOCs, with some of them showing catalytic rates and stabilities that might make their application in technical devices feasible [51–55]. Structural analyses have revealed that in particular layered manganese oxides belonging to the naturally occurring mineral families of birnessites and buserites show good catalytic performances [26, 56]. Interestingly, especially good reaction rates were observed when – as in the OEC – Ca^{2+} was incorporated into the oxide structure. Figure 7.3 shows a section of an oxide layer of such a layered calcium manganese oxide.

A second class of true manganese-based WOCs could be prepared if (otherwise catalytically inactive) Mn_2 complexes were adsorbed on hydrophilic surfaces, especially clay minerals [26, 57–60]. Detailed investigations of the phenomenon have revealed that the clay surfaces facilitate the formation of oxido-bridged oligomers from the dimeric precursors, which in solution only occurs under special conditions. These oligomers (Figure 7.4 shows a model of a possible structure) then seem to be the active species for water oxidation.

Details of the catalytic properties of either of the two classes of catalyst would fill a chapter of their own. However, they both deal with catalysis on the surfaces of solids,

Figure 7.3 *Model of a section of the highly disordered structure of calcium birnessites. Color code: Ca in green, Mn in purple, O in red. See color plate*

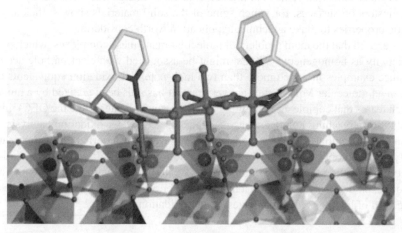

Figure 7.4 *Possible arrangement of an Mn$_4$ species formed by the aggregation of two Mn$_2$ complexes on the surface of a clay mineral. Reproduced with permission from [29]. Copyright © 1987, American Chemical Society. See color plate*

and this topic should thus not be addressed in great length in this book with its focus on homogeneous WOCs. Nevertheless, the results seem to indicate some important points in support of the picture of molecular manganese compounds discussed before:

- Manganese-based water oxidation catalysis can be observed outside the special protein environment provided by the PSII enzyme.

- Again, manganese centers of +III and +IV oxidation states in an oxygen-rich environment are the key players. This agrees with the data from the Mn Pourbaix diagram, in which Mn(III,IV)-oxides are ideally positioned for water oxidation at low overpotentials.
- Because more than two Mn centers are electronically coupled in both oxides and clay hybrids, the accumulation of the four holes necessary for water oxidation seems feasible.
- Both the layered manganese oxides and the Mn oligomers on surfaces represent highly disordered, open structures. The metal centers are thus accessible for substrate water.

7.8 Conclusion

This chapter presented a brief overview of the syntheses, structures, and redox chemistries of dinuclear manganese complexes for water oxidation catalysis. Research in this field can build on two strong foundations: the well-studied basics concerning the aqueous redox chemistry of manganese and a very detailed knowledge of natural water oxidation catalysis by the OEC within the PSII enzyme. A large number of Mn_2 complexes have been synthesized in the last few decades, many of which were investigated in detail concerning their redox properties in the presence of water. These studies showed that the fundamental steps of any likely catalytic cycle for water oxidation can be observed for such Mn_2 species in solution: multiple electron transfer steps, the coordination of water, the deprotonation of aqua ligands to yield μ-oxido ligands, and the catalysis of O–O bond formation. However, no molecular dimanganese species studied so far acts as a convincing catalyst for homogeneous water oxidation. This is in marked contrast to manganese oxides and manganese oxido clusters on surfaces, for which some of the solid materials show sufficiently good catalytic properties to make a technical application worth considering.

It was argued that the main problem of molecular manganese complexes, which prevents their activity as homogeneous WOCs, might be associated with electron hole accumulation, since examples in which more than two high-potential oxidation equivalents can be stored in an molecular Mn_2 species are very rare. It has thus been realized by a number of research teams that complexes containing *four* manganese centers (like the OEC) might be the solution to reaching homogeneous manganese-based water-oxidation catalysis. Some very interesting initial examples along this line have been reported, and some catalytic activity has been observed from Mn_4 units in solution [61–63]. It will be the exciting but also very demanding task of the coming years to study the reactions of these and related compounds in detail. Thus, despite a history of over 30 years, many discoveries lie ahead for the field of homogeneous manganese water oxidation chemistry.

Acknowledgements

The author thanks the Deutsche Forschungsgemeinschaft, the Fonds der Chemischen Industrie, and the German federal states of Schleswig-Holstein and Baden-Württemberg for generous financial support of his research efforts during the last years. The graphics shown as Figures 7.1, 7.3, and 7.4 were kindly provided by Mathias Wiechen. The outline and contents of Scheme 7.4 were heavily influenced by a drawing in reference [8] by the author's close collaboration partners of the Dau research team at Freie Universität Berlin.

References

[1] Hopkins, E. F. (1930) Science, 72: 609–610.

[2] Pirson, A. (1937) Z. Botan., 31: 193–267.

[3] Kessler, E. (1955) Arch. Biochem. Biophys., 59: 527–529.

[4] Sauer, K. (1980) Acc. Chem. Res., 13: 249–256.

[5] Barber, J. (2009) Chem. Soc. Rev., 38: 185–196.

[6] Blankenship, R. E. (2002) Molecular Mechanisms of Photosynthesis. Blackwell Science, Oxford, UK.

[7] Dau, H., Haumann, M. (2008) Coord. Chem. Rev., 252: 273–295.

[8] Dau, H., Limberg, C., Reier, T., Risch, M., Roggan, S., Strasser, P. (2010) ChemCatChem, 2: 724–761.

[9] Ferreira, K. N., Iverson, T. M., Maghlaoui, K., Barber, J., Iwata, S. (2004) Science, 303: 1831–1838.

[10] Loll, B., Kern, J., Saenger, W., Zouni, A., Biesiadka, J. (2005) Nature, 438: 1040–1044.

[11] McEvoy, J. P., Brudvig, G. W. (2006) Chem. Rev., 106: 4455–4483.

[12] Umena, Y., Kawakami, K., Shen, J.-R., Kamiya, N. (2011) Nature, 473: 55–60.

[13] Yano, J., Kern, J., Sauer, K., Latimer, M. J., Pushkar, Y., Biesiadka, J., Loll, B., Saenger, W., Messinger, J., Zouni, A., Yachandra, V. K. (2006) Science, 314: 821–825.

[14] Lubitz, W., Reijerse, E. J., Messinger, J. (2008) Energy Environ. Sci., 1: 15–31.

[15] Greenwood, N. N., Earnshaw, A. (1998) Chemistry of the Elements. Butterworth Heinemann, Oxford, UK.

[16] Holleman, A. F., Wiberg, E. (2007) Lehrbuch der Anorganischen Chemie. de Gruyter, Berlin, Germany.

[17] Brookins, D. G. (1988) Eh-pH Diagrams for Geochemistry. Springer, Berlin, Heidelberg, Germany.

[18] Takeno, N. (2005) Atlas of Eh-pH Diagrams. Geological Survey of Japan. Published online, http://www.fssm.ucam.ac.ma/biblioadmin/opac_css/chimie/Atlas_Eh-pH_diagrams.pdf (accessed December 13, 2013).

[19] Post, J. E. (1999) Proc. Natl. Acad. Sci. U. S. A., 96: 3447–3454.

[20] Spiro, T. G., Bargar, J. R., Sposito, G., Tebo, B. M. (2010) Acc. Chem. Res., 43: 2–9.

[21] Helm, L., Merbach, A. E. (1999) Coord. Chem. Rev., 187: 151–181.

[22] Mukhopadhyay, S., Mandal, S. K., Bhaduri, S., Armstrong, W. H. (2004) Chem. Rev., 104: 3981–4026.

[23] Wu, A. J., Penner-Hahn, J. E., Pecoraro, V. L. (2004) Chem. Rev., 104: 903–938.

[24] Cady, C. W., Crabtree, R. H., Brudvig, G. W. (2008) Coord. Chem. Rev., 252: 444–455.

[25] Mullins, C. S., Pecoraro, V. L. (2008) Coord. Chem. Rev., 252: 416–443.

[26] Wiechen, M., Berends, H. M., Kurz, P. (2012) Dalton Trans., 41: 21–31.

[27] Diril, H., Chang, H. R., Zhang, X. H., Larsen, S. K., Potenza, J. A., Pierpont, C. G., Schugar, H. J., Isied, S. S., Hendrickson, D. N. (1987) J. Am. Chem. Soc., 109: 6207–6208.

[28] Limburg, J., Vrettos, J. S., Liable-Sands, L. M., Rheingold, A. L., Crabtree, R. H., Brudvig, G. W. (1999) Science, 283: 1524–1527.

[29] Sheats, J. E., Czernuszewicz, R. S., Dismukes, G. C., Rheingold, A. L., Petrouleas, V., Stubbe, J., Armstrong, W. H., Beer, R. H., Lippard, S. J. (1987) J. Am. Chem. Soc., 109: 1435–1444.

[30] Wieghardt, K., Bossek, U., Bonvoisin, J., Beauvillain, P., Girerd, J. J., Nuber, B., Weiss, J., Heinze, J. (1986) Angew. Chem. Int. Ed., 25: 1030–1031.

[31] Bard, A. J., Faulkner, L. R. (2000) Electrochemical Methods: Fundamentals and Applications. John Wiley and Sons, Ltd, Hoboken, NJ, USA.

[32] Wieghardt, K., Bossek, U., Zsolnai, L., Huttner, G., Blondin, G., Girerd, J. J., Babon-neau, F. (1987) J. Chem. Soc. Chem. Comm.: 651–653.

[33] Collomb, M. N., Deronzier, A. (2009) Eur. J. Inorg. Chem.: 2025–2046.

[34] Berends, H. M., Manke, A. M., Näther, C., Tuczek, F., Kurz, P. (2012) Dalton Trans., 41: 6215–6224.

[35] Parent, A. R., Crabtree, R. H., Brudvig, G. W. (2013) Chem. Soc. Rev., 42: 2247–2252.

[36] Dubois, L., Pecaut, J., Charlot, M. F., Baffert, C., Collomb, M. N., Deronzier, A., Latour, J. M. (2008) Chem. Eur. J., 14: 3013–3025.

[37] Larson, E. J., Pecoraro, V. L. (1991) J. Am. Chem. Soc., 113: 7809–7810.

[38] Dunand-Sauthier, M. N. C., Deronzier, A., Piron, A., Pradon, X., Menage, S. (1998) J. Am. Chem. Soc., 120: 5373–5380.

[39] Beckmann, K., Uchtenhagen, H., Berggren, G., Anderlund, M. F., Thapper, A., Messinger, J., Styring, S., Kurz, P. (2008) Energy Environ. Sci., 1: 668–676.

[40] Kurz, P., Berggren, G., Anderlund, M. F., Styring, S. (2007) Dalton Trans.: 4258–4261.

[41] Shevela, D., Koroidov, S., Najafpour, M. M., Messinger, J., Kurz, P. (2011) Chem. Eur. J., 17: 5415–5423.

[42] Poulsen, A. K., Rompel, A., McKenzie, C. J. (2005) Angew. Chem. Int. Ed., 44: 6916–6920.

[43] Tagore, R., Chen, H. Y., Zhang, H., Crabtree, R. H., Brudvig, G. W. (2007) Inorg. Chim. Acta, 360: 2983–2989.

[44] Shimazaki, Y., Nagano, T., Takesue, H., Ye, B. H., Tani, F., Naruta, Y. (2004) Angew. Chem. Int. Ed., 43: 98–100 (2004).

[45] Siegbahn, P. E. (2009) Acc. Chem. Res., 42: 1871–1880.

[46] Romain, S., Vigara, L., Llobet, A. (2009) Acc. Chem. Res., 42: 1944–1953.

[47] Li, X. C., Chen, G. J., Schinzel, S., Siegbahn, P. E. M. (2011) Dalton Trans., 40: 11 296–11 307.

[48] Betley, T. A., Wu, Q., Van Voorhis, T., Nocera, D. G. (2008) Inorg. Chem., 47: 1849–1861.

[49] Najafpour, M. M., Moghaddam, A. N., Allakhverdiev, S. I., Govindjee, (2012) Biochim. Biophys. Acta, 1817: 1110–1121.

[50] Shafirovich, V. Y., Khannanov, N. K., Shilov, A. E. (1981) J. Inorg. Biochem., 15: 113–129.

[51] Gorlin, Y., Jaramillo, T. F. (2010) J. Am. Chem. Soc., 132: 13 612–13 614.

[52] Harriman, A., Pickering, I. J., Thomas, J. M., Christensen, P. A. (1988) J. Chem. Soc. Faraday Trans. I, 84: 2795–2806.

[53] Hocking, R. K., Brimblecombe, R., Chang, L. Y., Singh, A., Cheah, M. H., Glover, C., Casey, W. H., Spiccia, L. (2011) Nature Chem., 3: 461–466.

[54] Najafpour, M. M., Ehrenberg, T., Wiechen, M., Kurz, P. (2010) Angew. Chem. Int. Ed., 49: 2233–2237.

[55] Najafpour, M. M., Pashaei, B., Nayeri, S. (2012) Dalton Trans., 41: 4799–4805.

[56] Zaharieva, I., Najafpour, M. M., Wiechen, M., Haumann, M., Kurz, P., Dau, H. (2011) Energy Environ. Sci., 4: 2400–2408.

[57] Yagi, M., Narita, K. (2004) J. Am. Chem. Soc., 126: 8084–8085.

[58] Yagi, M., Narita, K., Maruyama, S., Sone, K., Kuwabara, T., Shimizu, K. (2007) Biochim. Biophys. Acta, 1767: 660–665.

[59] Berends, H.-M., Homburg, T., Kunz, I., Kurz, P. (2011) Appl. Clay Sci., 53: 174–180.

[60] Li, G., Sproviero, E. M., Snoeberger, R. C. III, Iguchi, N., Blakemore, J. D., Crabtree, R., Brudvig, G., Batista, V. S. (2009) Energy Environ. Sci., 2: 230–238.

[61] Gao, Y. L., Crabtree, R. H., Brudvig, G. W. (2012) Inorg. Chem., 51: 4043–4050.

[62] Kanady, J. S., Tsui, E. Y., Day, M. W., Agapie, T. (2011) Science, 333: 733–736.

[63] Karlsson, E. A., Lee, B. L., Åkermark, T., Johnston, E. V., Kärkäs, M. D., Sun, J. L., Hansson, Ö., Bäckvall, J. E., Åkermark, B. (2011) Angew. Chem. Int. Ed., 50: 11715–11718.

8

Molecular Water Oxidation Catalysts from Iron

W. Chadwick Ellis[1], Neal D. McDaniel[2], and Stefan Bernhard[3]

[1] Department of Chemistry and Chemical Biology, Baker Laboratory, Cornell University, Ithaca, NY, USA

[2] Phillips 66, Bartlesville, OK, USA

[3] Department of Chemistry, Carnegie Mellon University, Pittsburgh, PA, USA

8.1 Introduction

Conversion between water and oxygen is a fundamental step in natural photosynthesis and respiration. Originally of interest to the electrolysis community, water oxidation has a rich history both as an academic exploration and as an industrial pursuit [1]. The reverse reaction, oxygen reduction, is of growing interest in the fields of fuel cells and oxygen sensors [2]. The transformation between these two deceptively simple molecules proceeds by four energetic electron transfers. The mechanism of transformation is therefore relatively complex, despite the simplicity of reagents and products. Additionally, in the endothermic direction of water oxidation, reaction complexity is exacerbated by the need for controlled energy input, rendering the catalysis highly problematic [3].

This chapter highlights the catalysis of water oxidation by iron. Iron is an especially attractive transition-metal catalyst due to its natural abundance, ecological compatibility, and wide range of available oxidation states. While nature has a highly complicated set of requirements for catalysts in biological systems, the performance metrics of water oxidation in the lab are chiefly two: rate and stability. In the field of alkaline electrolysis, iron-based catalysts are known to exhibit both.

It has been noted in electrochemical studies that iron can be exceptionally efficient or extremely inefficient when used as an anode in alkaline electrolysis. When anodized in

Molecular Water Oxidation Catalysis: A Key Topic for New Sustainable Energy Conversion Schemes,
First Edition. Edited by Antoni Llobet.

raw form, iron passivates quickly, and the resulting Fe_2O_3 layer is unable to assist water oxidation [4]. Stainless steel, on the other hand, is an efficient, stable anode found in commercial alkaline electrolyzers [5]. It has been widely reported that alloys of iron, especially those that resist passivation, are active towards the oxygen evolution reaction [6].

In one study, Merrill and Dougherty report a specific mixture of nickel and iron to catalyze the anodic reaction with a performance very near the thermodynamic efficiency limit; that is, very nearly zero overpotential [7]. The authors note a wide range of oxidation states present in the catalyst they prepared, as well as poor crystallinity in the mixed oxide they prescribe. It stands to reason that the oxidation state of iron in such a catalyst is more amenable to change than that of pure hematite (due to lattice energy), and this freedom to accommodate a varying oxidation state may account for much of the observed activity enhancement.

Other work by Tilley and coworkers illuminates a different problem in the use of hematite as an electrocatalyst: poor electrical conductivity [8]. The device targeted by these authors is photoelectrochemical in nature, requiring a highly crystalline hematite thin film. As just discussed, hematite is relatively inactive in the catalysis of water oxidation, so the authors impregnate the surface of their electrodes with iridium to assist the reaction. The issue of poor conductivity is addressed by adjusting the length scale of the hematite film. On reducing the dimensions of the film, the authors report dramatic performance enhancement.

Combined, these electrochemical examples point to reduction in iron oxide crystallinity and domain size being key to both unlocking iron's catalytic performance and removing the conductivity constraint. When put in these terms, it is expected that the extreme case of a molecular iron complex should be significantly more active than anything achievable on the bulk scale of an electrode. While there is significant precedence for molecular species capable of efficiently catalyzing the water oxidation reaction [9, 10], the vast majority of previous work utilizes ruthenium, iridium, and manganese. Only a handful of molecular iron water oxidation catalysts (WOCs) have been reported to date.

8.2   Fe-Tetrasulfophthalocyanine

Elizarova and coworkers reported oxygen evolution from a number of metallophthalocyanine complexes in 1981 [11]. In the experiments described, trisbipyridyl ruthenium(III) was used as a sacrificial oxidant at an unbuffered pH of 2. The reduced form of this oxidant, namely trisbipyridyl ruthenium(II), is a commonly used light harvester in photochemical water splitting endeavors, and is known to undergo oxidative quenching. The use of this oxidized ruthenium(III) species to drive water oxidation is therefore particularly relevant for photocatalytic applications. Under acidic conditions, this oxidant possesses sufficient potential to oxidize water, but it requires a catalyst to effect the reaction at a measurable rate. The tests described in Elizarova's article were initiated at alkaline pH (9–10), achieved by addition of sodium hydroxide, and were reported to terminate at a final pH that was slightly acidic (3–4).

The catalysts being investigated by these authors included phthalocyanine complexes of aluminum, chromium, manganese, iron, cobalt, nickel, copper, and zinc. Upon addition of trisbipyridyl ruthenium(III), evolution of oxygen and acid was reported to occur to the extent indicated in Table 8.1.

Table 8.1 *Catalysts studied by Elizarova and coworkers [11] and corresponding oxygen yields*

Catalyst	O_2 Yield
Disulfodihydrophthalocyanine	0%
Al- Disulfochlorophthalocyanine	20%
Cr- Tetrasulfochlorophthalocyanine	5%
Mn- Tetrasulfophthalocyanine	14%
Fe- Tetrasulfophthalocyanine	47%
Co- Tetrasulfophthalocyanine	50%
Co- Disulfotetrahydrophthalocyanine	24%
Co- Disulfotetraaminophthalocyanine	54%
Co- Disulfotetranitrophthalocyanine	65%
Co- Disulfooctabromophthalocyanine	62%
Co- Tetra(sulfophenyl)porphyrin	65%
Ni- Tetrasulfophthalocyanine	18%
Cu- Disulfophthalocyanine	21%
Zn- Disulfophthalocyanine	25%

The authors concluded that phthalocyanines of iron and cobalt were efficient catalysts towards the oxidation of water. They also posited that the reaction of interest was preceded by removal of two electrons from the phthalocyanine complexes. Rationale for the surprising performance of unlikely candidates (especially the aluminum, chromium, copper, and zinc complexes) was not offered directly, but it was implied that the role of the metals in these complexes might involve electronic structure tuning of the phthalocyanine. The publication presents neither a detailed description of the synthetic protocols that were used to synthesize the phtalocyanines nor details of the kinetics of the observed O_2 evolution. It is noteworthy that the authors observed no O_2 evolution from water oxidation reactions utilizing sulfonated Fe(II) and Cu(II) phthalocyanine complexes with ceric ammonium nitrate (CAN) (pH = 0.7).

8.3 Fe-TAML

In the 1980s, Collins and coworkers initiated a program of iterative ligand design to develop a homogeneous metalloredox-active oxidant [12]. The class of compounds known as iron-centered tetraamido macrocyclic ligands **1–5** (Fe-TAMLs, Figure 8.1), some of which are in commercial deployment for environmental remediation purposes, resulted from this program [13, 14]. The macrocycle equatorially coordinates an Fe^{III} atom using four deprotonated amide groups, with Na^+ as a typical cation. In solution, labile aqua ligands coordinate the iron center residing above and below the plane of the macrocycle. Their lability yields sites for exchange with and activation of dioxygen and reduced oxygen species. Collins and coworkers engineered Fe-TAMLs to resist intramolecular oxidative degradation and hydrolysis [15].

Over the last decade, extensive kinetic characterization of Fe-TAMLs has shown them to function as peroxidase mimics [16]. They typically catalyze the destructive oxidation of substrates using peroxides in water at neutral to alkaline pH. Oxidation rates were shown to be optimal between pH 8 and 11, a range dictated by the pK_a of the axial aqua ligands. This pK_a is controlled by the Lewis acidity of the metal complex, which can be tuned by the electron-donating character of the ligand through the appropriate selection of ligand substituents, as exemplified by **1–5** [17]. This relation between ligand substituents and catalyst reactivity should play a strong role in future ligand design strategy.

After optimizing the Fe-TAML system, Collins and coworkers applied it to the purification of water from recalcitrant organic pollutants under conditions native to the waste stream. A few examples of contaminants vulnerable to oxidation by the Fe-TAML/peroxide system include chlorophenols, phosphate pesticides, estrogens, pharmaceuticals, and dyes [18]. Product characterization in many cases shows extensive mineralization of the pollutant.

In attempts to understand the mechanism of these unique oxidation catalysts, a number of high-valent Fe–oxo intermediates have been isolated and characterized. Fe-TAMLs activate dioxygen in noncoordinating solvents such as dichloromethane to form an $Fe^{IV} \mu$-oxo dimer [19]. In water, this dimer exists in a pH-dependent equilibrium with a monomeric Fe^{IV}–oxo species, which can be irreversibly oxidized by one electron electrochemically [20]. Using cryogenic conditions, the Fe^V–oxo can be isolated and characterized spectroscopically [21]. Recently, the reactivity of both the Fe^{IV}– and the Fe^V–oxo species has been characterized in elegant work [22–24].

Upon the exciting realization that this rich Fe–oxo chemistry could allow for entry into the catalytic cycle of water oxidation, Ellis and coworkers investigated the potential for **1–5** to oxidize water using CAN and sodium periodate ($NaIO_4$) as sacrificial oxidants (Figure 8.1) [25]. CAN is a commonly used single-electron oxidant in the field of homogeneous water oxidation with experimental advantages including its large oxidation potential,

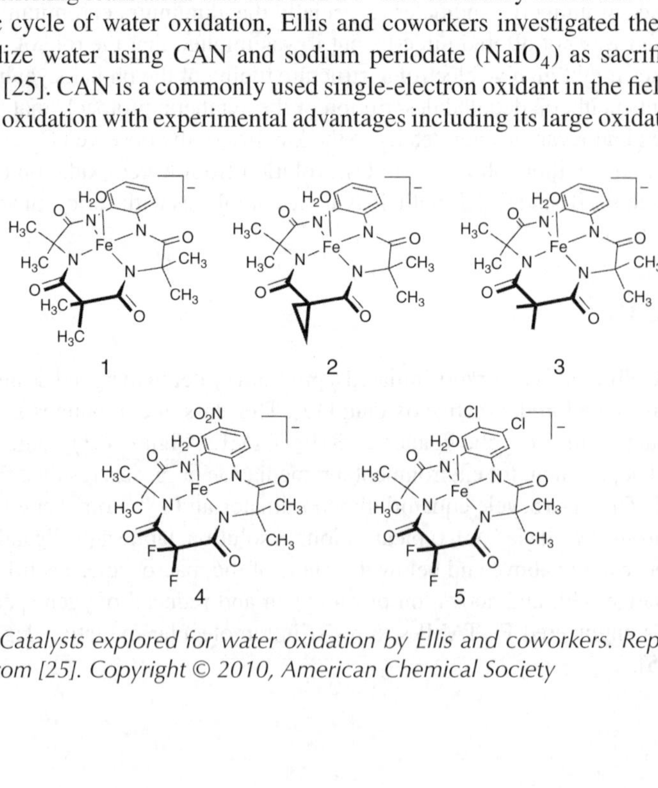

Figure 8.1 *Catalysts explored for water oxidation by Ellis and coworkers. Reproduced with permission from [25]. Copyright © 2010, American Chemical Society*

its slow oxidation of water in the absence of a catalyst, and its inability to yield O_2 via disproportionation (CAN may not, however, be altogether innocent when it is used in water oxidation catalysis [26]). Two drawbacks to its use are its intense yellow/orange color and its highly acidic nature. As an alternative, periodate was used in situations that called for solution transparency or elevated pH. The utility of periodate in water oxidation was recently validated by Crabtree, and it was shown to function with catalysts of low overpotential [27].

Compound **1** was found to be inactive towards water oxidation using CAN, presumably due to its acid sensitivity [13]. However, **2–5** showed fast liberation of O_2 upon addition of CAN. Meanwhile, simple iron salts in combination with tetra-amido macrocyclic ligand (TAML) fragments exhibited no activity. The turnover frequency (TOF) increases as the substituents on the ligand are made more electronegative, with **5** affording an extremely fast TOF in excess of 1.3 s^{-1}. A definitively catalytic turnover number (TON) of 16 was achieved.

Dioxygen evolution was noted to be first-order in [5], thereby undermining claims that intermolecular catalysis results in the formation of an O–O bond. In addition, the *trans* ligation sites suggest that an intramolecular O–O bond-forming process is unlikely. (It is noteworthy that this catalyst is active with *trans* ligation sites, while the catalysts studied by Filliol and coworkers function with *cis* sites, *vide infra*.) Therefore, given the precedent for Fe^V–oxo speciation, one could surmise that an Fe^V–oxo reacts with water to generate an equivalent of hydrogen peroxide, which then quickly decomposes into O_2 gas. The authors note that O_2 evolution was observed to follow two separate rate profiles, with an extremely fast but short-lived ($\sim$ 10 seconds) rate of evolution giving way to a much slower, but longer-lived ($\sim$ 2 hours) evolution.

Ellis and coworkers note that **5** (in various oxidation states) is stable in the presence of either acid or oxidant (periodate), but not both simultaneously. This observation suggests that one or both of the high-valence FeIV and FeV states, which are only achieved in the presence of a strong oxidant, is far more susceptible to decomposition in acid than the catalyst's reduced state. In order to improve the performance of Fe-TAMLs as WOCs, future attention should be devoted to: (i) the origin of the dichotomous rate behavior; (ii) the mechanism of catalyst deactivation and enhancement of catalyst longevity; and (iii) the mechanism of Fe-TAML-catalyzed water oxidation and the nature of the active species. The decomposition pathways of molecular WOCs have recently been reviewed [28], and the use of periodate in conjunction with ultraviolet/visible (UV/vis) techniques should yield future insight into mechanism and catalyst speciation. Theoretical investigations can also yield valuable insight into catalyst function; for example, Ertem and coworkers have described a plausible mechanism for water oxidation by Fe-TAMLs based on quantum calculations [29].

8.4 Fe-mcp

Fillol and coworkers reported significant activity towards water oxidation from several iron complexes of modular, tetradentate nitrogen-based ligands [30]. The sacrificial oxidants CAN and sodium periodate were both employed in the study of these complexes. The pH of reaction was strongly acidic in all cases where oxygen evolution was detected. The most active of these complexes, shown in Figure 8.2, is reported to generate over 1050 turnovers

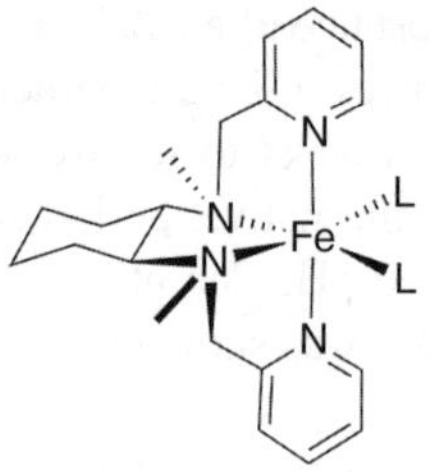

Figure 8.2   *One of several active iron complexes studied for water oxidation by Fillol and coworkers. Reproduced with permission from [30]. Copyright © 2011, Rights Managed by Nature Publishing Group*

using periodate as the oxidant, or a maximum TOF of 838 hr^{-1} with cerium(IV) as the oxidant. These values mark a significant milestone in homogeneous water oxidation from earth-abundant materials.

Other iron complexes were explored as a control group, for which either (i) the tetradentate ligands forced the position of the two monodentate ligands to be a *trans* configuration instead of *cis* or (ii) a pentadentate ligand was used, leaving room for only a single monodentate ligand. In both cases, no significant activity was observed, leading the authors to conclude that a *cis* arrangement of monodentate ligands was a prerequisite to oxygen evolution activity.

The authors additionally point out the formation of a visible absorbance peak for their complexes at a wavelength near 775 nm, which they attributed to the formation of an iron(IV) oxo species. This species is believed to be the resting state of iron in the strongly oxidizing cerium(IV) solution. Kinetic analyses revealed a first-order rate dependence on both [iron] and [cerium], suggesting that the rate-determining step is a further oxidation by cerium(IV) to form an iron(V) intermediate. It is postulated that this intermediate then undergoes fast reaction with a nearby water molecule to generate a peroxide bond, following which decomposition to oxygen is relatively straightforward.

8.5 Fe_2O_3 as a Microheterogeneous Catalyst

The Lau group very recently reported water oxidation activity from a number of iron-containing compounds, which they attribute ultimately to the uncontrolled formation of Fe_2O_3 particles from their various precursors [31]. Unlike the previous two studies, the work of Chen and coworkers takes place in an alkaline medium (pH 7.5–9.0). In one set of reactions, trisbipyridyl ruthenium(III) is prepared and used as an oxidant in combination with the set of catalysts shown in Table 8.2. Abbreviations for the employed ligands are: mcp = N, N*ı*-dimethyl-N, N*ı*-bis(2-pyridylmethyl)cyclohexane-1, 2-diamine, bpy = 2, 2′-bipyridine, tpy = 2, 2′ : 6′, 2″-terpyridine, cyclen = 1, 4, 7, 10-tetraazacyclodecane, and tmc = 1, 4, 8, 11-tetramethyl-1, 4, 8, 11-tetraazacyclotetradecane.

Simple iron salts are found to catalyze the reaction much more effectively than complexes of multidentate ligands. In the case of the tetradentate mcp ligand (the same complex studied by Fillol and coworkers, shown in Figure 8.2), only background-level oxygen production is observed. However, when a similar experiment is conducted at acidic pH (using

Table 8.2 *Iron-catalyzed water oxidation driven by [Ru(bpy)$_3$](ClO$_4$)$_3$ at pH 8.5 in borate buffer. Reproduced with permission from [31]. Copyright © 2013 WILEY-VCH Verlag GmbH & Co. KGaA, Weinheim*

Catalyst	Yield[a]	TOF, (s^{-1})	TON[b]
No catalyst	8%	-	-
Fe(mcp)Cl$_2$	9%	-	-
[Fe(bpy)$_2$Cl$_2$]Cl	49%	3.6	95
[Fe(tpy)$_2$]Cl$_2$	16%	1.5	19
[Fe(cyclen)Cl$_2$]Cl	54%	4.4	108
Fe(tmc)Br$_2$	48%	4.6	93
Fe(ClO$_4$)$_3$	71%	9.6	147

[Catalyst] = 0.8 µM, [Ru(bpy)$_3$$^{3+}$] = 750 µM, pH = 8.5 (15 mM borate buffer), 23 °C, time = 120 s
[a] Determined by Clark Electrode
[b] (mol O$_2$ − blank)/(mol catalyst)

cerium(IV) as the oxidant), the mcp complex shows activity similar to that observed by Fillol and coworkers [30].

In a second set of experiments, trisbipyridyl ruthenium(II) is used as a photosensitizer, with persulfate (S$_2$O$_8$$^{2-}$) as the corresponding sacrificial oxidant. It is postulated that a photon is absorbed by the ruthenium photosensitizer, exciting the complex into a relatively long-lived triplet metal–ligand charge transfer (MLCT) state. At this point the excited photosensitizer is quenched oxidatively by persulfate to generate sulfate, a sulfuryl radical, and an oxidized photosensitizer molecule. The latter two of these species are each kinetically and thermodynamically capable of withdrawing an electron from a WOC, whereas minimal reaction is observed directly between persulfate and the catalyst. Additional control experiments are reported to confirm that all three components (catalyst, photosensitizer, and oxidant) are necessary for oxygen generation, although it should be noted that, in agreement with literature precedence [32], the authors do observe some oxygen evolution from decomposition of persulfate in the absence of catalyst or light. The results of this second set of experiments are reproduced in Table 8.3.

Table 8.3 *Visible light-driven water oxidation by iron catalysts. Reproduced with permission from [31]. Copyright © 2013 WILEY-VCH Verlag GmbH & Co. KGaA, Weinheim*

Entry	Catalyst	O$_2$(µmol)	%Yield[a]	TON[b]
1	No catalyst	0.4	5	-
2	Fe(mcp)Cl$_2$	2.0	24	194
3	[Fe(bpy)$_2$Cl$_2$]Cl	1.7	21	157
4	[Fe(tpy)$_2$]Cl$_2$	3.5	42	376
5	[Fe(cyclen)Cl$_2$]Cl	3.8	46	412
6	Fe(tmc)Br$_2$	3.4	41	364
7	Fe(ClO$_4$)$_3$	4.0	48	436

Note: [Catalyst] = 1.0 µM, [Ru(bpy)3Cl2] = 0.2 mM, [Na$_2$S$_2$O$_8$] = 2 mM in 15 mM borate buffer (pH 8.5) at 23 °C, λ > 420 nm
[a] Determined by GC, yield = (mol of O$_2$)/(1/2 × mol of Na$_2$S$_2$O$_8$);
[b] TON = (mol of O$_2$ after subtracting the blank)/(mol of catalyst), error around 10%

On monitoring their reaction mixtures with dynamic light scattering (DLS; particle size analysis), the authors report clear evidence of particle formation on the timescale of minutes following initiation of the reaction. Agglomeration of smaller particles into micron-size clusters is observed to coincide with loss of catalytic activity. In summary, the authors attribute the observed oxygen evolution activity at high pH to a colloidal iron oxide species that forms *in situ*. Meanwhile, their experimental results at acidic pH concur with the conclusions of other groups [25, 30].

8.6   Conclusion

The labile nature of first-row transition-metal complexes renders their use in most catalytic applications difficult. Demetalation is a probable and common deactivation pathway for

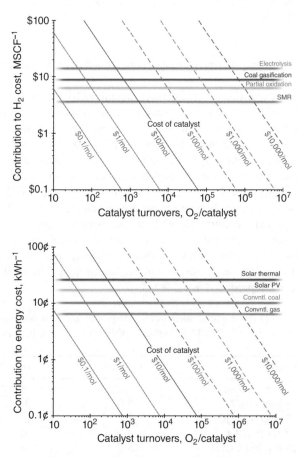

Figure 8.3   *The cost of artificially photosynthesized hydrogen is shown as a function of water oxidation catalyst robustness and price, in units of hydrogen cost (top) and corresponding power cost (bottom). Reference lines have been added to give context where available. The unit of hydrogen normalization, MSCF, is a thousand standard cubic feet. SMR refers to steam methane reforming, which accounts for > 90% of all hydrogen produced today [33, 34]*

these catalysts, documented by the behavior observed in the iron-based WOCs described in this chapter. It is evident that a strongly chelating ligand sphere is a critical prerequisite for these catalysts. The high-valent intermediates commonly detected in water oxidation cycles will be even more likely to decomplex and liberate the central iron cation by either demetallation or ligand oxidation pathways. The chelate effect underscores the need for polydentate ligands, and experimental evidence validates that poly-anionic ligands better stabilize these high-oxidation states. Future work on these catalysts should continue to develop these design principles in order to obtain active catalysts with stable ligand environments. The relationship between a catalyst's robustness and its ultimate contribution to the process cost of artificial photosynthesis is dramatic. Figure 8.3 highlights this critical relationship on a log scale. It has to be an objective of future studies to discover materials with activity across a wide pH range, with TONs comparable to complexes using second- and third-row transition metals. The development of techniques for establishing the nature of the catalytically active species is critical to these iron-based complexes. Then photocatalytic water oxidations that are thermodynamically uphill must be explored in order to enable the deployment of these reactions in full water-splitting systems powered by sunlight.

In conclusion, the initial work on iron-based WOCs has offered insights into these fast and surprisingly robust molecular systems. The inexpensive, biologically innocuous, and abundantly available iron ion renders these materials extremely appealing. The established feasibility of stable and robust first-row WOCs will multiply the efforts in this field, which will, hopefully, lead to the utilization of such complexes in future energy solutions, thus enabling the Great Energy Transition [35].

References

[1] Hall, D. E. (1985) J. Electrochem. Soc., 132(2): 41C–48C.

[2] Song, C., Zhang, J. (2008) In PEM Fuel Cell Electrocatalysts and Catalyst Layers (ed. Zhang, J.). Springer, Berlin, Germany, p. 89.

[3] Dau, H., Limberg, C., Reier, T., Risch, M., Roggan, S., Strasser, P. (2010) Chem. Cat. Chem., 2(7): 724–761.

[4] Pourbaix, M. (1974) Atlas of Electrochemical Equilibria in Aqueous Solutions, 2nd English Ed. National Association of Corrosion Engineers, Houston. Cebelcor, Brussels.

[5] Companies that sell alkaline electrolyzers include: Norsk Hydro, Lurgi, De Nora, Bamag, Teledyne, Electrolyzer Corp.

[6] Ma, B., Yang, J., Han, H., Wang, J., Zhang, X., Li, C. (2010) J. Phys. Chem. C, 114: 12818–12822.

[7] Merrill, M. D., Dougherty, R. C. (2008) J. Phys. Chem., 112: 3655–3666.

[8] Tilley, S. D., Cornuz, M., Silvula, K., Grätzel, M. (2010) Ang. Chem. Int. Ed., 49(36): 6405–6408.

[9] Yagi, M., Kaneko, M. (2001) Chem. Rev., 101: 21–35.

[10] Wasylenko, D. J., Palmer, R. D., Berlinguette, C. P. (2013) Chem. Commun., 49(3): 218–227.

[11] Elizarova, G. L., Matvienko, L. G., Lozhkina, N. V., Maizlish, V. E., Parmon, V. N. (1981) React. Kinet. Catal. Lett., 16(2–3): 285–288.

[12] Collins, T. J. (1994) Acc. Chem. Res., 27: 279–285.

[13] Collins, T. J. (2002) Acc. Chem. Res., 35: 782–790.

[14] GreenOx Catalysts, www.greenoxcatalysts.com (accessed December 13, 2013).

[15] Ryabov, A. D., Collins, T. J. (2009) Adv. Inorg. Chem., 61: 471–521.

[16] Ghosh, A., Mitchell, D. A., Chanda, A., Ryabov, A. D., Popescu, D.-.L., Upham, E. C., Collins, G. J., Collins, T. J. (2008) J. Am. Chem. Soc., 130: 15 116–15 126.

[17] Ellis, W. C., Tran, C. T., Roy, R., Rusten, M., Fischer, A., Ryabov, A. D., Blumberg, B., Collins, T. J. (2010) J. Am. Chem. Soc., 132: 9774–9781.

[18] Collins, T. J., Khetan, S. K., Ryabov, A. D. (2009) In Handbook of Green Chemistry (Anastas, P., Crabtree, R. eds). Wiley-VCH, Weinheim, Germany, p. 39.

[19] Ghosh, A., Tiago de Oliveira, F., Yano, T., Nishioka, T., Beach E. S., Kinoshita, I., Münck, E., Ryabov, A. D., Horwitz, C. P., Collins, T. J. (2005) J. Am. Chem. Soc., 127: 2505–2513.

[20] Chanda, A., Shan, X., Chakrabarti, M., Ellis, W. C., Popescu, D.-.L., Tiago de Oliveira, F., Wang, D., Que, L., Jr., Collins, T. J., Muenck, E., Bominaar, E. L. (2008) Inorg. Chem., 47: 3669–3678.

[21] Tiago de Oliveira, F., Chanda, A., Banerjee, D., Shan, X., Mondal, S., Que, L. Jr., Bominaar, E. L., Münck, E., Collins, T. J. (2007) Science, 315: 835–838.

[22] Popescu, D.-.L., Vrabel, M., Brausam, A., Madsen, P., Lente, G., Fabian, I., Ryabov, A. D., van Eldik, R., Collins, T. J. (2010) Inorg. Chem., 49: 11 439–11 448.

[23] Kundu, S., Thompson, J. V. K., Ryabov, A. D., Collins, T. J. (2011) J. Am. Chem. Soc., 133: 18 546–18 549.

[24] Kundu, S., Annavajhala, M., Kurnikov, I. V., Ryabov, A. D., Collins, T. J. (2012) Chem. Eur. J., 18: 10 244–10 249.

[25] Ellis, W. C., McDaniel, N. D., Bernhard, S., Collins, T. J. (2010) J. Am. Chem. Soc., 132: 10 990–10 991.

[26] Stull, J. A., Britt, R. D., McHale, J. L., Knorr, F. J., Lymar, S. V., Hurst, J. K. (2012) J. Am. Chem. Soc., 134: 19 973–19 976.

[27] Parent, A. R., Brewster, T. P., De Wolf, W., Crabtree, R. H. Brudvig, G. W. (2012) Inorg. Chem., 51: 6147–6152.

[28] Limburg, B., Bouwman, E., Bonnet, S. (2012) Coord. Chem. Rev., 256: 1451–1467.

[29] Ertem, M. Z., Gagliardi, L., Cramer, C. J. (2012) Chem. Sci., 3: 1293–1299.

[30] Fillol, J. L., Codolá, Z., Garcia-Bosch, I., Gómez, L., Pla, J. J., Costas, M. (2011) Nat. Chem., 3: 807–813.

[31] Chen, G., Chen, L., Ng, S.-M., Man, W.-L., Lau, T.-C. (2013) Ang. Chem. Int. Ed., 52: 1789–1791.

[32] Beylerian, N. M., Vardanyan, L. R., Harutyunyan, R. S., Vardanyan, R. L. (2002) Macromol. Chem. Phys., 203(1): 212–218.

[33] Annual Energy Outlook 2012. US Energy Information Administration.

[34] Basye, L., Swaminathan, S. (1997) Hydrogen Production Costs: A Survey. US Department of Energy.

[35] Oreskes, N., Conway, E. M. (2013) Dædalus, 142(1): 40–58.

9

Water Oxidation by Co-Based Oxides with Molecular Properties

Marcel Risch[1,2], Katharina Klingan[1], Ivelina Zaharieva[1], and Holger Dau[1]

[1]*Department of Physics, Free University of Berlin, Berlin, Germany*
[2]*Electrochemical Energy Laboratory, Massachusetts Institute of Technology, Cambridge, MA, USA*

9.1 Introduction

For decades, intense research has been directed towards the synthesis of structurally well-defined molecular water oxidation catalysts (WOCs) based on inexpensive first-row transition metals. However, their application in technological systems is seriously hampered by turnover numbers (TONs) that are several orders of magnitude below the threshold ($10^7 – 10^{12}$) needed for use in devices of technological relevance. Although degradation (corrosion) remains a crucial issue, inorganic solid-state materials carry the promise of superior robustness. Therefore, the (re)discovery in 2008 of electrodeposited amorphous Co oxides as WOCs [1] has spurred a wave of investigations into this material class. In the meantime, the electrodeposition of reasonably efficient amorphous WOCs based on Ni [2] and Mn [3] has also been reported; these materials share striking structural similarities [3–5], as too do water-oxidizing Mn–Ca oxide particles [6, 7] and amorphous Co oxide nanoparticles [8]. Structurally and functionally, these oxide materials are remarkable. The similarity of the Co catalyst (CoCat) to a molecular catalyst appears to be stronger than that to a classical heterogeneous catalyst. Indeed, this class of amorphous and hydrated (water-containing) catalyst may be viewed best as nonclassical solid-state materials with molecular properties.

In this chapter, we will summarize early and recent developments relating to the use of amorphous Co-based oxides for water oxidation. These oxides can be surprisingly

Molecular Water Oxidation Catalysis: A Key Topic for New Sustainable Energy Conversion Schemes, First Edition. Edited by Antoni Llobet.
© 2014 John Wiley & Sons, Ltd. Published 2014 by John Wiley & Sons, Ltd.

complex. They often consist of Co ions, redox-inactive cations (K, Na, Ca, etc.), anions that can act as proton-accepting bases (phosphate, borate, chloride, etc.), and (layers) of water molecules, as detailed further in the text. For simplicity, we will denote these oxide catalysts as CoCat, NiCat, or MnCat, according to their respective redox-active metal. We will use these expressions irrespective of whether the material was electrodeposited as a thick film (50–3000 nm) on a planar electrode or synthesized in particle form.

9.2   CoCat Formation

The first report of electrodeposition of an amorphous Co oxide and its electrochemical characterization may likely be attributed to Coehn and Gläser [9]. Over a century ago, the ancestral CoCat was already discussed in the context of water oxidation and hydrogen evolution. A few reports about related bulk oxides (as opposed to surface oxides) and their characterization followed in the first decades of the 20th century [10–13].

El Wakkad and Hickling [14] investigated various Co oxide films deposited galvanostatically. The resulting films were tested in a variety of electrolytes, among which was a phosphate buffer (pH 6.8). They found that the presence of phosphate stopped the dissolution of Co, which was observed with other alkaline (NaOH, $NaCO_3$) and acidic (CH_3OONa/H) electrolytes. Moreover, initial dissolution was reverted in phosphate buffer, indicating that the films possessed a self-repair mechanism there. However, the application of these films for water oxidation was not discussed.

Benson and coworkers [15, 16] deposited Co oxide films on platinum and nickel. Galvanostatic deposition in potassium hydroxide yielded a black film (like the CoCat) with sufficient crystallinity to resolve the unit-cell parameters (hexagonal lattice; $a = 6.75$ Å, $c = 5.36$ Å) [15]. Oxygen evolution was detected, but the origin of the oxygen was not discussed [16]. A discussion of the oxygen evolution of an early Co oxide film may be found in the patent application of Osamu [17]. Electrodeposition of Co hydroxides by cyclic voltammetry was later extensively characterized in alkaline and neutral solutions [18, 19]. Tseung and coworkers published an extensive series of reports on reactive galvanostatic deposition of Co oxides [20–27]. The surface morphology of their electrodes [17] showed the same nodules as the initially reported CoCat [1] (see Figure 9.1b).

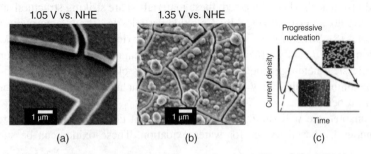

Figure 9.1   Surface of the Co oxide electrodeposited at (a) 1.05 and (b) 1.35 V versus normal hydrogen electrode (NHE). Adapted with permission from [27]. Copyright ©, 2009 American Chemical Society. (c) Illustration of film formation by progressive nucleation. Adapted with permission from [37]. Copyright ©, 2012 American Chemical Society.

The CoCat for electrochemical water oxidation reported by Kanan and Nocera [1] has been studied widely due to its efficiency at neutral pH, its self-repair mechanism, and its electrochemical self-assembly from low-cost materials. While the catalytic material may have been similar to previously reported Co oxides, Kanan and Nocera were the first to see its enormous potential for future energy applications. Their work, published in 2008, has triggered a wave of in-depth investigations into basic and applied aspects of the CoCat. In reference [1], the CoCat was produced by electrodeposition on indium tin oxide (ITO) substrates from buffer solution (pH 7) of 0.1 M potassium phosphate (KP$_i$) containing 0.5 mM Co(OH$_2$)$_6$(NO$_3$)$_2$. More recently, the following modifications have been reported:

(i)   Deposition on other semiconductor and metal substrates, such as glassy carbon, carbon felt, fluorinated tin oxide (FTO) [28], high-surface-area nickel foam [29], and various photoanodes (Section 9.5), either by electrodeposition (this section), photodeposition (Section 9.5), or sputtering [30].
(ii) Exchange of the electrolyte; specifically, substitution of potassium for group 1 and group 2 cations and of phosphate for phosphinate, borate, fluorides, amines, acetates, or chlorides, with extension of the range of operation and formation from pH 3.5 to 14.0 [31–35].

The significance of (2) is that neither potassium nor phosphate is essential for catalytic activity; they are likely not an essential part of the catalytic unit itself (see Section 9.3 for a detailed discussion).

The CoCat forms on a variety of conducting substrates, such as those already discussed. Surface morphology and film thickness depend on the deposition time, the composition of the electrolyte, and the applied potential (Figure 9.1a,b) [1, 28, 29, 36]. Nucleation of the catalytic film is progressive (Figure 9.1c) [37]. Catalyst operation and formation do not require deionized reagent-grade water: operation in brine and river waters has been demonstrated [29], suggesting that other ions, especially chloride, do not inhibit oxygen evolution. Self-repair of the oxide has been found for pH greater than ~ 6 [37]. Under these conditions, leaching of both the redox-inert cations and the anions is faster than that of the Co ions, as shown by radioactive labeling experiments under open-circuit conditions (no voltage applied to CoCat electrode) [34].

Amorphous deposits with properties and structures related to the CoCat have been proposed to form during water oxidation on the surfaces of various solid precursor materials, such as LiCoO$_2$ and LiCoPO$_4$ [38] (common battery materials), as well as CoWO$_4$ [39], Co$_3$O$_4$ [40], Co-MOF71 [41], and Ba$_{0.5}$Sr$_{0.5}$Co$_{0.8}$Fe$_{0.2}$O$_3$ [42, 43]. In addition, it has been reported that Co-doped nanocrystalline ruthenium oxide contains Co-oxo clusters akin to the CoCat [44]. A further interesting class of materials that share the Co-oxo structure of the CoCat have been recently described as "unsupported nanoparticles" [8, 45].

Although not a bulk oxide catalyst, a recent molecular catalyst is also worth mentioning, as its tetranuclear Co-oxo core may model the smallest Co oxide unit needed for water oxidation. Hill and coworkers [46] reported an efficient WOC consisting of a soluble, carbon-free heteronuclear molecule: a Co polyoxometalate ([Co$_4$(H$_2$O)$_2$ (α-PW$_9$O$_{34}$)$_2$]$^{10-}$, Co-POM). There is an ongoing debate as to whether the molecular [Co$_4$(μ-O)$_6$O$_{10}$] unit may represent the catalytically active part [50b], or whether the polyoxometalate undergoes a hydrolytic decomposition to form CoCat fragments that oxidize water [50a, 51]. Possibly the mechanisms of water oxidation in the Co-POM and the CoCat are identical. Further, molecules with Co_4O_4 cubane cores have been studied as models for the active site of the CoCat [52–57].

9.3   Structure and Structure–Function Relations

Conventional diffraction analysis is not applicable to the elucidation of the structure of electrodeposited CoCat, due to its amorphous character. On the other hand, X-ray absorption spectroscopy (XAS) is well suited to the investigation of such materials. The analysis of X-ray absorption and the pair distribution function (PDF) [58] from high-energy X-ray scattering can provide information on interatomic distances. XAS and PDF suggest that CoCat films consist of octahedrally coordinated Co, which forms oligomeric cobaltates of molecular dimensions.

In Figure 9.2, the Fourier transform (FT) of the extended X-ray absorption fine structure (EXAFS) is shown for the CoCat and a crystalline reference material, $LiCoO_2$. The similarity in their spectra suggests a close structural correspondence. The peak positions in the FT relate to the average bond distance between the X-ray-absorbing Co atom and groups of backscattering atoms [59, 60]. The reduced distances of the x-axis in Figure 9.2 are about 0.3 Å lower than the real interatomic distances; the latter can be determined precisely by

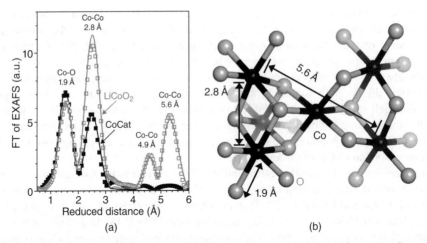

Figure 9.2 *(a) Fourier transform (FT) of the Co K-edge extended X-ray absorption fine structure (EXAFS) for the amorphous CoCat (■) and crystalline $LiCoO_2$ (□), as well as simulated spectra (solid lines; see [5, 62] for further details). Extensive edge sharing is characteristic of both the CoCat and $LiCoO_2$. Adapted with permission from [45]. Copyright ©, 2009 IOP. (b) Illustration of basic structural motifs deduced from the simulations in panel (a). The edge-sharing octahedra in the CoCat can be arranged as incomplete $Co_3(\mu\text{-}O)_4$ cubanes or as complete $Co_4(\mu\text{-}O)_4$ cubanes. The structural unit containing seven Co ions and the 1 : 1 ratio between complete and incomplete Co-oxo cubanes are used for illustrative purposes only. The extension of a contiguous Co-oxo cluster likely depends on the type of anion present in the CoCat and has been estimated to be close to nine Co atoms in the phosphate–CoCat [35]. Adapted with permission from [5]. Copyright © 2009, American Chemical Society.*

EXAFS simulations. The amplitude of the peaks in the FT is a rough measure of the average number of atoms at the respective distances and the width of their distance distributions. The precision of EXAFS simulations may be as high as 0.01 Å for the distances and about 15% for the number of backscattering atoms [61].

EXAFS and PDF analysis of the local structure have been performed on CoCat films deposited either electrochemically [5, 35, 62, 63] or photochemically [64] from phosphate, acetate, and chloride electrolytes on ITO, glassy carbon, and titania substrates; Co–O bond lengths between 1.89 and 1.91 Å were obtained. This agrees well with the average bond length for octahedral Co^{III} of 1.90 ± 0.03 Å [65]. EXAFS analysis further suggests ligation by six oxygen atoms [5, 63, 64]. Both results corroborate the prevalence of Co^{III} octahedra in the CoCat. The oxidation state of +3 of the Co ion and its octahedral coordination have also been confirmed by analysis of the X-ray absorption near-edge structure (XANES) [5, 63, 64], electron paramagnetic resonance (EPR) [22, 66], and X-ray photoelectron spectroscopy (XPS) [32].

The Co atoms in CoCat films are connected by di-μ-oxo bridges, as suggested by Co–Co distances of 2.81–2.85 Å. These distances are characteristic of edge-sharing octahedra in layered oxides, such as $Li_x CoO_2$ [67], $Na_x CoO_2$ [68], $H_x CoO_2$ [69], and complexes with cubane cores, such as $Co_4(\mu-O)_4$ [70–72]. The reported number of Co ions neighboring the X-ray-absorbing Co ion ranges between 3.3 and 5.3 [5, 63, 64], which implies the presence of only a few interconnected incomplete or complete cubane units ($Co_{3/4}(\mu-O)_4$; see Figure 9.2b). The presence of $Co_4(\mu_3-O)_4$ cubane units is neither excluded nor proven by the presently available experimental [5, 58, 63, 64] and theoretical [73, 74] results. Longer Co–Co distances of 4.9 and 5.6 Å can also be resolved. However, these peaks are clearly more pronounced in crystalline materials, such as $LiCoO_2$ (Figure 9.2a), which further supports the molecular dimensions of the Co-oxo fragments.

Due to the absence of atomic coordinates obtained from diffraction data, structural models of the amorphous CoCat are based on the atomic distances and number of neighboring atoms, as estimated from XAS and PDF data. The visualization of a representative model for the structure is further complicated by the absence of detectable periodicity in the Co-oxide plane beyond ~ 6 Å in EXAFS analysis and ~ 13 Å in PDF analysis. Therefore, clusters of a few Co atoms and an appropriate number of oxygens are chosen to represent plausible atomic structures. For electrodeposition of the CoCat in phosphate, the size of these clusters in the bulk ranges from 9 to 16 Co atoms in EXAFS analysis [5, 63], which compares favorably with the 13–14 atoms found by PDF analysis [58].

All proposed models for the CoCat are related to the crystalline parent structures being either layered $ACoO_2$ (all incomplete cubanes) or spinel ACo_2O_4 (all complete cubanes), where A can be any cation. Models based on the layered $ACoO_2$ structure (Figure 9.3a) match the experimental EXAFS and PDF spectra better than those based *exclusively* on the spinel ACo_2O_4 structure. Models with mixed characteristics are conceivable; indeed, the cubane motif can be formed in crystalline $ACoO_2$ by interlayer octahedra. In addition to such a "cubane defect," vacancies in the Co octahedra layer should be expected for an amorphous material (Figure 9.3b). Furthermore, the cluster models are usually circular to ellipsoidal in shape, for which there is no support in any experimental data. A multitude of more complicated shapes that can include vacancies and cubane defects are equally well described by the available EXAFS data. For the sake of simplicity, we only show models of the CoCat that do not contain any of these defects here; the presence of minor cubane or

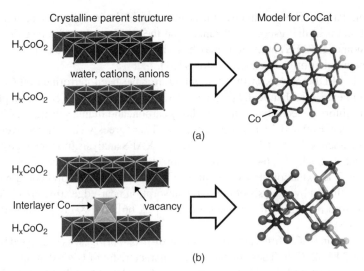

Figure 9.3   (a) Ideal layered oxide/hydroxide and analogous simple layer-like CoCat model. (b) Comparison of a layered oxide/hydroxide with interstitial octahedra and Co-layer vacancies versus an analogous CoCat model with cubane and vacancy defects

vacancy defects would likely not seriously influence the conclusions drawn on the grounds of these simplified models.

Redox-inert cations (A = Li, K, Na, or Ca) are incorporated into CoCat films if they are present in the electrolyte during deposition [1, 30, 33, 35]. Early reports of the molar ratio of ~ 3 : 1 for Co : K [1, 33] led to the hypothesis that the CoCat contains cubanes of type $ACo_3(\mu-O)_4$. Manganese analogs of this motif are present in the natural paragon of water oxidation, the $CaMn_3(\mu-O)_4$ core of photosystem II (PSII; Figure 9.4a) [75–78] and in amorphous Mn–Ca oxides [6, 79]. These motifs are difficult (Na, K, Ca) or impossible (Li) to resolve using EXAFS spectroscopy at the Co K-edge as the scattering probability decays rapidly with decreasing atomic number [80]. Based on Co K-edge EXAFS spectra, Risch and coworkers [5] tentatively proposed the presence of a Co–K distance of approximately 3.8 Å for CoCat deposited in KP_i. Subsequently, Kanan and coworkers [63] found no differences exceeding the noise level in their EXAFS analysis of films made with K^+ and Na^+. They concluded that their data were inconsistent with a significant number of Co–K motifs. A more direct measurement at the potassium K-edge strongly suggests that potassium (and likely other monovalent cations) binds largely unspecifically with the incomplete Co cubanes of the CoCat [35], in analogy to layered oxides where the cations are located between the Co-containing layers (Figure 9.3). On the other hand, a sizable fraction of bivalent calcium may participate in the formation of $CaCo_3(\mu-O)_4$ in the CoCat (Figure 9.4b) [35]. EXAFS simulations yield Ca–Co distances of 3.46 Å: this is very close to the 3.49 Å reported for the $CaCo_3(\mu-O)_4$ cubane in a synthetic complex [72] and compares favorably to the Ca–Mn distance in the photosynthetic Mn complex [77, 78, 81, 82].

The macroscopic CoCat can be composed of a large number of Co-oxo clusters separated by redox-inert cations (e.g. K^+), anions (e.g. HPO_4^{2-}), and water, in analogy to crystalline-layered double hydroxides (Figure 9.5) [83]. We emphasize that it would not

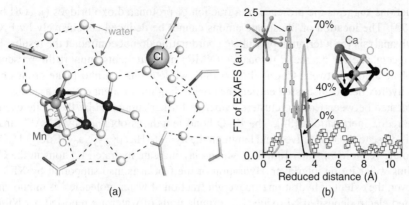

(a)   (b)

Figure 9.4 *(a) Crystallographic model of the water-oxidizing CaMn$_3$(μ–O)$_4$ core of PSII and its ligand environment, based on the crystallographic model of [77]. (b) Fourier transform (FT) of the Ca K-edge EXAFS for a CoCat deposited in CaCl$_2$. Open squares indicate experimental data and lines show simulations for the indicated abundance (per cent) of the cubane motif in the CoCat (see [35] for further details). The structural motifs assigned to selected peaks are displayed schematically. Adapted with permission from [35]. Copyright © 2012 WILEY-VCH Verlag GmbH & Co. KGaA, Weinheim.*

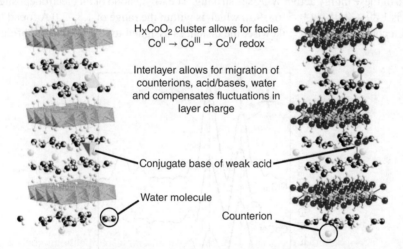

Figure 9.5 *Hypothetical arrangement of cations (light gray), anions (conjugate weak base), and water molecules (oxygen atoms in black) in the vicinity of Co-oxo clusters comprising the CoCat (left: polyhedral representation; right: ball-and-stick representation). We note that in the CoCat the extent of order in the layers of edge-sharing CoO$_6$ octahedra is clearly lower than suggested by this scheme. Reproduced with permisson from [31]. Copyright © American Chemical Society, 2011.*

be in conflict with experimental data if some of the clusters were loosely interconnected or not aligned in parallel. There is no direct experimental evidence of protonated Co-oxo clusters, but a recent Pourbaix diagram suggests that Co hydroxides are stable in the usual operating range of the CoCat [31] and a combination of EXAFS analysis and *ab initio* calculations suggests the presence of a fraction of protonated oxo bridges (μ_2-OH bridging) [73]. The location of phosphate anions cannot be deduced conclusively by EXAFS or PDF analysis [58]; terminal phosphate coordination do not contradict the results. However, recent nuclear magnetic resonance (NMR) measurements support the presence of phosphate anions between CoCat clusters [84], while excluding phosphate groups bound to Co hydroxide. As discussed earlier, the nonbinding monovalent cations are most likely also located between Co-oxo clusters, which is further supported by the facile exchange of the redox-inert cations [34]. The K–O bond length, as obtained by EXAFS analysis [35], is close to the respective bond length in the hydrate $H_{0.19}K_{0.25}Na_{0.06}CoO_2 \cdot H_2O$ [85] and solute K^+ ions in water [86, 87], which may indicate hydrated K^+ ions in the CoCat and thus water in the interlayer. Hydration of the CoCat is also supported by XPS [32]. However, the extent of hydration (the weight fraction of water molecules) is unknown. For a related electrodeposited Ni oxide, 1.5 formula units of water are reported per Ni ion in dried films [2].

Since the initial discussion of the CoCat by Kanan and Nocera [1], electrodeposited films of Ni (NiCat) [2] and, more recently, Mn (MnCat) [3] with comparable activity have been reported. Figure 9.6 compares the EXAFS spectra of these catalytically active metal oxides to each other and to the spectrum of the $Mn_4Ca(\mu\text{-}O)_5$ catalyst of photosynthetic water oxidation, which is bound to the proteins of PSII [77, 89]. The similarity in the EXAFS spectra of these highly active WOCs is striking. The M–O bond of all electrodeposited catalysts is 1.89 ± 0.01 Å [3–5, 63, 90], which is within the range of 1.8–2.0 Å found for the Mn–Ca core of PSII [91]. Likewise, the distances assigned to di-μ-oxo metal bridges are

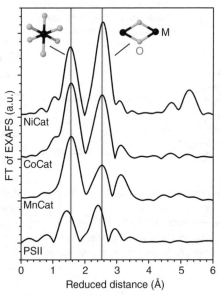

Figure 9.6 *FT of the transition metal K-edge EXAFS of NiCat [4], CoCat [35], and MnCat [3]. For comparison, the EXAFS of the Mn–Ca core of the protein-bound Mn_4Ca catalyst of photosynthetic water oxidation is shown [76]. The structural motifs assigned to selected peaks are displayed schematically*

2.81 ± 0.01 Å for the NiCat and CoCat [4, 5, 63, 90]. For the MnCat and PSII, the bridging distances are 2.86 Å [3] and 2.70 Å [91], respectively. Dau and coworkers [92] previously concluded that O(H)-bridged metal centers are a candidate for a common structural motif in multinuclear WOCs. The O(H) bridges may have an essential role in reducing the overpotential for water oxidation by redox-potential leveling; that is, the maintenance of an approximately constant redox potential for the four sequential oxidation steps assumed to occur before O_2 formation, as discussed in more detail in reference [93–95]. This redox leveling could be facilitated on the atomic scale by coupling of μ-OH deprotonation to the oxidation step, possibly in close analogy to processes in PSII [91]. The important role of the μ-oxo bridges in efficient water oxidation has also been discussed for synthetic Mn–Ca oxides [6]. Experiments addressing changes in the bridging configuration of these electrodeposited oxides represent a major challenge, but the results could provide crucial insight into the mechanism of water oxidation.

For the CoCat, Risch and coworkers [35] have shown that the catalytic activity may be determined by the extension of the Co-oxo clusters, which was modulated by the choice of anion in the electrolyte during electrodeposition. All investigated CoCat films were characterized by extensive di-μ-oxo bridging between Co^{III} ions, resulting in more extended (deposition in $CaCl_2$) or less extended (deposition in KP_i) Co-oxo structures consisting of edge-sharing CoO_6 octahedra (Figure 9.7). The data in Figure 9.8 provide a link between the structural information (derived from EXAFS analysis) and the functional

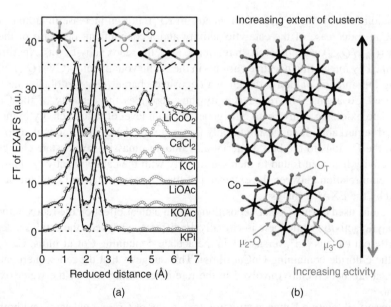

Figure 9.7 (a) FT of Co K-edge EXAFS spectra deposited in the indicated electrolytes, with crystalline $LiCoO_2$ as a reference. Symbols indicate experimental data and lines indicate EXAFS simulations (see [35] for further detail). (b) Hypothetical clusters of different sizes that relate to CoCat films of low order (bottom; KP_i-CoCat) and increased order (top; KCl-CoCat). Adapted with permission from [35]. Copyright © 2012 WILEY-VCH Verlag GmbH & Co. KGaA, Weinheim.

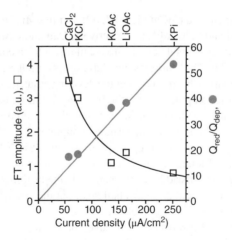

Figure 9.8 *Relation between current density, redox activity (○), and a qualitative measure of structural order (□), the latter obtained from the FT amplitude of the peaks related to the double μ-oxo bridges in Figure 9.7a. The lines were calculated under the assumption of linearly increasing redox activity and hyperbolically decreasing structural order and are shown to guide the eye. Reproduced with permission from [35]. Copyright © 2012 WILEY-VCH Verlag GmbH & Co. KGaA, Weinheim*

information (obtained by cyclic voltammetry in KP_i). For an increase in atomic order of the CoCat, a decrease of the catalytic activity and a concomitant decrease of the redox activity (Q_{red}/Q_{dep}) was observed. In reference [35], the redox activity was experimentally determined by integrating negative currents in the cyclic voltammetry (CV) (Q_{red}) and then dividing this integral value by the known deposited charge (Q_{dep}) to obtain the fraction of reduced Co atoms. The redox activity thus describes the capacity of the CoCat to accumulate oxidizing equivalents. The generally high level of redox activity in the CoCat is indeed remarkable. It suggests not only that a surface layer of Co ions is redox-active but that the Co ions in the bulk material also undergo massive redox-state changes. The redox activity has been found to increase linearly with the catalytic activity (Figure 9.8). Moreover, a relation between the extension of the Co-oxo cluster and catalytic activity is implied by the EXAFS analysis.

The anions used during CoCat deposition exhibit a clear effect on the redox activity and the catalytic activity, while there is virtually no difference between the catalysts deposited with different cations (Li^+ versus K^+ in the acetate-containing CoCat films; Ca^{+2} versus K^+ in the chloride-containing CoCat films). This suggests that the redox-inert cations of the CoCat are not crucially involved in the rate-determining step of the water oxidation reaction.

The observed relation between structure (extension of Co-oxo clusters) and both redox and catalytic activity can be rationalized by the following hypothesis [95b]. The active sites of the catalyst facilitate both redox and catalytic activity and are located at the periphery of Co-oxo clusters. They comprise the peripheral Co and oxygen atoms (emphasized by a darker hue in Figure 9.7b), as well as (possibly) ions of the electrolyte. Consequently, the fraction of redox-active Co ions (relative to the total number of Co ions, Q_{dep}) and the water oxidation activity decrease with increasing cluster size. It is therefore likely that the terminal oxygen atoms and the oxygen atoms in the μ_2-O bridges of the metal oxide play a pivotal role in both the local accumulation of oxidizing equivalents and the O–O bond-formation chemistry.

9.4   Functional Characterization

The CoCat function, or the mechanism of electrocatalytic water oxidation, has been studied predominantly by electrochemical methods, such as CV, potentiostatic polarization, and galvanostatic polarization, all of which may be utilized to construct Tafel plots. Ideally, a straight line is obtained, whose slope (the Tafel slope) depends on the nature of the rate-limiting step [96]. Table 9.1 shows a collection of Tafel slopes for different electrodeposited

Table 9.1   *Tafel slopes for select first-row transition metal catalysts*

Type of catalyst	Tafel slope (mV/decade)	Deposition conditions	Tafel data conditions
CoO_x [15]	60	Co wire, 1 M H_2SO_4	6 M NaOH,
CoO_x [75]	62	0.1 M KP_i, pH 7	0.1 M KP_i, pH 7
	170	0.1 M KP_i, pH 7, Pt RDE	0.1 M $NaClO_4$, pH 8, 2000 rpm
	59		0.1 M KP_i, pH 8, 2000 rpm
CoO_x [21]	100	RF sputtering 800 nm	0.1 M KP_i, pH 7
	110	0.1 M KP_i, pH 7	
	60	0.1 M KF, pH 3.4	1 M KF, pH 3.4
	113	0.1 M KF, pH 3.4	0.1 M KF, pH 3.4
CoO_x [22]	65	0.1 M P_i, pH 7	0.1 M P_i, pH 7
	70	1 M P_i, pH 7	1 M P_i, pH 7
	35		1 M KOH, pH 14, mQH_2O
	103	Ni foil, 1 M KOH, pH 14	1 M KOH, pH 14, riverwater
NiO_x [20]	116	2 mA/cm^2, 5 min	1 M KOH, pH 14, seawater
CoO_x [20]	50	0.1 M KB_i, pH 9.2	1 M KB_i, pH 9.2, mQH_2O
	54		1 M KB_i, pH 9.2, riverwater
	82		1 M KB_i, pH 9.2, seawater
NiO_x [2]	59	0.1 M B_i, pH 9.2	0.2 M B_i, pH 9.2, after 12 hours' electrolysis
	120		0.2 M B_i, pH 9.2
MnO_x [3]	80	Deionized water	0.1 M KP_i, pH 7
CoO_x [78]	42		
NiO_x [78]	29	Spin coating	1 M KOH, pH 14
MnO_x [78]	49		
Co_3O_4 [79]	49	103 °C	1 M KOH, pH 14
	36	50 °C	
$LiCoO_2$ [29]	60	–	0.1 M KOH, pH 13
$LiCoPO_4$ [29]	60	–	
CoO_x [29]	60	0.1 M KP_i, pH 7,	
NiO_x [80]	30	0.1 M KB_i, pH 9.2	0.5 M KB_i, 1.75 M KNO_3, pH 9.2

films containing Co, Ni, and Mn, as wells crystalline Co compounds in a variety of electrolytes and at various pH values.

The Tafel slope of thin CoCat films in the buffering KP_i electrolyte has been reported to be 59 mV/decade (of current) at pH 7 [28, 97], which reflects a pre-equilibrium followed by a rate-limiting O–O bond-formation step. Furthermore, the Tafel slope and pH dependence of the catalytic current (about 1 decade per pH unit) are compatible with the hypothesis of a proton-coupled electron transfer (PCET) in the formation of the pre-equilibrium for water oxidation [97]. While Tafel slopes close to 60 mV/decade have been reported in 0.1 M KPi (pH 7) for up to 60 mC/cm^2 deposited Co [97], values of $\sim$ 100 mV/decade have also been reported for (presumably very thick) CoCat films [30]. In 1.0 M borate (B_i, pH 9.2) [31], the Tafel slope is $\sim$ 50 mV/decade for film thicknesses ranging from 12 to 384 mC/cm^2 deposited Co (Table 9.1). The surface morphology and porosity of the CoCat depend on the conditions during deposition (Figure 9.1) and might cause the variance in reported values for CoCat-like deposits (denoted CoO_x in Table 9.1) [98].

Electrodeposited catalysts based on Ni (NiCat) and Mn (MnCat) have been reported to share structural motifs with the CoCat [3, 4, 90]. In order to discuss the question of whether the mechanism of water oxidation is the same in these related first-row transition metal oxides, Tafel slopes for these catalyst materials are also included in the survey of Table 9.1. A Tafel slope close to 60 mV/decade is not consistently observed for either the CoCat, the NiCat, or the MnCat.

The CoCat has been studied by pH titration in galvanostatic measurements. The observed slope of −64 mV/pH unit suggests an inverse first-order dependence of current on proton activity [97]. The buffering species (P_i) may serve as the proton acceptor, transferring the rapid one-electron–one-proton step from Co^{III}–OH to Co^{IV}=O.[75] Symes and coworkers [52] investigated the latter mechanistic step with an electrochemical and stopped-flow NMR study of a molecular $Co_4(\mu_3\text{-}O)_4$ cubane model and concluded that the unidirectional PCET (electron and proton transferred to the same cubane) includes a concerted self-exchange reaction between two Co ions of the cubane. These results may provide a model for charge transfer through CoCat films.

The proton-acceptor capacity of the electrolyte plays a significant role in the performance of the WOC [33]. The CoCat has similar Tafel slopes and a Faradaic efficiency close to 100% (O_2 evolved) in buffering electrolytes, such as KP_i and B_i, while nonbuffering electrolytes (SO_4^{2-}, ClO_4^-, NO_3^-) reduce the current density in the oxygen-evolving regime by more than one order of magnitude [33]. The dependence of the buffer concentration above 30 mM is of zero order [97]. Natural buffer systems such as seawater and riverwater increase the overpotential of the CoCat [29].

Several hypotheses for the catalytic cycle of the CoCat have been reported. Initially, it was hypothesized that the CoCat cycled through solid and liquid phases and that these phases were in equilibrium during catalytic turnover [28]. For buffering electrolytes, such as KP_i at pH 7, this mode of catalysis is no longer favored [29–31, 37, 97]. More recently, a two-step reaction scheme has been proposed based on electrokinetic experiments [97]. The authors assigned a reversible PCET between Co^{III}–OH and Co^{IV}=O to the first step in the mechanistic sequence and a two-electron Co reduction accompanying O–O bond formation to the second step. They hypothesized that the latter step was rate-limiting, but its nature remained unclear. Density functional theory (DFT) calculations on the rate-determining O–O bond formation in a complex with a cubane Co_4O_4 core support a mechanistic model, in which two terminal oxygen atoms couple directly. The kinetic barrier of O–O bond formation was very low after the formation of two adjacent $Co^{IV}=O$ on the cubane [99], which may contradict the mechanistic proposals based on Tafel slopes, where the bond formation is assumed to be rate-limiting for the CoCat.

The analysis of EPR and Tafel data within a wide pH range (0–14) was interpreted in terms of two different mechanisms: one for buffered systems (heterogeneous catalysis) and one for acidic conditions (homogenous catalysis) [31]. For buffered conditions (pH ≈ 3.5–14.0), a Tafel slope of 60 mV/decade was detected. It was suggested that the rate-limiting O–O bond-formation step involved a reaction between a $Co^{IV}=O$ species and a water molecule. The resting state of the CoCat was proposed to involve $Co^{IV}=O$ and Co^{III}–OH motifs, with the latter being deprotonated to $Co^{IV}=O$ upon Co oxidation. In the rate-determining step, a Co^{IV} intermediate reacted with water from the electrolyte to form two times the (Co^{III}–OH) motif. Under acidic conditions (pH 0–3.5), the Tafel slope was 120 mV/decade. The suggested nonbuffering mechanism involved an adsorption/deprotonation equilibrium between Co^{II} in solution and Co^{II}–OH, Co^{III}–OH on the electrode surface.

In summary, the CoCat has been thoroughly characterized by the collection of electrokinetic data. Due to a lack of information on the electronic and atomic structure of reaction intermediates, the corresponding mechanistic proposals have remained highly hypothetical.

9.5 Directly Light-Driven Water Oxidation

So far we have discussed CoCat-catalyzed water oxidation, which was driven by an anodic potential created by an external source of electrical power. Another and clearly more tempting alternative is utilization of solar light energy to drive water oxidation directly. For this approach, three requirements should be fulfilled: (i) the simple conducting electrode should be replaced by a material system that facilitates both light absorption and charge separation; (ii) the light-generated charges should live long enough to be transferred to the CoCat with high efficiency; and (iii) this should be achieved at a reasonably low final cost on a large scale, excluding the use of rare chemical elements [103].

A classical example of a light-absorbing–charge-separating system that can drive water oxidation is a molecular system based on ruthenium dyes. Often $[Ru(bipy_3)]^{2+}$ (bipy = 2, 2′-bipyridine) is used as a photosensitizer and $[Ru(bipy_3)]^{3+}$, created by light-induced charge separation, as a carrier of the cation radical. In this system, persulfate ($S_2O_8^{2-}$) is commonly used as an electron acceptor. The $[Ru(bipy_3)]^{3+}$ has a potential of about 1.3 V versus normal hydrogen electrode (NHE), which is positive enough to oxidize water (E_m around 0.8 V at pH 7). A serious drawback of this system, when used together with CoCat films deposited on an electrode, is the low probability of electron transfer between the surface of the solid CoCat and $[Ru(bipy_3)]^{3+}$ in solution. For that reason, $[Ru(bipy_3)]^{3+}$ has been used as an oxidant together with CoCat only when the CoCat is present in the form of a suspension of nanoparticles. In such a system, Shevchenko and coworkers [45] detected oxygen formation at a rate of 0.2 mol O_2/s/mol Co upon illumination of a suspension of Co^{2+} ions, $[Ru(bipy_3)]^{2+}$, and $S_2O_8^{2-}$ (conditions that result in self-assembly of CoCat particles). If methylenediphosphonate (M2P) is added to the illuminated solution, it does not change the CoCat structure at the atomic level, but binds at the periphery of the oxide particles, preventing their agglomeration upon light oxidation. After appropriate chemical modification, these M2P ligands can be used to directly connect the photosensitizer to the catalytically active particles, thereby building a partial system for the direct utilization of solar energy (Figure 9.9) [8, 45]. A similar approach was used in a recent work in which Co^{2+} ions were adsorbed on SiO_2 nanoparticles and oxidized chemically by addition of

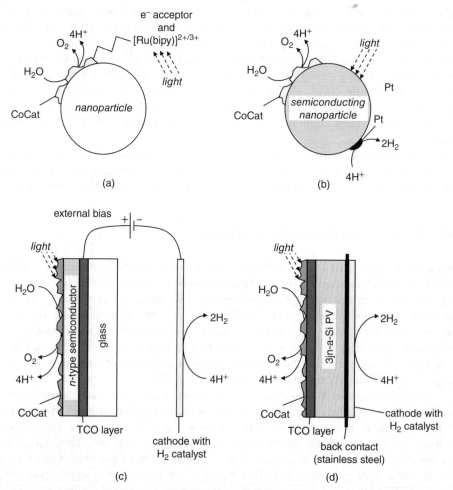

Figure 9.9 *Possible device configurations for directly light-driven water oxidation involving water oxidation catalysis by the CoCat. (a) Hypothetical device combining the colloidal CoCat systems reported in references [8, 104]. (b) Colloidal system for complete water splitting based on light-absorbing semiconducting materials [105]. (c) "Classical photoanode" – CoCat-covered and combined with a hydrogen-producing cathode – for operation with an external bias voltage. (d) "Artificial leaf" geometry [106]. TCO indicates transparent conducting oxide (most commonly FTO or ITO). The system presented in (a) drives water oxidation but not hydrogen formation, while the systems in (b), (c), and (d) facilitate total light-driven water splitting*

$[Ru(bipy_3)]^{3+}$ to $Co(OH)_2$ [104]. The detected turnover frequency (TOF) per Co ion of $\sim 300\,s^{-1}$ was astoundingly high, which can possibly be explained by the small size of the Co–oxide clusters (1–2 nm) formed by adsorption on the SiO_2 nanoparticles.

The use of CoCat with M2P ligands and the use of $Co(OH)_2$ on SiO_2 in combination with $[Ru(bipy_3)]^{3+}$ are examples of quasihomogeneous catalytic systems. More prominent is the approach of depositing CoCat films on a semiconducting anode consisting of a light-absorbing/charge-separating material (water oxidation at photoanodes). A variety of semiconductor-based photoanodes provide a reasonable overpotential that might be sufficient for oxidation of water without additional catalysts (Figure 9.10). The efficiency, however, is restricted by two factors: (i) short hole-diffusion length, a problem that can be solved by use of nanostructured materials, and (ii) strong electron–hole recombination, which can be decreased by rapid use of the holes for water oxidation promoted by a (co)catalyst bound to the photoanode, such as the CoCat [107].

CoCat formation has been achieved by electrodeposition directly on to the semiconductor surface [109, 110]. Moreover, the CoCat can be photodeposited on the surface by illuminating the photoactive semiconductor in the presence of Co^{2+} ions. One advantage of the photodeposition is that the CoCat will be formed faster at locations where the holes are especially accessible [64, 111–115]. In a recent study, it has been shown that the atomic structures of the CoCat photodeposited on TiO_2 nanoparticles and of the electrodeposited CoCat are identical [64].

TiO_2 was the first semiconductor demonstrated to be capable of oxidizing water and driving proton reduction at a Pt cathode upon ultraviolet (UV) irradiation (in 1972) [116]. TiO_2 is a wide-bandgap semiconductor (3.2 eV); that is, it absorbs only a minor fraction of the solar spectrum. In order to optimize the absorption of visible light, it has been coupled with molecular dye sensitizers [109]. Bledowski and coworkers [117] reported a polyheptazine-coated nanocrystaline TiO_2 in which a CoCat was deposited. This catalyzed water oxidation under visible-light illumination and moderate bias potential and improved photocatalytic water oxidation by photo-excitation with UV light [64, 118]. The mechanism of this improvement is still debated, however. Bledowski and coworkers [117] suggested that it was the result of the improved photooxidation kinetics in the presence of the CoCat cocatalyst; another alternative is that the passivation of TiO_2 surface states resulted in increased hole lifetime and thus improved energy conversion efficiency [118].

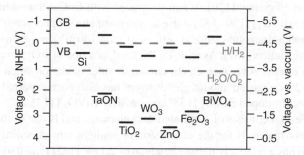

Figure 9.10 *Band edges of the conduction band (CB) and valence band (VB) for semiconducting photoanodes used in combination with the CoCat. Data taken from [108]*

Tungsten oxide (WO_3) is another n-type semiconductor (Figure 9.10). Its bandgap of 2.7 eV is smaller than that of TiO_2 [119]. It absorbs photons in the visible spectrum and can generate a potential of about 3 V versus NHE, which may make it a promising photoanode for solar-driven water oxidation [120]. The overpotential is so high that peroxide formation is also thermodynamically feasible (E_m around 1.4 V versus NHE at pH 7) [121]. Peroxide formation is typically kinetically more favorable than water oxidation, however, and thus dominant [122]. The accumulation of peroxo species on the WO_3 surface has been suggested to cause loss of photoactivity [123]. In this system, deposition of the CoCat as a cocatalyst suppresses the formation of surface-bound peroxo species and increases photoinduced O_2 formation by improving the kinetics of the O_2 evolution reaction and preventing photodissolution of WO_3 [122]. Similarly, in a ZnO photoanode the valence band edge is located at highly positive potentials (Figure 9.10). Moreover, due to its wide bandgap of 3.2 eV, which is similar to WO_3 and TiO_2, it absorbs mostly in the UV range. Further, in ZnO, when the CoCat is (photo)deposited on to the ZnO surface, a general enhancement of the anodic photocurrent is observed, likely due to decreased electron–hole recombination [111].

Tantalum oxynitride (TaON) has band levels that could be appropriate for both water oxidation and proton reduction (Figure 9.10) and a narrow bandgap (< 3 eV) that allows for visible-light absorption [124]. The photocurrent generated decreases rapidly on irradiation due to self-oxidative deactivation, however, which proceeds competitively with water oxidation [125]. In this system, deposition of the CoCat as a cocatalyst results in stable and efficient generation of O_2 under visible-light irradiation at a relatively low applied potential [126]. Moreover, the stability of the CoCat/TaON system has been found to be superior to that of IrO_x/TaON system, which is explained by the more homogeneous distribution of the Co-oxide nanoparticles on the TaON surface, which reduces the self-oxidation and retards charge recombination [126].

Bismuth vanadate ($BiVO_4$, also called bismuth yellow) is another promising photoanode material for solar water-splitting applications. It has a bandgap of around 2.4 eV [127], which ensures absorption in the visible region. CoCat deposition on the surface of undoped $BiVO_4$ increases the photocurrent [105], and CoCat may also remove the kinetic limitation of slow oxygen evolution (at high light intensities) [128, 129]. For $BiVO_4$ modified with CoCat, poor electron transport has been suggest to be the main performance bottleneck [128]. Doping may improve the overall photoelectrochemical water oxidation efficiencies of $BiVO_4$ photoanodes, improving the electron transport in the $BiVO_4$ and thus increasing charge-separation efficiency [130]. When interfaced with CoCat, tungsten-doped $BiVO_4$ photoanodes (W : $BiVO_4$) [130–132] cause almost complete suppression of surface charge recombination and a dramatic cathodic shift of around 440 mV in the onset potential for sustained water oxidation [131]. As with TaON, the deposition of IrO_x, known as an excellent WOC, does not improve photoelectrochemical water oxidation when combined with W : $BiVO_4$ [132]. Enhanced anodic photocurrent has also been reported when a CoCat was deposited on Mo-doped $BiVO_4$ [133] and on SiO_2-$BiVO_4$ [114].

Hematite (α-Fe_2O_3) consists of earth-abundant elements and is one of the most promising semiconductor materials for the conversion of sunlight into chemical fuels by water splitting [134]. It has a relatively narrow bandgap of 2.1 eV [135] (and thus absorbs a large portion of the solar spectrum) and a valence band edge of around 1.8 V versus NHE, which makes it suitable for photoinduced water oxidation. However, other drawbacks remain, especially strong charge recombination and low conductivity. In 2006, Grätzel and coworkers had already shown that if Co oxide nanoparticles were deposited in a certain way on the hematite surface, they increase the photocurrent by introducing an 80 mV cathodic shift of the onset potential [136]. Following Nocera's report in *Science* in 2008 [1], the CoCat was deposited on α-Fe$_2$O$_3$ [137, 138], and the external bias required to drive water oxidation using this catalyst at 1 mA/cm^2 (1 sun, AM 1.5 solar irradiation) has since decreased by more than 0.5 V [113, 139]. Based on the results of transient absorption spectroscopy [140] and impedance spectroscopy [141], it has been suggested that the enhanced photochemical activity of CoCat/α-Fe$_2$O$_3$ for water oxidation is caused by reduced electron–hole recombination. This effect has been attributed either to improved water oxidation kinetics with the CoCat [115, 141] or to an increase in the lifetime of photogenerated holes in the hematite photoanode, due to formation of an inorganic heterojunction [140]. While the first explanation suggests water oxidation by the CoCat, the latter implies water oxidation by the hematite itself [141, 142]. A Zn^{2+} treatment to form a zinc ferrite (ZnFe$_2$O$_4$) layer between the CoCat and α-Fe$_2$O$_3$ resulted in changed valence and conducting band levels, situating the conducting band edge very close to the H$^+$ reduction potential of the second half-reaction of water splitting [143]. This treatment has also been found to enhance electron–hole separation at the Fe$_2$O$_3$/ZnFe$_2$O$_4$ interface [143]. In order to improve the conductivity of α-Fe$_2$O$_3$, Si [136, 137, 139] and Nb [140] were also used as dopants.

In all of the photoanodes described in this chapter, the semiconductor material is deposited on a conducting electrode – in most cases a transparent conducting oxide (mostly FTO) on a glass plate. There is one report in which a CoCat was not deposited on a macroscopic electrode but instead on semiconducting nanoparticles (yttrium-doped BiVO$_4$) operating in solution. In this system, water oxidation was directly driven by light, without any external oxidizing agents [105]. Pt clusters also deposited on the semiconducting nanoparticles served as a cocatalyst for proton reduction. Aside from this system of dissolved nanoparticles, whenever CoCat-based composite photoanodes are used as part of a complete system for overall water splitting, it is essential to apply an external bias between the photoanode and counter electrode as the conduction band of the semiconducting oxides is too positive for H$_2$ evolution [134]. The external potential also prevents charge recombination between water-oxidation intermediates and conduction-band electrons [141]. The external bias voltage can result from photovoltaic [134] or dye-sensitized solar cells, possibly integrated within a single device.

A common problem for all photoanodes is the gradual loss of activity caused by anodic photocorrosion. As an alternative to the semiconducting oxides, Nocera and coworkers used a silicon solar cell to absorb light and create the potential needed to split water [106, 112, 144]. The photovoltaic cell was covered with ITO to protect its surface against photoinstability. In this case, the water oxidation was catalyzed entirely by the CoCat, and the solar cell was used to generate the necessary anode–cathode potential difference upon illumination with visible light. The photovoltaic part of the device was a triple-junction amorphous silicon (3-jn-a-Si) solar cell, which created a potential difference sufficient to drive hydrogen formation at the cathode, employing an appropriate catalyst for proton reduction. The cathode can be connected directly to the back of the device to form a prototype of a wireless "artificial leaf" (Figure 9.9d) [106]. This "artificial leaf" design may be the most straightforward and robust concept to date for direct solar-to-hydrogen conversion. There remain serious drawbacks, however, especially the use of relatively expensive triple-junction photovoltaic cells of moderate efficiency and the clearly insufficient stability of the device function. Nonetheless, this "artificial leaf" design can serve as a proof of concept that technologically relevant devices are within reach.

References

[1] Kanan, M. W., Nocera, D. G. (2008) Science, 321: 1072–1075.

[2] Dincă, M., Surendranath, Y., Nocera, D. G. (2010) Proc. Natl. Acad. Sci. U.S.A., 107: 10337–10341.

[3] Zaharieva, I., Chernev, P., Risch, M., Klingan, K., Kohlhoff, M., Fischer, A., Dau, H. (2012) Energy Environ. Sci., 5: 7081–7089.

[4] Risch, M., Klingan, K., Heidkamp, J., Ehrenberg, D., Chernev, P., Zaharieva, I., Dau, H. (2011) Chem. Commun., 47: 11912–11914.

[5] Risch, M., Khare, V., Zaharieva, I., Gerencser, L., Chernev, P., Dau, H. (2009) J. Am. Chem. Soc., 131: 6936–6937.

[6] Zaharieva, I., Najafpour, M. M., Wiechen, M., Haumann, M., Kurz, P., Dau, H. (2011) Energy Environ. Sci., 4: 2400–2408.

[7] Najafpour, M. M., Ehrenberg, T., Wiechen, M., Kurz, P. (2010) Angew. Chem. Int. Ed., 49: 2233–2237.

[8] Risch, M., Shevchenko, D., Anderlund, M. F., Styring, S., Heidkamp, J., Lange, K. M., Thapper, A., Zaharieva, I. (2012) Int. J. Hydrogen Energy, 37: 8878–8888.

[9] Coehn, A., Gläser, M. (1902) Z. Anorg. Chem., 33: 9–24.

[10] Skirrow, F. W. (1902) Z. Anorg. Chem., 33: 25–30.

[11] Siemens, A. (1904) Z. Anorg. Chem., 41: 249–275.

[12] Müller, E., Spitzer, F. (1906) Z. Anorg. Chem., 50: 321–354.

[13] Hüttig, G. F., Kassler, R. (1929) Z. Anorg. Allg. Chem., 184: 279–288.

[14] El Wakkad, S. E. S., Hickling, A. (1950) Trans. Faraday Soc., 46: 820–824.

[15] Benson, P., Briggs, G. W. D., Wynne-Jones, W. F. K. (1964) Electrochim. Acta, 9: 275–280.

[16] Benson, P., Briggs, G. W. D., Wynne-Jones, W. F. K. (1964) Electrochim. Acta, 9: 281–288.

[17] Osamu, S., Masao, T., Tomio, F., Kuboyama, J. (1968) United States Patent 3399966.

[18] Burke, L. D., Lyons, M. E., Murphy, O. J. (1982) J. Elelectroanal. Chem., 132: 247–261.

[19] Chen, Y.-W. D., Noufi, R. N. (1984) J. Electrochem. Soc., 131: 1447–1451.

[20] Jiang, S. P., Tseung, A. C. C. (1991) J. Electrochem. Soc., 138: 1216–1222.

[21] Jiang, S. P., Tseung, A. C. C. (1990) J. Electrochem. Soc., 137: 3387–3393.

[22] Jiang, S. P., Tseung, A. C. C. (1990) J. Electrochem. Soc., 137: 3381–3386.

[23] Jiang, S. P., Cui, C. Q., Tseung, A. C. C. (1991) J. Electrochem. Soc., 138: 3599–3605.

[24] Jiang, S. P., Chen, Y. Z., You, J. K., Chen, T. X., Tseung, A. C. C. (1990) J. Electrochem. Soc., 137: 3374–3380.

[25] Cui, C. Q., Jiang, S. P., Tseung, A. C. C. (1992) J. Electrochem. Soc., 139: 60–66.

[26] Cui, C. Q., Jiang, S. P., Tseung, A. C. C. (1992) J. Electrochem. Soc., 139: 1276–1282.

[27] Cui, C. Q., Jiang, S. P., Tseung, A. C. C. (1992) J. Electrochem. Soc., 139: 1535–1544.

[28] Kanan, M. W., Surendranath, Y., Nocera, D. G. (2009) Chem. Soc. Rev., 38: 109–114.

[29] Esswein, A. J., Surendranath, Y., Reece, S. Y., Nocera, D. G. (2011) Energy Environ. Sci., 4: 499–504.

[30] Young, E. R., Nocera, D. G., Bulovic, V. (2010) Energy Environ. Sci., 3: 1726–1728.

[31] Gerken, J. B., McAlpin, J. G., Chen, J. Y. C., Rigsby, M. L., Casey, W. H., Britt, R. D., Stahl, S. S. (2011) J. Am. Chem. Soc., 133: 14 431–14 442.

[32] Gerken, J. B., Landis, E. C., Hamers, R. J., Stahl, S. S. (2010) ChemSusChem, 3: 1176–1179.

[33] Surendranath, Y., Dinca, M., Nocera, D. G. (2009) J. Am. Chem. Soc., 131: 2615–2620.

[34] Lutterman, D. A., Surendranath, Y., Nocera, D. G. (2009) J. Am. Chem. Soc., 131: 3838–3839.

[35] Risch, M., Klingan, K., Ringleb, F., Chernev, P., Zaharieva, I., Fischer, A., Dau, H. (2012) ChemSusChem, 5: 542–549.

[36] Nocera, D. G. (2009) Inorg. Chem., 48: 10 001–10 017.

[37] Surendranath, Y., Lutterman, D. A., Liu, Y., Nocera, D. G. (2012) J. Am. Chem. Soc., 134: 6326–6336.

[38] Lee, S. W., Carlton, C., Risch, M., Surendranath, Y., Chen, S., Furutsuki, S., Yamada, A., Nocera, D. G., Shao-Horn, Y. (2012) J. Am. Chem. Soc., 134: 16 959–16 962.

[39] Jia, H. F., Stark, J., Zhou, L. Q., Ling, C., Sekito, T., Markin, Z. (2012) RSC Advances, 2: 10 874–10 881.

[40] Yeo, B. S., Bell, A. T. (2011) J. Am. Chem. Soc., 133: 5587–5593.

[41] Miles, D. O., Jiang, D., Burrows, A. D., Halls, J. E., Marken, F. (2013) Electrochem. Commun., 27: 9–13.

[42] May, K. J., Carlton, C. E., Stoerzinger, K. A., Risch, M., Suntivich, J., Lee, Y.-L., Grimaud, A., Shao-Horn, Y. (2012) J. Phys. Chem. Lett., 3: 3264–3270.

[43] Risch, M., Grimaud, A., May, K. J., Stoerzinger, K. A., Chen, T. J., Mansour, A. N., Shao-Horn, Y. (2013) J. Phys. Chem. C, 117(17): 8628–8635, doi: 10.1021/jp3126768.

[44] Petrykin, V., Macounová, K., Okube, M., Mukerjee, S., Krtil, P. (2013) Catal. Today, 202: 63–69.

[45] Shevchenko, D., Anderlund, M. F., Thapper, A., Styring, S. (2011) Energy Environ. Sci., 4: 1284–1287.

[46] Yin, Q., Tan, J. M., Besson, C., Geletii, Y. V., Musaev, D. G., Kuznetsov, A. E., Luo, Z., Hardcastle, K. I., Hill, C. L. (2010) Science: 342–345.

[47] Geletii, Y. V., Yin, Q., Hou, Y., Huang, Z., Ma, H., Song, J., Besson, C., Luo, Z., Cao, R., O'Halloran, K. P., Zhu, G., Zhao, C., Vickers, J. W., Ding, Y., Mohebbi, S., Kuznetsov, A. E., Musaev, D. G., Lian, T., Hill, C. L. (2011) Isr. J. Chem., 51: 238–246.

[48] Huang, Z., Luo, Z., Geletii, Y. V., Vickers, J. W., Yin, Q., Wu, D., Hou, Y., Ding, Y., Song, J., Musaev, D. G., Hill, C. L., Lian, T. (2011) J. Am. Chem. Soc., 133: 2068–2071.

[49] Lieb, D., Zahl, A., Wilson, E. F., Streb, C., Nye, L. C., Meyer, K., Ivanović-Burmazović, I. (2011) Inorg. Chem., 50: 9053–9058.

[50a] Natali, M., Berardi, S., Sartorel, A., Bonchio, M., Campagna, S., Scandola, F. (2012) Chem. Commun., 48: 8808–8810.

[50b] Schiwon, R., Klingan, K., Dau, H., Limberg, C. (2014) Chem. Commun., 50: 100–102.

[51] Stracke, J. J., Finke, R. G. (2011) J. Am. Chem. Soc., 133: 14 872–14 875.

[52] Symes, M. D., Surendranath, Y., Lutterman, D. A., Nocera, D. G. (2011) J. Am. Chem. Soc., 133: 5174–5177.

[53] McAlpin, J. G., Stich, T. A., Ohlin, C. A., Surendranath, Y., Nocera, D. G., Casey, W. H., Britt, R. D. (2011) J. Am. Chem. Soc., 133: 15 444–15 452.

[54] Berardi, S., La Ganga, G., Natali, M., Bazzan, I., Puntoriero, F., Sartorel, A., Scandola, F., Campagna, S., Bonchio, M. (2012) J. Am. Chem. Soc., 134: 11 104–11 107.

[55] La Ganga, G., Puntoriero, F., Campagna, S., Bazzan, I., Berardi, S., Bonchio, M., Sartorel, A., Natali, M., Scandola, F. (2012) Faraday Discuss., 155: 177–190.

[56] McCool, N. S., Robinson, D. M., Sheats, J. E., Dismukes, G. C. (2011) J. Am. Chem. Soc., 133: 11 446–11 449.

[57] Symes, M. D., Lutterman, D. A., Teets, T. S., Anderson, B. L., Breen, J. J., Nocera, D. G. (2013) ChemSusChem, 6: 65–69.

[58] Du, P. W., Kokhan, O., Chapman, K. W., Chupas, P. J., Tiede, D. M. (2012) J. Am. Chem. Soc., 134: 11 096–11 099.

[59] Dau, H., Liebisch, P., Haumann, M. (2003) Anal. Bioanal. Chem., 376: 562–583.

[60] Teo, B. (1986) EXAFS: Basic Principles and Data Analysis. Springer Verlag, Berlin, Germany.

[61] Li, G. G., Bridges, F., Booth, C. H. (1995) Phys. Rev. B, 52: 6332–6348.

[62] Risch, M., Ringleb, F., Khare, V., Chernev, P., Zaharieva, I., Dau, H. (2009) J. Physics: Conf. Series, 190: 012167.

[63] Kanan, M. W., Yano, J., Surendranath, Y., Dinca, M., Yachandra, V. K., Nocera, D. G. (2010) J. Am. Chem. Soc., 132: 13 692–13 701.

[64] Khnayzer, R. S., Mara, M. W., Huang, J., Shelby, M. L., Chen, L. X., Castellano, F. N. (2012) ACS Catalysis, 2: 2150–2160.

[65] Wood, R. M., Palenik, G. J. (1998) Inorg. Chem., 37: 4149–4151.

[66] Casey, W. H., McAlpin, J. G., Surendranath, Y., Dinca, M., Stich, T. A., Stoian, S. A., Nocera, D. G., Britt, R. D. (2010) J. Am. Chem. Soc., 132: 6882–6883.

[67] Delmas, C., Fouassier, C., Hagenmuller, P. (1980) Physica B+C, 99: 81–85.

[68] Fouassier, C., Matejka, G., Reau, J.-M., Hagenmuller, P. (1973) J. Solid State Chem., 6: 532–537.

[69] Delaplane, R. G., Ibers, J. A., Ferraro, J. R., Rush, J. J. (1969) J. Chem. Phys., 50: 1920–1927.

[70] Dimitrou, K., Folting, K., Streib, W. E., Christou, G. (1993) J. Am. Chem. Soc., 115: 6432–6433.

[71] Ama, T., Rashid, M. M., Yonemura, T., Kawaguchi, H., Yasui, T. (2000) Coord. Chem. Rev., 198: 101–116.

[72] Ama, T., Yonemura, T., Morita, S., Yamaguchi, M. (2010) Acta Cryst. E, 66: M483–U1364.

[73] Mattioli, G., Risch, M., Bonapasta, M. A., Dau, H., Guidoni, L. (2011) Phys. Chem. Chem. Phys., 13: 15 437–15 441.

[74] Hu, X. L., Piccinin, S., Laio, A., Fabris, S. (2012) ACS Nano, 6: 10 497–10 504.

[75] Ferreira, K. N., Iverson, T. M., Maghlaoui, K., Barber, J., Iwata, S. (2004) Science, 303: 1831–1838.

[76] Grundmeier, A., Dau, H. (2012) Biochim. Biophys. Acta, 1817: 88–105.

[77] Umena, Y., Kawakami, K., Shen, J.-R., Kamiya, N. (2011) Nature, 473: 55–60.

[78] Dau, H., Grundmeier, A., Loja, P., Haumann, M. (2008) Phil. Trans. R. Soc. B, 363: 1237–1244.

[79] Wiechen, M., Zaharieva, I., Dau, H., Kurz, P. (2012) Chem. Sci., 3: 2330–2339.

[80] Koningsberger, D. C., Mojet, B. L., van Dorssen, G. E., Ramaker, D. E. (2000) Top. Catal., 10: 143–155.

[81] Cinco, R. M., Holman, K. L. M., Robblee, J. H., Yano, J., Pizarro, S. A., Bellacchio, E., Sauer, K., Yachandra, V. K. (2002) Biochemistry, 41, 12 928–12 933.

[82] Müller, C., Liebisch, P., Barra, M., Dau, H., Haumann, M. (2005) Phys. Scr., T115: 847–850.

[83] Wang, Q., O'Hare, D. (2012) Chem. Rev., 112: 4124–4155.

[84] Harley, S. J., Mason, H. E., McAlpin, J. G., Britt, R. D., Casey, W. H. (2012) Chemistry – A European Journal, 18: 10 476–10 479.

[85] Butel, M., Gautier, L., Delmas, C. (1999) Solid State Ionics, 122: 271–284.

[86] Fulton, J. L., Heald, S. M., Badyal, Y. S., Simonson, J. M. (2003) J. Phys. Chem. A, 107: 4688–4696.

[87] Glezakou, V. A., Chen, Y. S., Fulton, J. L., Schenter, G. K., Dang, L. X. (2006) Theor. Chem. Acc., 115: 86–99.

[88] McAlpin, J. G., Stich, T. A., Casey, W. H., Britt, R. D. (2012) Coord. Chem. Rev., 256: 2445–2452.

[89] Dau, H., Zaharieva, I., Haumann, M. (2012) Curr. Opin. Chem. Biol., 16: 3–10.

[90] Bediako, D. K., Lassalle-Kaiser, B., Surendranath, Y., Yano, J., Yachandra, V. K., Nocera, D. G. (2012) J. Am. Chem. Soc., 134: 6801–6809.

[91] Haumann, M., Müller, C., Liebisch, P., Iuzzolino, L., Dittmer, J., Grabolle, M., Neisius, T., Meyer-Klaucke, W., Dau, H. (2005) Biochemistry, 44: 1894–1908.

[92] Dau, H., Limberg, C., Reier, T., Risch, M., Roggan, S., Strasser, P. (2010) ChemCatChem, 2: 724–761.

[93] Magnuson, A., Liebisch, P., Hogblom, J., Anderlund, M. F., Lomoth, R., Meyer-Klaucke, W., Haumann, M., Dau, H. (2006) J. Inorg. Biochem., 100: 1234–1243.

[94] Dau, H., Haumann, M. (2008) Coord. Chem. Rev., 252: 273–295.

[95] Dau, H., Haumann, M. (2005) Photosynth. Res., 84: 325–331.

[95b] Klingan, K., Ringleb, F., Zaharieva, I., Heidkamp, J., Chernev, P., Gonzalez-Flores, D., Risch, M., Fischer, A., Dau, H. ChemSusChem, doi: 10.1002/cssc.201301019.

[96] Bockris, J. O. M. (1956) J. Phys. Chem., 24: 817.

[97] Surendranath, Y., Kanan, M. W., Nocera, D. G. (2010) J. Am. Chem. Soc., 132: 16 501–16 509.

[98] Banham, D. W., Soderberg, J. N., Birss, V. I. (2009) J. Phys. Chem. C, 113: 10 103–10 111.

[99] Wang, L.-P., Van Voorhis, T. (2011) J. Phys. Chem. Lett., 2: 2200–2204.

[100] Trotochaud, L., Ranney, J. K., Williams, K. N., Boettcher, S. W. (2012) J. Am. Chem. Soc., 134: 17253–17261.

[101] Koza, J. A., He, Z., Miller, A. S., Switzer, J. A. (2012) Chem. Mater., 24: 3567–3573.

[102] Bediako, D. K., Surendranath, Y., Nocera, D. G. (2013) J. Am. Chem. Soc., 135: 3662–3674.

[103] Lewis, N. S., Nocera, D. G. (2006) Proc. Natl. Acad. Sci. U. S. A., 103: 15729–15735.

[104] Zidki, T., Zhang, L., Shafirovich, V., Lymar, S. V. (2012) J. Am. Chem. Soc., 134: 14275–14278.

[105] Wang, D., Li, R., Zhu, J., Shi, J., Han, J., Zong, X., Li, C. (2012) J. Phys. Chem. C, 116: 5082–5089.

[106] Reece, S. Y., Hamel, J. A., Sung, K., Jarvi, T. D., Esswein, A. J., Pijpers, J. J. H., Nocera, D. G. (2011) Science, 334: 645–648.

[107] Li, Z., Luo, W., Zhang, M., Feng, J., Zou, Z. (2013) Energy Environ. Sci., 6: 347–370.

[108] Chen, S., Wang, L.-W. (2012) Chem. Mater., 24: 3659–3666.

[109] Young, K. J., Martini, L. A., Milot, R. L., Snoeberger, R. C. III, Batista, V. S., Schmuttenmaer, C. A., Crabtree, R. H., Brudvig, G. W. (2012) Coord. Chem. Rev., 256: 2503–2520.

[110] Yamazaki, H., Shouji, A., Kajita, M., Yagi, M. (2010) Coord. Chem. Rev., 254: 2483–2491.

[111] Steinmiller, E. M. P., Choi, K. S. (2009) Proc. Natl. Acad. Sci. U. S. A., 106: 20633–20636.

[112] Young, E. R., Costi, R., Paydavosi, S., Nocera, D. G., Bulovic, V. (2011) Energy Environ. Sci., 4: 2058–2061.

[113] Zhong, D. K., Cornuz, M., Sivula, K., Grätzel, M., Gamelin, D. R. (2011) Energy Environ. Sci., 4: 1759–1764.

[114] Pilli, S. K., Deutsch, T. G., Furtak, T. E., Turner, J. A., Brown, L. D., Herring, A. M. (2012) Phys. Chem. Chem. Phys., 14: 7032–7039.

[115] McDonald, K. J., Choi, K. S. (2011) Chem. Mater., 23: 1686–1693.

[116] Fujishima, A., Honda, K. (1972) Nature, 238: 37–38.

[117] Bledowski, M., Wang, L., Ramakrishnan, A., Bétard, A., Khavryuchenko, O. V., Beranek, R. (2012) ChemPhysChem, 13: 3018–3024.

[118] Hoang, S., Guo, S., Hahn, N. T., Bard, A. J., Mullins, C. B. (2012) Nano Lett., 12: 26–32.

[119] Santato, C., Odziemkowski, M., Ulmann, M., Augustynski, J. (2001) J. Am. Chem. Soc., 123: 10639–10649.

[120] Bak, T., Nowotny, J., Rekas, M., Sorrell, C. C. (2002) Int. J. Hydrogen Energy, 27: 991–1022.

[121] Pourbaix, M. (1974) Atlas of Electrochemical Equilibria in Aqueous Solutions, 2nd English Ed. National Association of Corrosion Engineers, Houston, TX, USA.

[122] Seabold, J. A., Choi, K. S. (2011) Chem. Mater., 23: 1105–1112.

[123] Augustynski, J., Solarska, R., Hagemann, H., Santato, C. (2006) Proc. SPIE 6340, Solar Hydrogen and Nanotechnology, 6340: U140–U148.

[124] Chun, W.-J., Ishikawa, A., Fujisawa, H., Takata, T., Kondo, J. N., Hara, M., Kawai, M., Matsumoto, Y., Domen, K. (2003) J. Phys. Chem. B, 107: 1798–1803.

[125] Hitoki, G., Takata, T., Kondo, J. N., Hara, M., Kobayashi, H., Domen, K. (2002) Chem. Commun., (16): 1698–1699.

[126] Higashi, M., Domen, K., Abe, R. (2012) J. Am. Chem. Soc., 134: 6968–6971.

[127] Tokunaga, S., Kato, H., Kudo, A. (2001) Chem. Mater., 13: 4624–4628.

[128] Abdi, F. F., van de Krol, R. (2012) J. Phys. Chem. C, 116: 9398–9404.

[129] Ding, C., Shi, J., Wang, D., Wang, Z., Wang, N., Liu, G., Xiong, F., Li, C. (2013) Phys. Chem. Chem. Phys, 15(13): 4589–4595.

[130] Abdi, F. F., Firet, N., van de Krol, R. (2012) ChemCatChem, 5: 490–496.

[131] Zhong, D. K., Choi, S., Gamelin, D. R. (2011) J. Am. Chem. Soc., 133: 18370–18377.

[132] Ye, H., Park, H. S., Bard, A. J. (2011) J. Phys. Chem. C, 115: 12464–12470.

[133] Pilli, S. K., Furtak, T. E., Brown, L. D., Deutsch, T. G., Turner, J. A., Herring, A. M. (2011) Energy Environ. Sci., 4: 5028–5034.

[134] Sivula, K., Le Formal, F., Grätzel, M. (2011) ChemSusChem, 4: 432–449.

[135] Ahmed, S. M., Leduc, J., Haller, S. F. (1988) J. Phys. Chem., 92: 6655–6660.

[136] Kay, A., Cesar, I., Grätzel, M. (2006) J. Am. Chem. Soc., 128: 15714–15721.

[137] Zhong, D. K., Sun, J. W., Inumaru, H., Gamelin, D. R. (2009) J. Am. Chem. Soc., 131: 6086–6087.

[138] Hong, Y.-R., Liu, Z., Al-Bukhari, S. F. B. S. A., Lee, C. J. J., Yung, D. L., Chi, D., Hor, T. S. A. (2011) Chem. Commun., 47: 10653–10655.

[139] Zhong, D. K., Gamelin, D. R. (2010) J. Am. Chem. Soc., 132: 4202–4207.

[140] Barroso, M., Cowan, A. J., Pendlebury, S. R., Grätzel, M., Klug, D. R., Durrant, J. R. (2011) J. Am. Chem. Soc., 133: 14868–14871.

[141] Klahr, B., Gimenez, S., Fabregat-Santiago, F., Bisquert, J., Hamann, T. W. (2012) J. Am. Chem. Soc., 134: 16693–16700.

[142] Gamelin, D. R. (2012) Nat. Chem., 4: 965–967.

[143] McDonald, K. J., Choi, K.-S. (2011) Chem. Mater., 23: 4863–4869.

[144] Pijpers, J. J. H., Winkler, M. T., Surendranath, Y., Buonassisi, T., Nocera, D. G. (2011) Proc. Natl. Acad. Sci. U. S. A., 108: 10056–10061.

10

Developing Molecular Copper Complexes for Water Oxidation

Shoshanna M. Barnett, Christopher R. Waidmann, Margaret L. Scheuermann, Jared C. Nesvet, Karen Goldberg and James M. Mayer
Department of Chemistry, University of Washington, Seattle, WA, USA

10.1 Introduction

The field of molecular water oxidation catalysis has grown dramatically in recent years, as demonstrated throughout this volume. This chapter is an account of our efforts to develop the first copper water oxidation catalysts (WOCs). We were inspired originally by the beautiful biomimetic chemistry of copper complexes with dioxygen, developed in a number of laboratories [1–3]. In particular, Tolman and coworkers developed Cu(I) complexes that react with O_2 at low temperatures to form a mixture of dicopper(III)-μ-oxo and dicopper(II)-μ-peroxo complexes, and made the remarkable discovery that these are in rapid equilibrium [4] (see Scheme 10.1). We felt that this oxo–peroxo equilibrium was relevant to water oxidation because it is one of the very rare cases in solution chemistry where the formation and cleavage of an O–O bond is facile and reversible. At the time when we started down this path, roughly 8 years ago, we and many others believed that O–O bond formation was the kinetic bottleneck for water oxidation. This view has changed in recent years, as described elsewhere in this volume. Still, a system with a low barrier to reversible O–O bond formation is a good starting point for the pursuit of new WOCs.

This chapter tells our story chronologically. The initial studies used biomimetic compounds developed by Kitajima, Tolman, Stack, Warren, and others [1–3]. While this effort was ultimately unsuccessful, it was quite instructive. We approached water oxidation as a proton-coupled electron transfer (PCET) process that requires both oxidant and base to remove e^- and H^+, respectively. Section 10.2.1 discusses the thermochemistry of this approach, and the challenges of combining strong oxidants and bases (as described in more

Molecular Water Oxidation Catalysis: A Key Topic for New Sustainable Energy Conversion Schemes,
First Edition. Edited by Antoni Llobet.

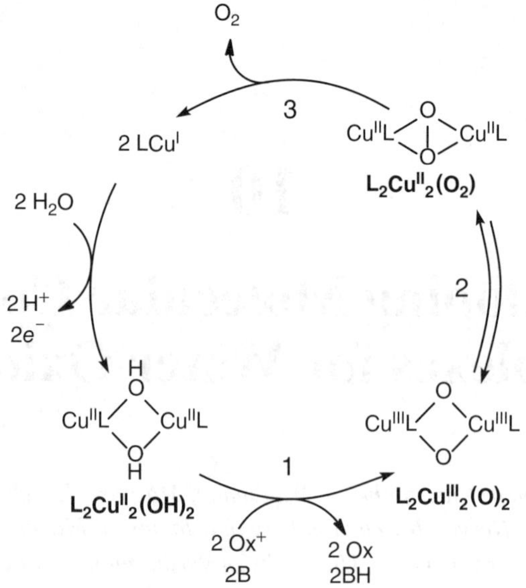

Scheme 10.1 *Envisioned catalytic cycle for water oxidation involving dimeric copper complexes*

detail in reference [5]). This story also illustrates some of the challenges of working under the strongly oxidizing conditions required for water oxidation. Section 10.3 then describes our successful development of a water oxidation electrocatalyst [6]. These studies were also motivated by the biomimetic chemistry, although recent experiments indicate that the electrocatalysis has a mechanism different from the path originally envisioned.

Prior to our work, we believe that the only discussion of molecular copper catalysts for water oxidation was from Elizarova and coworkers, who reported a few turnovers by $CuCl_2$, [Cu(bpy)$_2$]Cl$_2$, and [Cu(bpy)$_3$]Cl$_2$, using [Ru(bpy)$_3$]$^{3+}$ as the terminal oxidant [7–9]. There are also some controversial reports of mechanochemical water splitting involving a copper oxide [10–12, 21–23]. Since our initial report there has been renewed interest in the use of copper for water oxidation, starting with a study of electrocatalysis by copper salts in carbonate and phosphate buffers [13] and one with a peptide ligand [14]. As these emerging studies indicate, the foundation of biomimetic copper chemistry plus copper's low cost and high abundance make it a promising metal to explore for use in water oxidation catalysis.

10.2   A Biomimetic Approach

Our initial efforts aimed to use known transformations of copper species with nitrogen ligands (L) [1–3] to develop a mechanistically well-defined catalytic cycle for water oxidation. Our inspiration came from reported biomimetic reactions of LCu^I and O_2 to form peroxo or oxo intermediates, eventually leading to dicopper-bis(μ-hydroxo) complexes ($L_2Cu^{II}_2(\mu$-OH)$_2$). We hypothesized that oxidants and bases could be used to drive this chemistry in the reverse direction, as summarized in the envisioned water oxidation cycle in Scheme 10.1. Removal of two electrons and two protons from $L_2Cu^{II}_2(\mu\text{-}OH)_2$ would give the corresponding Cu^{III}_2-bis(μ-oxo) species, indicated as step 1 in the Scheme. The reverse of this step, oxidation of substrates by $L_2Cu^{III}_2(\mu\text{-}O)_2$, is well precedented [1–3]. We expected the $2e^-/2H^+$ oxidation to be challenging, but we felt we could apply new understanding of related kinds of PCET reactions, developed in our lab and others [15–18]. The resulting $L_2Cu^{III}_2(\mu\text{-}O)_2$ complex could then isomerize to a $Cu^{II}_2\text{-}\mu(\eta^2\text{:}\eta^2\text{-peroxo})$ species, with formation of the O–O bond. Step 3, the release of O_2 (or H_2O_2) from a $L_2Cu^{II}_2(O_2)$ species, has also been previously observed [1–3, 19–23]. The cycle is then completed by reaction of Cu^I with H_2O, with removal of H^+ and e^-. An attractive feature of this approach is that the key intermediates, $L_2Cu^{III}_2(\mu\text{-}O)_2$ and $L_2Cu^{II}_2(\mu\text{-}O_2)$, can be generated independently from O_2 and in many cases had already been well characterized. Thus, if step 1 is possible, Scheme 10.1 could be an unusual, mechanistically well-defined cycle for the oxidation of water to O_2. A similar cycle has recently been proposed for the four-electron *reduction* of dioxygen to water by a linked dicopper complex with bis[2-(2-pyridyl)ethyl]amine) ligands [24].

10.2.1 Thermochemistry: Developing Oxidant/Base Combinations as PCET Reagents

The extraction of 2 e^- and 2 H^+ from $L_2Cu_2(\mu\text{-}OH)_2$ (step 1 of Scheme 10.1) effectively removes two hydrogen atoms. We envisioned this occurring via two sequential concerted proton–electron transfer (CPET) steps, removing $1e^-/1H^+ \equiv H\cdot$ in each step. From this perspective, step 1 is the cleavage of two CuO–H bonds. Therefore, the relevant thermochemical parameter is the average bond dissociation free energy (BDFE) of these bonds. The average BDFE will indicate the oxidant/base combinations that are required, as described in this section.

The CuO–H BDFEs can be roughly estimated from reactions of these compounds. A number of $L_2Cu^{III}_2(\mu\text{-}O)_2$ compounds can remove a hydrogen atom from phenols, with the simplest example being the conversion of 2, 4, 6-$^tBu_3(C_6H_2)OH$ (abbreviated $^tBu_3PhOH$) to the phenoxyl radical $^tBu_3PhO\cdot$ (Equation 10.1) [25, 26]. For reaction 10.1 to be exoergic ($\Delta G° < 0$), the average BDFE of the CuO–H bonds in the $L_2Cu^{II}_2(\mu\text{-}OH)_2$ dimer that are being formed must be larger than the BDFE of $^tBu_3PhO$–H. This is 77 kcal/mol in most organic solvents [27]. The reaction of the bis(μ-hydroxide) with hydrogen peroxide [21] provides another limit, as shown by Equation 10.2. Cleavage of the two CuO–H bonds must require less free energy than $2H\cdot + H_2O_2 \rightarrow 2H_2O$, which has $\Delta G° = -183$ kcal/mol (in the gas phase[1]). Equation 10.2 thus indicates a rough upper limit of ~ 92 kcal/mol for the average CuO–H BDFE. Overall, the analysis gives $77 < \text{BDFE(CuO–H)} < 92$ kcal/mol, or $\text{BDFE(CuO–H)} \cong 84 \pm 7$ kcal/mol. This is in the range needed for water oxidation, which has a per-hydrogen energy requirement of 86 kcal/mol (Equation 10.3; the value is for aqueous solutions and is equivalent to $E° = 1.23$ V) [27]. This is very close to the BDFE for aqueous tyrosine, 87.8 kcal/mol [27], which is consistent with the redox equivalents in the water-oxidizing complex in photosystem II (PSII) passing through the tyrosine/tyrosyl

---

[1] Data from NIST Chemsitry Webbook, NIST Standard Reference Database Number 69, June 2005 Release. The gas-phase values therein have been corrected to the solution standard state of 1 mol/L (a correction of 1.9 kcal/mol per species).

radical couple.

$$L_2Cu^{III}_2(\mu\text{-}O)_2 + 2\,{}^tBu_3PhOH \longrightarrow L_2Cu^{II}_2(\mu\text{-}OH)_2 + 2\,{}^tBu_3PhO^\bullet \tag{10.1}$$

$$H_2O_2 + 2H^\bullet \longrightarrow 2H_2O \qquad \Delta G^\circ \sim -183\ \text{kcal mol}^{-1}$$

$$L_2Cu^{II}_2(\mu\text{-}OH)_2 \longrightarrow L_2Cu^{II}_2(\mu\text{-}O)_2 + 2H^\bullet \qquad \Delta G^\circ = 2 \times \text{BDFE (CuO–H)}$$

$$L_2Cu^{II}_2(\mu\text{-}OH)_2 + H_2O_2 \longrightarrow L_2Cu^{II}_2(\mu\text{-}O)_2 + 2H_2O \qquad \Delta G^\circ < 0 \tag{10.2}$$

$$2\,H_2O \longrightarrow O_2 + 4H^\bullet \qquad \Delta G^\circ_{(aq)} = 344\ \text{kcal mol}^{-1} \tag{10.3}$$

The phenoxyl radical ${}^tBu_3PhO^\bullet$ is one of the strongest isolable organic hydrogen atom abstracting reagents reported [28], so we concluded that oxidation of copper bis(hydroxo) dimers would likely require more thermodynamic driving force than can be provided by typical hydrogen atom abstractors. Thus we developed combinations of outer-sphere oxidants and organic bases to deliver the required driving force [5]. An oxidant/base combination has a thermodynamic affinity for a hydrogen atom that can be calculated from the E° of the oxidant and pK_a of the base (Equation 10.4) [5, 27]. This value is an effective BDFE, even though no X–H bond is being homolytically formed or cleaved. The C term in Equation 10.4 depends on the solvent and the reference potential for the redox couple [27]. In principle, there are a very large number of oxidant/base combinations that could be used, but in practice combinations that have high effective BDFEs often involve incompatible reagents. Bases are generally electron rich and thus typically susceptible to oxidation. Oxidants are electron poor and are therefore often susceptible to nucleophilic attack by bases. Still, our survey identified relatively stable combinations of oxidants and organic bases, with effective BDFEs from 71 to at least 98 kcal/mol in acetonitrile [5, 29]. Pyridine bases and triarylaminium radical cations $(NAr_3^{\bullet+})$ have been the most valuable. $NAr_3^{\bullet+}$ generally reacts only as a simple outer-sphere electron transfer agent (although a counterexample is discussed later). It has innocuous NAr_3 byproducts and an intense, characteristic ultraviolet/visible (UV/vis) spectrum that is useful for monitoring reactions.

$$-\text{BDFE}(Ox^+/:B) = 23.1\ E^\circ(Ox^{+/0}) + 1.37pK_a(HB^+) + C_G(\text{in kcal/mol}) \tag{10.4}$$

10.2.2 Copper Complexes with Alkylamine Ligands

We explored three copper systems with neutral amine ligands, containing 1,4,7-trimethyl-1,4,7-triazacyclononane (Me$_3$TACN) [30–36], N,N'-diethyl-N,N'-dimethyl-1,2-ethylenediamine ((MeEt)$_2$en)^2 [31, 37–39], and N,N'-di-*tert*-butyl-1,2-ethylenediamine (${}^tBu_2$en) [29]. Me$_3$TACN offers well-characterized $[L_2Cu_2^{II}(\mu\text{-}OH)_2]^{2+}$ and $[L_2Cu^{III}_2(\mu\text{-}O)_2]^{2+}$ complexes, and (MeEt)$_2$en was chosen for its steric properties and unusually stable bis(μ-oxo) complexes. These bis(μ-oxo) complexes have intense, characteristic UV/vis absorptions ($\epsilon \approx 15\,000\ \text{M/cm}$; see Figure 10.1) [1–3, 36, 48], which make them readily detectable even at low concentrations. They can be independently generated by adding O_2 to solutions of LCu^{I}X complexes in methylene chloride (CH$_2$Cl$_2$) at $-75\,°C$. In contrast, O_2 reacts with (${}^tBu_2$en)Cu^I(OTf)^3 to form a μ–(η^2:η^2-peroxo) complex, which

---

^{2} Note that oxygenation of more concentrated solutions of ((MeEt)$_2$en)Cu^I leads to formation of 3:1 Cu:O$_2$ species.

^{3} Attempts to isolate (${}^tBu_2$en)Cu^I(PF$_6$) yielded intractable viscous oils, so the triflate salt was used instead.

has a characteristic charge-transfer band at ca. 350 nm, $\epsilon \approx 3000\,M^{-1}\,cm^{-1}$ [40–43]. Successful oxidation of the bis(μ-hydroxo) derivative of ($^{t}Bu_2$en) should form an O–O bond, representing steps 1 and 2 of the proposed catalytic cycle in Scheme 10.1.

The dicopper(II)-*bis-μ-hydroxo* starting complexes were synthesized using two general methods, following literature procedures (Scheme 10.2) [29–31, 40–43]. In some cases, a copper salt can simply be treated with ligand, water, and a base. An alternative route is oxygenation of $LCu^{I}X$ followed by reduction with a sacrificial hydrogen atom donor such as dihydroanthracene or with a solvent such as THF. However, these bis(μ-hydroxo) species have been viewed as the "rust" of biomimetic copper chemistry, so their preparation and characterization have not received extensive attention and can be challenging [29].

The reactions of alkylamine Cu(II) bis(μ-hydroxo) complexes with various oxidant/base combinations have been surveyed. Table 10.1 gives representative examples, and more detail is available in reference [29]. Base and then oxidant were added to solutions of the bis-μ-hydroxo complexes in acetonitrile (CH_3CN) or methylene chloride, both quite oxidation-resistant solvents. Acetonitrile was chosen because its redox potential and pK_a data are available (at 25 °C), so that the oxidant/base thermodynamics are well defined, while methylene chloride allows for the use of lower temperatures [5].

Reactions were monitored via the disappearance of the intense absorbance of the $NAr_3^{\cdot+}$ oxidant. For instance, treatment of $[(Me_3TACN)_2Cu^{II}_2(\mu\text{-}OH)_2](OTf)_2$ in acetonitrile with $[N(2,4\text{-}Br_2\text{-}C_6H_3)_3]^{+\cdot}$ and pyridine or 2,6-Me_2pyridine showed rapid bleaching of the aminium oxidant, within $\sim$ 20 seconds at $-40\,°C$. With the less oxidizing $[N(4\text{-}Br\text{-}C_6H_4)_3]^{+\cdot}$, over an hour was required for the blue aminium color to disappear. In none of these – or any of the related reactions – was the spectrum of the anticipated bis(μ-oxo) species observed. Given the high absorbance of the μ-oxo material, even a 10% yield would have been evident. Bleaching of $[N(4\text{-}Br\text{-}C_6H_4)_3]^{+\cdot}$ was sometimes accompanied by growth of a broad absorption at 485 nm that disappeared on warming. Similar results were obtained with $[((MeEt)_2en)Cu^{II}(\mu\text{-}OH)]_2(PF_6)_2$.

Control experiments were performed in order to make sense of these negative results. $[N(4\text{-}Br\text{-}C_6H_4)_3]^{+\cdot}$ is fairly stable to pyridine and to $[(Me_3TACN)_2Cu^{II}_2(\mu\text{-}OH)_2](OTf)_2$ separately, even at 298 K, but decays much more rapidly when both the copper complex and the base are present (Figure 10.1). To test whether the bis(μ-oxo) species could be decomposing before it could be observed, it was generated independently from O_2 and $[(Me_3TACN)Cu^{I}(CH_3CN)]OTf$ and was shown to be stable to other species present in the reaction mixtures. In general, although the amine-ligated *bis(μ-oxo)* and $\mu\text{-}\eta^2{:}\eta^2$-peroxo compounds are not thermally stable, they are stable to the oxidants, bases, and acids used in this work for at least an hour at low temperatures. Thus, if *bis(μ-oxo)* and $\mu\text{-}\eta^2{:}\eta^2$-peroxo compounds had been formed under our reaction conditions, they would have been observed. Still, the difference in the rate of bleaching of $[N(2,4\text{-}Br_2\text{-}C_6H_3)_3]^{+\cdot}$ and $[N(4\text{-}Br\text{-}C_6H_4)_3]^{+\cdot}$ upon reaction with $[(Me_3TACN)_2Cu^{II}_2(\mu\text{-}OH)_2](OTf)_2$ and pyridine suggests that the reactions require an effective BDFE between roughly 90 and 98 kcal/mol. This would be consistent with the thermochemical data already presented if PCET from the copper hydroxide were the pathway for oxidant decay.

Reactions of oxidants and base with the $^{t}Bu_2$en derivative $[(^{t}Bu_2en)_2Cu^{II}_2(OH)_2](OTf)_2$ show similar results. Addition of the strong oxidant $[N(2,4\text{-}Br_2\text{-}C_6H_3)_3]^{+\cdot}$ to cold solutions of the copper bis(μ-hydroxide) and 2,6-Me_2 pyridine in acetonitrile or methylene chloride led to bleaching of the oxidant in less than 30 seconds. With the hexafluorophosphate salt of the copper complex under the same reaction conditions, bleaching of

Table 10.1 Representative reactions of $[L_2Cu^{II}_2(\mu\text{-}OH)_2]^{n+}$ with aminium oxidants and bases, and the effective hydrogen atom affinities, $\Delta G_{H•}$, of the oxidant/base combinations

Ligand[a]	X$^-$	Oxidant	Base	$\Delta G_{H•}$ (kcal/mol)	Solvent	Result
Me₃TACN	OTf$^-$	[N(2,4-Br₂-C₆H₃)₃]$^{•+}$	2,6-Me₂py	−101	CH₃CN	Ox. bleach in 20 seconds
Me₃TACN	OTf$^-$	[N(2,4-Br₂-C₆H₃)₃]$^{•+}$	py	−98	CH₃CN	Ox. bleach in 20 seconds
Me₃TACN	OTf$^-$	[N-(4-Br-C₆H₄)₃]$^{•+}$	2,6-Me₂py	−90	CH₃CN	Ox. bleach in 1 hour
Me₃TACN	OTf$^-$	[N-(4-Br-C₆H₄)₃]$^{•+}$	py	−87	CH₃CN	Ox. bleach in 1 hour
(MeEt)₂en	PF₆$^-$	[N(2,4-Br₂-C₆H₃)₃]$^{•+}$	py	−98	CH₃CN	Ox. bleach in 40 seconds
(MeEt)₂en	PF₆$^-$	[N-(4-Br-C₆H₄)₃]$^{•+}$	py	−87	CH₃CN	Ox. bleach in 1500 seconds
^{t}Bu₂en	OTf$^-$	[N(2,4-Br₂-C₆H₃)₃]$^{•+}$	2,6-Me₂py	−101	CH₂Cl₂	Ox. bleach in 30 seconds
^{t}Bu₂en	PF₆$^-$	[N(2,4-Br₂-C₆H₃)₃]$^{•+}$	2,6-Me₂py	−101	CH₂Cl₂	Ox. bleach in 200 seconds
^{t}Bu₂en	PF₆$^-$	[N-(4-Br-C₆H₄)₃]$^{•+}$	2,6-Me₂py	−90	CH₂Cl₂	Ox. bleach in 3000 seconds
^{t}Bu₂en	PF₆$^-$	[N-(4-Me-C₆H₄)₃]$^{•+}$	2,6-Me₂py	−83	CH₂Cl₂	No reaction
Tp^{iPr2}	-	[N-(4-Br-C₆H₄)₃]$^{•+}$	py	−87	CH₃CN	Ox. bleach in 500 seconds
Tp^{iPr2}	-	[N-(4-Br-C₆H₄)₃]$^{•+}$	py + 2,6-Me₂py	−90	CH₂Cl₂	Ox. bleach in 5000 seconds
Tp^{iPr2}	-	[N(2,4-Br₂-C₆H₃)₃]$^{•+}$	py + 2,6-Me₂py	−98	CH₂Cl₂	Ox. bleach in <10 seconds
MeL^{Me2}	-	[N-(4-Me-C₆H₄)₃]$^{•+}$	4-NH₂-py	−88	CH₂Cl₂	No reaction
MeL^{Me2}	-	[N-(4-Me-C₆H₄)₃]$^{•+}$	py	−81	CH₂Cl₂	No reaction
Orange[b] MeL^{Me2}	-	[N-(4-Me-C₆H₄)₃]$^{•+}$	^{t}BuOK/py	−119	CH₂Cl₂	Gives bis(μ-oxo)
Orange[b] MeL^{Me2}	-	[N-(4-Me-C₆H₄)₃]$^{•+}$	4-NH₂-py	−88	CH₂Cl₂	Gives bis(μ-oxo)
Orange[b] MeL^{Me2}	-	Fe(C₅H₄COMe)(Cp)$^+$	4-NH₂-py	−85	CH₂Cl₂	Gives bis(μ-oxo)
Orange[b] MeL^{Me2}	-	[N-(4-Me-C₆H₄)₃]$^{•+}$	py	−80	CH₂Cl₂	No reaction
CF₃L^{Me2}	-	[N-(4-Br-C₆H₄)₃]$^{•+}$	py	−87	CH₂Cl₂	No reaction
CF₃L^{Me2}	-	[N-(4-Me-C₆H₄)₃]$^{•+}$	4-NH₂-py	−88	CH₂Cl₂	No reaction
MeL^{Cl2}	-	[N-(4-Br-C₆H₄)₃]$^{•+}$	py	−87	CH₂Cl₂	Ox. bleach in 200 seconds
MeL^{Cl2}	-	[N-(4-Me-C₆H₄)₃]$^{•+}$	py	−80	CH₂Cl₂	No reaction

Note: Reactions in CH₃CN were performed at −40°C, while reactions in CH₂Cl₂ were performed at −75°C
[a]For ligand abbreviations, see Section 10.3;
[b]See text

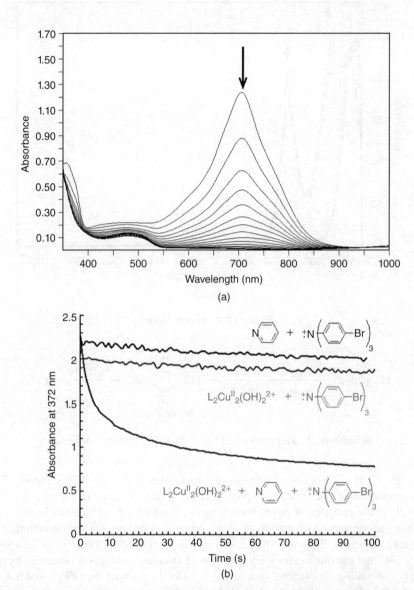

Figure 10.1 (a) Optical spectra of the reaction of $[(Me_3TACN)_2Cu^{II}_2(OH)_2](OTf)_2$ with $[N(4\text{-}Br\text{-}C_6H_4)_3]^{+\bullet}$ and pyridine in acetonitrile at 298 K, showing the decay of the aminium oxidant ($L = Me_3TACN$). (b) Changes in the absorbance at 372 nm for the first reaction (blue, bottom) and for control experiments omitting the copper (black, top) or the pyridine (red, middle). (c) (next page) Optical spectra of the reaction of $[LCu^I(CH_3CN)]OTf$ with O_2 at $-75\,°C$ in methylene chloride, showing the growth of the μ-oxo complex $[L_2Cu^{III}_2(\mu\text{-}O)_2]^{2+}$. See color plate

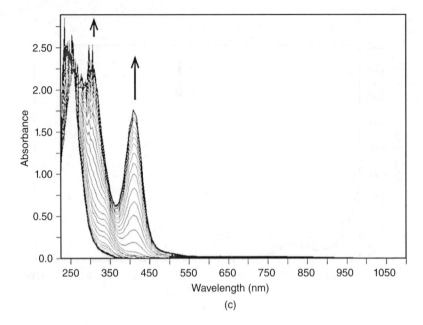

Figure 10.1   *(continued)*

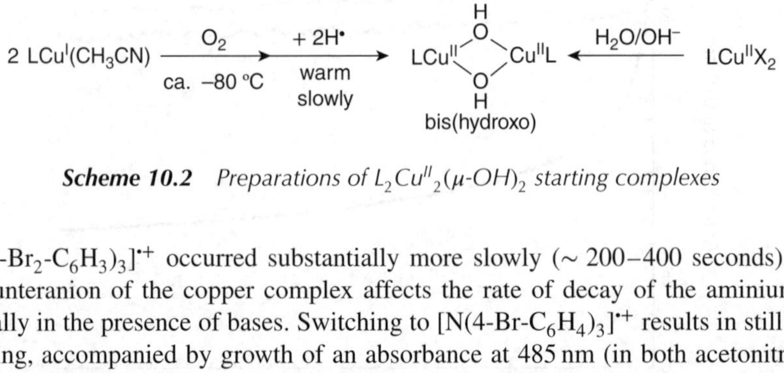

Scheme 10.2   *Preparations of $L_2Cu^{II}{}_2(\mu\text{-}OH)_2$ starting complexes*

$[N(2,4\text{-}Br_2\text{-}C_6H_3)_3]^{\bullet+}$ occurred substantially more slowly ($\sim 200\text{-}400$ seconds). Thus, the counteranion of the copper complex affects the rate of decay of the aminium ions, especially in the presence of bases. Switching to $[N(4\text{-}Br\text{-}C_6H_4)_3]^{\bullet+}$ results in still slower bleaching, accompanied by growth of an absorbance at 485 nm (in both acetonitrile and methylene chloride). This absorbance is very likely related to Eberson and Larsson's reports [44] that intermolecular decomposition of aminium oxidants is promoted by more strongly ion-pairing counterions such as BF_4^- and OTf^-, rather than PF_6^- and $SbCl_6^-$. In particular, decomposition of $[N(4\text{-}Br\text{-}C_6H_4)_3]^{+\bullet}$ is reported to give a characteristic absorbance at 485 nm, just as is observed here. The observed oxidant decay is thus likely due to the reaction of $NAr_3^{\bullet+}$ with the base and the counteranion of the copper complex, not to oxidation of the copper.

In sum, no Cu(III) bis(μ-oxo) or Cu(II) $\mu\text{-}\eta^2\text{:}\eta^2$-peroxo species were observed upon reaction of alkylamine Cu(II) bis(μ-hydroxo) compounds with oxidants and bases. This is in spite of our thermodynamic estimation that a number of the oxidant/base combinations have more than sufficient driving force for the desired reactions, as summarized in Table 10.1. This emphasis on thermochemistry has the implicit assumption that the PCET reactions have small kinetic barriers, as is typical for cases in which the proton moves between electronegative atoms [17]. Yet one pair of reactions may suggest a significant barrier. The mixture of $[(Me_3TACN)_2Cu^{II}_2(OH)_2](OTf)_2$, $[N(4\text{-}Br\text{-}C_6H_4)_3]^{+\bullet}$, and pyridine in acetonitrile decays over an hour at $-40\,^\circ C$ (and does not give the bis(μ-oxo) product), yet in the reverse direction, independently generated $[(Me_3TACN)_2Cu^{III}_2(\mu\text{-}O)_2]^{2+}$ does not appear to react with $[N(4\text{-}Br\text{-}C_6H_4)_3]$ and pyH$^+$, as only a 30% decrease in the bis(μ-oxo) complex is observed over an hour, and no formation of the aminium radical cation is observed. Since the reaction must be favorable in one direction or the other, the lack of reaction in both directions implies at least a modest barrier. We note that rapid addition of e^- and H$^+$ has been reported for the copper bis(μ-oxo) complexes $(^iPr_3TACN)_2Cu^{III}_2(O)_2$ and $(Bz_3TACN)_2Cu^{III}_2(O)_2$ using ferrocene and HBF$_4$ [45], although this combination has significantly more driving force. Still, the major problem with this chemistry appears not to be kinetic bottlenecks but rather the facile decomposition of the aminium oxidants under the reaction conditions.

10.2.3   Copper Complexes with Anionic Ligands

To avoid some of the problems just described, we turned to complexes with mono-anionic chelating nitrogen ligands. The more electron-donating nature of these ligands should facilitate oxidation of the copper, although the lower outer-sphere redox potential is likely offset in part by the hydroxo protons being less acidic. The $L_2Cu_2(OH)_2$ complexes with anionic ligands are neutral, thus eliminating the counterions that can contribute to oxidant decomposition.

Our studies started with Kitajima's complexes containing the hydrotris(3,5-diisopropylpyrazolyl)borate ligand [21, 46]. This Tp^{iPr2} ligand supports a fairly stable Cu$^{II}_2$-μ-(η^2:η^2-peroxo) complex that releases O$_2$ upon treatment with CO [22]. We later examined several β-diketiminate (NacNac) complexes that had been developed by Tolman [47–51], Warren [52], and others [53–55] (Scheme 10.3). These ligands reduce the Cu$^{II/I}$ reduction potential by ca. 0.5 V versus complexes with neutral bidentate ligands [49]. The Cu(III) bis(μ-oxo) complexes of MeL^{Cl2} (Scheme 10.3) are remarkably stable, persisting for hours at room temperature in toluene [56].

The bis(μ-hydroxo) compound $(MeL^{Me2})_2Cu^{II}_2(OH)_2$ was obtained via oxygenation of $(MeL^{Me2})Cu^I(CH_3CN)$ [50], as previously reported [52], but cleaner samples appeared to be obtained when 9,10-dihydroanthracene was added as the terminal reductant instead of the THF solvent. Professors William B. Tolman and Timothy H. Warren and their coworkers generously provided samples of the Cu(I)- and Cu(II)-hydroxo compounds of the CF$_3$L^{Me2} ligand [57], $(MeL^{Cl2})_2Cu^{II}_2(OH)_2$ (for synthesis see [58]) and $(Tp^{iPr2})_2Cu^{II}_2(OH)_2$. $(Tp^{iPr2})_2Cu^{II}_2(OH)_2$ was stored and handled in a glovebox, as it absorbs CO$_2$ from air [56].

Reactions of the β-diketiminate and tris-pyrazolylborate copper(II) bis(μ-hydroxo) compounds with oxidant/base combinations are included in Table 10.1. Reactions of $(Tp^{iPr2})_2Cu^{II}_2(OH)_2$ with the stronger oxidants $[N(2,4\text{-}Br_2\text{-}C_6H_3)_3]^{+\bullet}$, $[N(4\text{-}Br\text{-}C_6H_4)_2(4\text{-}MeO\text{-}C_6H_4)]^{+\bullet}$, and $[N(4\text{-}Br\text{-}C_6H_4)_3]^{+\bullet}$ all show a rapid bleaching of the oxidant. These experiments used a high concentration of pyridine to promote hydrogen bonding or proton transfer from the sterically encumbered copper hydroxides, with Me$_2$-pyridine

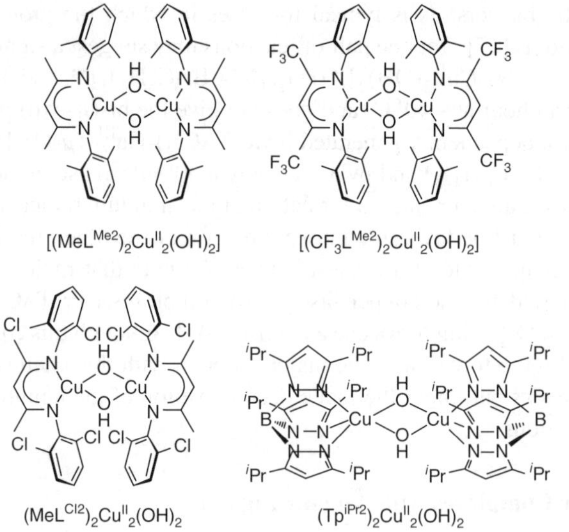

[(MeL^{Me2})$_2$Cu^{II}$_2$(OH)$_2$]   [(CF$_3$L^{Me2})$_2$Cu^{II}$_2$(OH)$_2$]

(MeL^{Cl2})$_2$Cu^{II}$_2$(OH)$_2$ (Tp^{iPr2})$_2$Cu^{II}$_2$(OH)$_2$

Scheme 10.3   *Copper bis(μ-hydroxo) complexes studied with oxidant/base combinations*

added to provide extra overall driving force, as discussed later. No bleaching of the weaker [N(4-Me-C$_6$H$_4$)$_3$]$^{+\bullet}$ oxidant was observed over 3 hours under these conditions. In all cases, no peaks characteristic of the Cu^{II}μ(η^2:η^2-peroxo) species were observed, although some of the UV/vis spectra suggest aminium decomposition, as described before.

In all of the reactions with the methyl-NacNac complex (MeL^{Me2})$_2$Cu^{II}$_2$(μ-OH)$_2$, again only bleaching of the aminium oxidant was observed, without formation of the intensely colored bis(μ-oxo) species (422 nm, 11 000 M^{-1} cm^{-1}) [29, 36, 48]. Control experiments showed, to our surprise, that the bis(μ-oxo) species is more reactive with oxidants than the bis(μ-hydroxide). (MeL^{Me2})$_2$Cu^{III}$_2$(μ-O)$_2$, generated from O$_2$ and (MeL^{Me2})Cu^{I}(CH$_3$CN) at −75 °C in THF, reacted instantaneously with [N(4-Br-C$_6$H$_4$)$_3$]$^{+\bullet}$, [N(4-Br-C$_6$H$_4$)$_2$(4-MeO-C$_6$H$_4$)]$^{+\bullet}$, and Mn^{III}(hfacac)$_3$ (but not with the milder oxidant [N(4-Me-C$_6$H$_4$)$_3$]$^{+\bullet}$), as evidenced by bleaching of the copper and oxidant absorbances (hfacac = CF$_3$C(O)CHC(O)CF$_3$). Since the Cu^{III} centers should be very difficult to oxidize, it is likely that the one-electron oxidants were reacting with the electron-rich NacNac ligand. Consistent with this hypothesis, the related complex with the less electron-rich CF$_3$-substituted ligand (CF$_3$L^{Me2})$_2$Cu^{III}$_2$(O)$_2$ reacted only with the strongly oxidizing [N(2, 4-Br$_2$-C$_6$H$_3$)$_3$]$^{+\bullet}$ but not with [N(4-Br-C$_6$H$_4$)$_3$]$^{+\bullet}$. Thus, replacing the backbone methyl groups with CF$_3$ groups makes (CF$_3$L^{Me2})$_2$Cu^{III}$_2$(O)$_2$ more resistant to oxidation than the Me-substituted analog, by 0.2–0.5 V. One set of problems appears to have been traded for another with the β-diketiminate ligand: the Cu(II) centers are easier to oxidize and the bis(μ-hydroxo) compound is uncharged, but the electron-rich ligand limits choice of oxidant.

Fortunately, the driving force for the PCET oxidation of L$_2$Cu^{II}$_2$(OH)$_2$ is the sum of the oxidant strength and the base strength (Equation 10.4). Thus the same driving force can be obtained with a weaker oxidant and a stronger base. This will favor the PCET oxidation over the apparent outer-sphere electron transfer oxidation of the Cu^{III}-bis(μ-oxo) compound. In this case, we used the stronger base 4-aminopyridine (4-NH$_2$py) with the weaker oxidant [N(4-Me-C$_6$H$_4$)$_3$]$^{+\bullet}$ to achieve an H$^\bullet$ affinity (minus effective BDFE) of -88 kcal/mol.

We were quite excited when mixing the [N(4-Me-C$_6$H$_4$)$_3$]$^{+\bullet}$/4-NH$_2$py combination with an "orange" batch of copper(II) starting material at $-75\,°$C in methylene chloride showed the growth of a strong absorption at 430 nm. This is consistent with the formation of significant yields of (MeL^{Me2})$_2$Cu^{III}$_2$(μ-O)$_2$. The strong absorbance at 430 nm bleaches quickly upon warming, at roughly the same rate of decay as independently prepared (MeL^{Me2})$_2$Cu^{III}$_2$(μ-O)$_2$ under similar conditions. While these data support the assignment of the product as the bis(μ-oxo) complex, and we refer to it that way here, attempts to confirm this by resonance Raman spectroscopy were unsuccessful due to interfering signals from the aminium oxidant.

Roughly 25% yields of (MeL^{Me2})$_2$Cu^{III}$_2$(μ-O)$_2$ were also observed using the less oxidizing mixed aminium [N(4-Br-C$_6$H$_4$)(4-MeO-C$_6$H$_4$)$_2$]$^{+\bullet}$ or the monoacetyl-ferrocenium cation in combination with 4-NH$_2$py as a base ($\Delta G_{H\bullet} = -86$ and -85 kcal/mol). However, no reaction was observed with the combinations [N(4-MeO-C$_6$H$_4$)$_3$]$^{+\bullet}$/4-NH$_2$py and [N(4-Me-C$_6$H$_4$)$_3$]$^{+\bullet}$/pyridine ($\Delta G_{H\bullet} = -83$ and -80 kcal/mol, respectively).[4] Similarly, no reaction was also observed with the 2,4,6-tri-*tert*-butylphenoxyl radical, which has an H-atom affinity of -77 kcal/mol. Thus the thermochemical BDFE analysis seems to be effective at predicting reactivity. One caveat is that the quoted $\Delta G_{H\bullet}$ values are for the oxidant/base pair in acetonitrile at room temperature, using the pK_as and $E°$s via Equation 10.4, and are thus only approximations to the values at $-75\,°$C in methylene chloride.

While the combination of [N(4-Me-C$_6$H$_4$)$_3$]$^{+\bullet}$ with pyridine alone does not oxidize the orange copper material, addition of the much stronger base potassium tert-butoxide (^{t}BuOK) leads to reduction of the oxidant and formation of the 430 nm absorption characteristic of the μ-oxo complex. The reaction does not proceed when ^{t}BuOK alone is used as the base, probably because of its very limited solubility and oligomeric nature in methylene chloride [59].[5] Apparently, pyridine is not a strong enough base in combination with [N(4-Me-C$_6$H$_4$)$_3$]$^{+\bullet}$ to perform stoichiometric PCET oxidation but it can act as a proton shuttle to the thermodynamically powerful but (under these conditions) kinetically slow ^{t}BuOK. These results illustrate that both the thermodynamics and the kinetics of proton transfer can play a key role. Intramolecular proton shuttles or relays have been shown to be highly effective in other PCET reactions, especially in hydrogenase catalysis [60]. They could be of particular value in water oxidation, because an insoluble non-oxidizable strong base such as Na$_3$PO$_4$ could be used in conjunction with a weaker but soluble base.

The likely formation of (MeL^{Me2})$_2$Cu^{III}$_2$(μ-O)$_2$ in these various reactions was very encouraging. The orange starting material was essentially unreactive with the oxidant or the base alone at $-75\,°$C in methylene chloride, indicating that oxidation is a PCET process. The PCET nature of the reactions is also indicated by oxidation occurring only with oxidant/base pairs above a certain effective BDFE. However, analyses of the orange starting complex indicated that it was a mixture of (MeL^{Me2})$_2$Cu^{II}$_2$(OH)$_2$ and another, unknown species, which despite a great deal of effort we were unable to characterize. Purification of the orange species and syntheses of purer samples of (MeL^{Me2})$_2$Cu^{II}$_2$(OH)$_2$

---

[4] No decay of the oxidant or growth of a peak at 430 nm was observed over 5500 seconds at $-75\,°$C: more than 10 times the time required for the other reactions to occur.

[5] ^{t}BuOK was added to the Cu/py/ CH$_2$Cl$_2$ solution as a concentrated solution in THF [29].

both showed that it was the second, orange species that was apparently being oxidized to the bis(μ-oxo) complex. Purified samples of the orange solid gave ca. 75–100% yields of the desired Cu(III) bis(μ-oxo) complex upon reaction with oxidant and base, based on the absorbance at 430 nm.

10.2.4   Lessons Learned: Thermochemical Insights and Oxidant/Base Compatibility

We were thus not successful in observing PCET oxidation of copper(II)-bis(μ-hydroxide) complexes to Cu(III)-bis(μ-oxo) or Cu(II)(μ-peroxo) species as the key step in a water oxidation cycle. In the NacNac systems, evidence for oxidation of Cu(II) to Cu(III)-bis(μ-oxo) complexes was obtained, but a specific reaction could not be proven. We have described these ultimately unsuccessful results here because we believe they have significant messages for water oxidation chemistry and for other demanding PCET processes.

First, the thermochemical implications of the need for both an oxidant to remove e^- and a base to remove H^+ are illuminating. In the aqueous media typically used for water oxidation, the pH describes the effective base strength of the buffer or hydroxide salt used, so discussions focus on the strength of the oxidant used (e.g. Ce^{IV} or $Ru(bpy)_3^{3+}$) and the thermodynamic potential of the $O_2 + 4H^+ + 4e^- \rightarrow 2H_2O$ half-reaction at that pH. Water oxidation in organic solvents, however, requires use of a specific base, whose thermochemical properties must be accounted for. Organic solvents have been suggested to be useful for water oxidation in some respects [61] and will likely prove valuable for some mechanistic studies. The work described in this section provides a thermochemical way of thinking about oxidant/base combinations and indicates that the affinity for a hydrogen atom (the ΔG_H°, or the negative effective BDFE) is a powerful way of analyzing such combinations [5].

The second lesson learned is that it is a substantial challenge to achieve compatibility between the metal complex catalyst, the oxidant, and the base under the strongly oxidizing conditions of water oxidation. We have found only a limited number of stable oxidant/base combinations with sufficient driving force for water oxidation [5]. However, the ability to achieve this driving force in various ways, from strong oxidants plus weak bases or weaker oxidants with stronger bases, can provide valuable approaches when problems arise. In the case just described, the absorption band likely resulting from $(MeL^{Me2})_2Cu^{III}_2(O)_2$ was only observed with weak oxidant/stronger base combinations, apparently because strong oxidants remove an electron from the NacNac ligand. Intermolecular proton relays to insoluble non-oxidizable bases could potentially be valuable in this regard.

10.3   An Aqueous System: Electrocatalysis with (bpy)Cu(II) Complexes

In light of the difficulties inherent in the biomimetic approach, with its requirement for oxidants and bases in organic solvents at low temperatures, we decided to try a different tactic. We moved to aqueous electrocatalysis, using hydroxide as the base and the electrode as the oxidant. Glassy carbon and indium tin oxide (ITO) electrodes are known to be poor WOCs. (There is, however, some degradation of carbon electrodes under strongly oxidizing and basic conditions, another example of oxidant–base incompatibility.) The initial choice of which copper complex to study is discussed in the next section.

10.3.1   System Selection: bpy + Cu

In selecting the best ligand for electrocatalysis, several criteria must be considered. A good WOC ligand should be robust in order to withstand highly oxidative conditions. It should be easily modified, so that electronic and steric effects can be studied. Given the potentially global application of water oxidation, the ligand should be inexpensive or easily synthesized. Finally, for our purposes there needed already to be some understanding of the chemistry of copper with the ligand, and some solubility of copper complexes, because we aimed to study molecular electrocatalysis.

The 2, 2'-bipyridine (bpy) family of ligands fits these criteria. Their aromatic structure makes them fairly robust for an organic ligand, although there are reports of bipyridine activation and oxidation in high-valent complexes [62–65]. Bpy and related ligands have been widely used in WOCs, starting from the blue dimer as described in Chapter 5. It is a ubiquitous ligand in inorganic chemistry and modified bpy ligands are commercially available or can be readily synthesized [66].

Dicopper(II) bis(μ-hydroxo) complexes with bipyridine and phenanthroline (phen) ligands have long been studied; they were first reported in 1968 in a paper by Woolliams [67]. These compounds are synthesized simply by dissolving a copper(II) salt in water, adding a single equivalent of ligand, and agitating the solution for 5–10 minutes until the ligand has fully dissolved (Equation 10.5). If the ligand is added as a solution in an organic solvent such as ethanol or acetonitrile, there is near-immediate formation of the bright-blue bpy/Cu(II) complex. The dominant species in acidic aqueous solutions is the bis(aquo) complex $[(bpy)Cu(H_2O)_2]^{2+}$ [78–80]. Upon raising the pH, as described by Woolliams, the dimeric, dicationic bis(μ-hydroxide) copper complex $[(bpy)_2Cu_2(\mu\text{-OH})_2]^{2+}$ is formed. With most starting copper(II) salts ($Cu^{II}X_2$), this complex precipitates as a dark-blue solid $[(bpy)_2Cu_2(\mu\text{-OH})_2]X_2$ upon addition of hydroxide.

$$2\,CuX_2 + 2\,bpy + 2\,OH^- \rightarrow [(bpy)_2Cu_2(\mu\text{-OH})_2]X_2 + 2\,X^- \qquad (10.5)$$

Of the various $[(bpy)_2Cu_2(\mu\text{-OH})_2]^{2+}$ salts surveyed, only the acetate one has substantial solubility in water (> 100 mM). The sulfate and triflate salts have modest solubility ($\leq$ 1 mM), while the nitrate and tetrafluoroborate are quite insoluble. The UV/vis spectra of the acetate, sulfate, and triflate salts are the same (0.5–1.0 mM Cu complex, pH 12.5), ruling out anion coordination to copper.

10.3.2   Observing Electrocatalysis

Cyclic voltammograms (CVs) of $[(bpy)_2Cu_2(\mu\text{-OH})_2](OAc)_2$ solutions, using a glassy carbon working electrode and aqueous NaOAc/NaOH electrolyte, show very large anodic (oxidation) currents in basic solutions (Figure 10.2). Peak currents of up to $700\,\mu A$ ($10\,\mu A/cm^2$) at 1 mM copper are much larger than expected for a typical one-electron redox couple. Scanning to negative potentials shows a more typical one-electron wave, with peak current $i_p \approx 10\,\mu A$, which we assign to the $Cu^{II/I}$ couple. The observation of peak currents ~ 70 times larger than the i_p for a $1e^-$ couple indicates electrocatalysis, as many more than one electron are removed from each copper in the vicinity of the electrode.

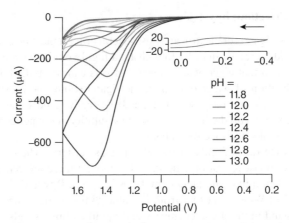

Figure 10.2   *CVs of a 0.5 mM solution of [(bpy)$_2$Cu$_2$(μ-OH)$_2$](OAc)$_2$ show increasing catalytic current as the solution becomes more alkaline. Arrow shows direction of potential sweep. Inset: The 1e$^-$ Cu$^{II/I}$ couple for the same solution, shown to scale. Conditions: 0.1 M electrolyte (NaOAc + NaOH), glassy carbon working electrode, platinum reference electrode, Ag/AgCl reference electrode (adjusted to normal hydrogen electrode, NHE). Reproduced with permission from [6]. Copyright © 2012, Rights Managed by Nature Publishing Group*

To test whether the electrocatalytic process observed by CV was water oxidation, we looked for the O$_2$ product upon controlled-potential bulk electrolysis (CPE) using high-surface-area reticulated vitreous carbon (RVC) or ITO electrodes. Since there are no standard procedures for the direct measurement of electrochemical oxygen production, we constructed a specialized electrochemical cell with a large Teflon cap (see Supplementary Information in reference [6]), which had air-tight ports for the reference electrode, an auxiliary chamber, and a fluorescence O$_2$ probe. A brass post was threaded through the cap so that the working electrode could be connected on the inside of the cell. The cap also had a gas inlet with a valve, so that the cell could be thoroughly purged with argon before electrolysis. Argon-sparged electrolyte was introduced by syringe to the cell after this purging, with the whole process being monitored by the oxygen probe. Typically, the dissolved oxygen concentration was monitored during electrolysis, with the headspace volume kept to a minimum in order to maximize the amount of oxygen in the aqueous phase. O$_2$ could also be detected in the headspace.

CPE at +1.35 V versus normal hydrogen electrode (NHE) (an overpotential of 860 mV) with the RVC electrode, with 1 mM [(bpy)$_2$Cu$_2$(μ-OH)$_2$](OAc)$_2$ at pH 12.5, showed a substantial increase in the concentration of dissolved oxygen, from 0 to ca. 550 μM. However, the Faradaic efficiency (with 4 e^- per O$_2$) was only 35–45%. Under these highly oxidizing conditions, the carbon electrode was consumed during CPE. We therefore switched the working electrode to unpolished ITO, which is oxidatively much more robust. However, there was still some background production of O$_2$ by the ITO at the potentials and pH used. This was taken into account through a simple subtraction of the current and O$_2$ production in the absence of catalyst. On ITO, a current of 0.5 mA/cm^2 was sustained for 30 minutes, with a Faradaic efficiency to O$_2$ of > 90%. It is possible that some of the "missing oxygen" escaped the cell or was in the headspace and therefore went unmeasured.

O_2 was observed electrochemically as well. CVs scanned through the O_2 production wave and then reversed to negative potentials showed a large current at ca. -0.2 V. This current is due to reduction of O_2 to O_2^- in basic solution. It is much larger than the $Cu^{II/I}$ wave in the inset of Figure 10.2, which is observed upon initial cathodic scan in O_2-free solutions. Scanning through the anodic wave and then sparging the solution with nitrogen removed the large cathodic current, confirming its assignment to the $O_2^{0/-}$ couple.

To test whether acetate oxidation might be taking place, as in the Kolbe process [68], we compared the CVs of the acetate salt in NaOAc/NaOH buffer with the sulfate and triflate salts in Na_2SO_4/NaOH and NaOTf/NaOH buffers. Very similar electrocatalysis was observed, with slight differences in total current, but with identical onset potentials. This indicates that the anion of the electrolyte is not intimately involved in the electrocatalysis, either by being oxidized (sulfate and triflate are not readily oxidized) or by acting as a ligand to Cu (triflate is a very weak ligand).

Thus $[(bpy)_2Cu_2(\mu\text{-OH})_2]^{2+}$ salts are electrocatalysts for water oxidation to O_2 in water at pH > 12. The onset of the water oxidation wave is at 1.1 V versus NHE and the half-peak potential ($E_{p/2}$) is at +1.3 V (potentials were recorded versus a Ag/AgCl reference and adjusted to NHE by addition of 0.199 V). High currents are observed, but the catalyst operates at somewhat high overpotentials. At pH 12.5, the thermodynamic potential for water oxidation is +0.493 V. This is 0.6 V below the 1.1 V for onset of catalysis, and ~ 0.8 V below the $E_{p/2}$. In general, the most efficient WOCs perform at overpotentials of 0.4–0.5 V [69, 70].

10.3.3 Catalyst Turnover Number and Turnover Frequency

Measuring benchmarks for soluble electrocatalysts is more complicated than catalysis with chemical oxidants. In homogeneous catalytic reactions, there is full mixing of substrate, catalyst, and oxidant, and the turnover number (TON) and turnover frequency (TOF) can be measured from the rate of appearance of products. Electrocatalysis, however, is a heterogeneous process, because only the catalyst molecules near the electrode are active. Therefore, the ratio of product to *total* catalyst in solution is much lower than the actual TON per *active* catalyst. To obtain such a lower limit, we performed water oxidation with $[(bpy)_2Cu_2(\mu\text{-OH})_2]^{2+}$ in a thin spectroelectrochemistry cell, in order to minimize the amount of inactive catalyst far from the ITO electrode. This experiment showed 60 catalyst turnovers (per Cu dimer) over 5.5 hours. The amount of generated oxygen was determined from the measured current during electrolysis and the Faradaic efficiency. Optical spectra of the solution showed a 15% decay of the catalyst during this experiment. However, we also observed substantial copper plating on to the auxiliary electrode (a platinum wire), since the thin cell did not allow for a membrane to separate the anode and cathode. This Cu reduction at the cathode may account for much of the decay observed, suggesting that the Cu/bpy system may be a quite robust water oxidation electrocatalyst. This conclusion is consistent with the rapid TOF observed by CV.

The TOF of an electrocatalyst can be determined from the current during CV [71–73]. However, the conditions of the CV are key to the accuracy of this calculation. The CVs must be in the "kinetic regime," where the rate of catalysis determines the maximum current, not the diffusion of the substrate. The kinetic regime is characterized by an S-shaped wave with a plateau current, and a lack of current dependence on scan rate (v). To determine the

TOF, or k_{cat}, it is common to divide the plateau current for the catalytic wave, i_c, by the peak CV current for the related noncatalytic redox process in the absence of substrate, i_p (Equation 10.6). In the kinetic regime, the ratio i_c/i_p is proportional to $(k_{cat}/v)^{1/2}$:

$$i_c/i_p = 2.24 n_c n_p^{-3/2} (k_{cat}RT)^{1/2} (Fv)^{-1/2} \tag{10.6}$$

where n_c is the number of electrons passed in each catalytic turnover (four for water oxidation), n_p is the number of electrons passed in the noncatalytic CV wave (typically one), and F is the Faraday constant.

For the Cu^{2+}/bpy catalyst, it is not possible to observe the noncatalytic anodic CV wave in the absence of substrate, since the substrate is hydroxide and high pH is needed to assemble the catalyst. We have not yet succeeded in obtaining bpy/Cu/hydroxide complexes in organic solvents. However, a good estimate of i_p is obtained from the quasi-reversible wave observed in initial cathodic scans of the Cu^{2+}/bpy catalyst (Figure 10.2 inset), which we assign as the one-electron $Cu^{II/I}$ couple. The magnitude of the peak current for this wave is close to that of ferricyanide at the same concentration.

The second challenge was reaching the kinetic regime needed to apply Equation 10.6. Under typical conditions, the CV waves are peaked rather than S-shaped (Figure 10.2), indicating that there is a diffusive contribution to the CV shape. This is most likely due to depletion of hydroxide near the electrode surface, such that the catalytic currents are limited not only by k_{cat} but also by the diffusion of OH^-. This has been suggested to occur in other systems with rapid TOFs [74]. We were able to reach the kinetic regime for this copper system by moving to low catalyst concentrations (≤ 0.1 mM in Cu) and low scan rates (≤ 10 mV/s).[6] The CV peak shapes are complicated by the background oxidation from the glassy carbon electrode under these conditions, but they are quite independent of scan rate, indicating the kinetic regime.

The electrochemical data at low scan rates indicate a TOF of $\sim 100\,s^{-1}$ per copper, using Equation 10.6. This is a relatively high value; it is consistent with the catalytic currents being much larger than $1e^-$ couples, even outside of the kinetic regime (as in Figure 10.3). As of this writing, there are only two homogeneous WOCs with similar or faster rates: a ruthenium catalyst at 800 s^{-1} and a cobalt catalyst at 80 s^{-1} [75, 76]. Previously reported WOCs were generally much slower at 5 s^{-1} or less [77].

At low copper concentrations (< 1 mM), in the kinetic regime, the catalytic currents, i_c, are linearly proportional to the [Cu(II)]. This is required by the equations that lead to Equation 10.6 (both i_c and i_p are proportional to [catalyst]). At copper concentrations higher than 1 mM, however, i_c is essentially independent of [Cu(II)]. This is likely due to catalysis being limited by substrate (hydroxide) availability at the electrode surface, which is why the high [Cu(II)] CVs are not in the kinetic regime.

The copper catalyst has only limited sensitivity to ligand substituents. We reacted several 4, 4'-disubstituted bipyridine derivatives with copper salts to yield the [(R$_2$-bpy)Cu]$^{2+}$ complexes, with R = MeO, Me, COO$^-$. We also examined the terpyridine and phenanthroline analogs. Most of the complexes performed water oxidation catalysis at roughly the same pH, potential, and current density as the unsubstituted bipyridine. For instance, the 4, 4'-dimethoxy-bpy complex performs water oxidation at the same applied potential as the

---

[6] At such slow scan rates, convection can be a concern (see: Constentin, C., Robert, M., Savéant, J. M. (2013) Chem. Soc. Rev., 42: 2423), but this does not appear to be an issue here due to the scan rate and [OH$^-$] independence.

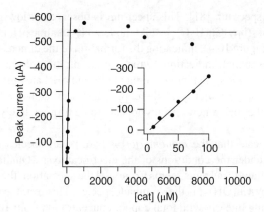

Figure 10.3 *Maximum current plotted versus total copper concentration. Inset: Linear regime at low catalyst concentration. Conditions: pH 12.5, 0.1 M electrolyte (NaOAc + NaOH), glassy carbon working electrode, platinum auxiliary electrode, Ag/AgCl reference electrode*

bpy complex. The 4, 4′-dicarboxy-bipyridine complex is one exception to this similarity, displaying catalysis at the same potential as other bpy complexes but with much lower currents. The terpyridine complex is another exception, with no observed catalysis in the potential range that was studied. It is tempting to ascribe this lack of activity to the tridentate nature of the terpyridine ligand, but a very recent report describes water oxidation by a copper complex of the *tetradentate* triglycylglycine (TGG^{4-}) ligand [14].

10.3.4 Catalyst Speciation: Monomer, Dimer, or Nanoparticles?

This copper system – in fact, any soluble electrocatalyst – is unlikely to be the technological solution to the challenge of water oxidation. Practical electrocatalysis is typically accomplished by catalytic sites that are part of or closely associated with an electrode. The goal of our studies is therefore to extract insight that could be used to develop better electrocatalysts. To that end, we include some mechanistic discussion here, although our current understanding is limited.

The first half of this chapter described how our inspiration for copper-mediated O_2 production was drawn from the elegant oxygen chemistry of copper dimers. In the aqueous studies, we started with isolated copper dimmers, but the speciation of the Cu(II)/bpy system is complex, as indicated by the various studies that are not in complete agreement [78–80, 83]. These papers all agree that in a solution of 1:1 Cu^{2+}:bpy under acidic conditions, the predominant species is the monomeric bis(aquo) complex [(bpy)Cu(H$_2$O)$_2$]$^{2+}$. Raising the pH causes dissociation of protons. One study suggests that at high pH, the dimeric [(bpy)$_2$Cu$_2$(μ-OH)$_2$]$^{2+}$ is the dominant species in solution [79], while an earlier report states that the monomeric bis(hydroxide) [(bpy)Cu(OH)$_2$] should predominate [80].

To establish the predominant form of the copper under our electrocatalysis conditions of 0.1 M electrolyte, pH $\sim$ 12.5, and $\sim$ 1 mM Cu(II), we turned to electron paramagnetic resonance (EPR) spectroscopy. The dimeric complex is effectively EPR-silent in solution at room temperature, due to antiferromagnetic coupling between the two copper ions [80]. The monomeric d^9 complexes are simple paramagnets and display the classic four-line copper(II) solution spectrum [81]. This spectrum is observed at low pH, consistent with the predominance of $[(bpy)Cu(H_2O)_2]^{2+}$. At moderately alkaline pH, no or very little EPR signal is observed (Figure 10.4), indicating the formation of the dimer. As the pH increases, so too does the EPR signal, indicating a greater concentration of the monomeric species $[(bpy)Cu(OH)_2]$. The EPR signal intensities at pH 13 and pH 4 are quite similar, indicating that at high pH, the large majority of the copper has been converted to the monomeric species. At even higher pH, a new species appears to be formed, likely the tetrahydroxide $[Cu(OH)_4]^{2-}$ [83].

These results indicate that the *monomeric* bis(hydroxide) species is the predominant species in solution under the conditions of the electrocatalysis. Qualitatively, the magnitude of the electrocatalytic currents seems to parallel the speciation: the largest currents are observed when $[(bpy)Cu(OH)_2]$ is the dominant species. This result, coupled with the catalytic currents varying linearly with total copper concentration from 10 to 100 μM, makes

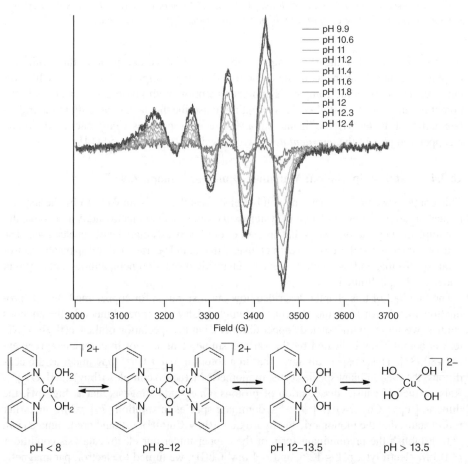

Figure 10.4 *Top: As the pH increases, the EPR signal corresponding to the (bpy)Cu(OH)₂ monomer increases, at 1 mM Cu(II)/bpy in 0.1 M electrolyte (NaOAc + NaOH). Bottom: Proposed speciation for the Cu(II)/bpy system under these conditions. See color plate*

it difficult to see how dimeric species can be involved in the catalytic cycle prior to the turnover-limiting step. This is contrary to our biomimetic motivation that Cu^{III}_2-bis(μ-oxo) and/or Cu^{II}_2-$\mu(\kappa^2:\kappa^2$-peroxo) would be key intermediates in copper water electrocatalysis.

One potentially valuable feature of this system is that all the copper(II) complexes appear to be in rapid equilibrium in aqueous media. Thus it is not necessary to use synthesized [(bpy)$_2$Cu$_2$(OH)$_2$]X$_2$ complexes, as was done for most of the studies previously discussed. In all of the cases we have tested, identical results were obtained with isolated [(bpy)$_2$Cu$_2$(μ-OH)$_2$]X$_2$ and with solutions of CuX$_2$ + bpy in the same buffer/electrolyte. This is consistent with the very high lability of most Cu(II) complexes.

A continuing concern in molecular catalysis of water oxidation is that the actual catalyst is not molecular, but rather a solid oxide/hydroxide deposit on the electrode or nanoparticle in solution [82]. While the Cu(II)/bpy system could potentially form such species at the high operating potentials and high pH of the electrocatalysis, this seems unlikely for a number of reasons: (i) CVs performed using ITO and glassy carbon electrodes were very similar, after correction for the significant background water oxidation by ITO itself; (ii) there is no initiation period before catalysis; and (iii) even after 5.5 hours of electrolysis, no discoloration of the electrode material is observed. The EPR and optical spectra show no evidence for the formation of significant solid or nanoparticulate material under electrocatalytic conditions. Furthermore, unlike related iridium, ruthenium, cobalt, and other systems, there is little evidence that solid copper oxide/hydroxides are good WOCs [9]. The recent report of water oxidation by simple copper salts in carbonate and phosphate buffers states that electrocatalysis is not observed at higher pHs due to formation of non-catalytic and insoluble Cu(OH)$_2$ [13]. In addition, a very recent paper on the use of a copper tripeptide for water oxidation presents evidence that the ligand is intact during catalysis [14].

The linearity of the catalytic currents with low copper concentrations is strong evidence for a molecular species. It is difficult to see how a nanoscale copper oxide/hydroxide, formed from $10-100\,\mu M$ [Cu^{2+}] under equilibrium conditions, could show this simple linear behavior. Finally, it is clear that the bipyridine ligand plays a key role in catalysis at pH > 13. At pH 13.5 in the absence of bpy, the dianionic copper(tetrahydroxide) [Cu(OH)$_4$]$^{2-}$ is the major species in solution [83, 84]. We do not observe any electrocatalysis with such solutions. The addition of 1 equiv. bpy causes a large increase in the current at the potential corresponding to water oxidation (Figure 10.5). Although it is nearly impossible to fully rule out catalysis by a heterogeneous species, these pieces of evidence strongly suggest a molecular catalyst.

Thus our mechanistic results at this point implicate a mono-copper mechanism, with [(bpy)Cu]$^{2+}$ the species undergoing the initial oxidation. A detailed mechanism for copper electrocatalysis of water oxidation was recently proposed by Zhang and coworkers for a triglycylglycine–copper electrocatalyst, on the basis of electrochemical results similar to those described here and the observation of an additional electrochemical reduction wave [14]. Their suggestion of water attack at a Cu(IV)-oxo/Cu^{III}(O$^\bullet$) species to form the O–O bond was by analogy with well-established ruthenium chemistry. Such a mechanism would be remarkable for copper as no Cu-terminal oxo complex has ever been observed, as opposed to the many well-characterized ruthenium–oxo compounds [85]. It therefore seems possible to us that the oxidation of water by copper complexes follows a path somewhat different from those observed by complexes of elements with lower d-electron counts.

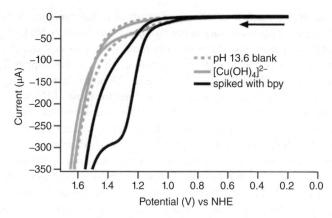

Figure 10.5 *CVs at pH 13.6 of: a solution of electrolyte only (grey, dashed); a solution of $Na_2[Cu(OH)_4]$ (grey, solid); a solution of $Na_2[Cu(OH)_4]$ with an added 1 equiv. bpy (black). The CV of the solution with $Na_2[Cu(OH)_4]$ is essentially the same as that of the blank without copper, while addition of bipyridine to form $(bpy)Cu(OH)_2$ gives a large increase in current at 1.2 V due to catalytic water oxidation. Conditions: 1 mM $Cu(OAc)_2$, 1 M NaOH, glassy carbon working electrode, platinum auxiliary electrode, Ag/AgCl reference electrode (adjusted to NHE)*

10.4   Conclusion

It has been a long, unpredictable, and rewarding process to add copper to the list of metals that can be molecular electrocatalysts for water oxidation. In the first half of our story, attempts to oxidize dimeric copper complexes to their well-characterized Cu^{III}_2-bis(μ-oxo) and/or Cu^{II}_2-μ-(κ^2:κ^2-peroxo) species were unsuccessful. Still, this pursuit brought new insights into the challenges of using organic solvents and organic ligands for water oxidation. In nonaqueous solvents, such PCET reactions require combinations of oxidants and bases whose thermochemistry has been defined and whose issues of incompatibility have been explored.

The second part of this chapter described our discovery that simple aqueous Cu^{2+}/bpy solutions are electrocatalysts for water oxidation at high pH. Catalysis is rapid and appears to be robust, although it occurs at somewhat high overpotentials. Recent reports from Meyer and coworkers show that other copper complexes, with carbonate, phosphate, or tripeptide ligands, are also water oxidation electrocatalysts [13, 14]. Work is thus ongoing in multiple laboratories to develop copper electrocatalysts, especially systems with lower overpotentials and operating pHs, and to unravel their mechanism(s). We are excited that copper is proving to be an exciting new avenue for water oxidation catalysis.

Acknowledgement

We thank the US National Science Foundation Center for Chemical Innovation, the Center for Enabling New Technology through Catalysis (CENTC), and its precursor, the Center for the Activation and Transformation of Strong Bonds (CATSB), for continuous funding despite discontinuous results. We also thank Professors William Tolman, Chris Cramer, and Tim Warren for valuable suggestions and insights, collaborative experiments, and chemical compounds.

References

[1] Lewis, E. A., Tolman, W. B. (2004) Chem. Rev., 104: 1047.

[2] Mirica, L. M., Ottenwaelder, X., Stack, T. D. P. (2004) Chem. Rev., 104: 1013.

[3] Solomon, E. I., Ginsbach, J. W., Heppner, D. E., Kieber-Emmons, M. T., Kjaergaard, C. H., Smeets, P. J., Tian, L., Woertink, J. S. (2011) Faraday Discuss., 148: 11.

[4] Halfen, J. A., Mahapatra, S., Wilkinson, E. C., Kaderli, S., Young, V. G. Jr., Que, L. Jr., Zuberbühler, A. D., Tolman, W. B. (1996) Science, 271: 1397.

[5] Waidmann, C. R., Miller, A. J. M., Ng, C. A., Scheuerman, M. L., Porter, T. R., Tronic, T. A., Mayer, J. M. (2012) Energy Environ. Sci., 5: 7771.

[6] Barnett, S. M., Goldberg, K. I., Mayer, J. M. (2012) Nat. Chem., 4: 498.

[7] Elizarova, G., Matvienko, L., Lozhkina, N., Maizlish, V., Parmon, V. (1981) React. Kinet. Catal. Lett., 16: 285.

[8] Elizarova, G., Matvienko, L., Lozhkina, N., Parmon, V., Zamaraev, K. (1981) React. Kinet. Catal. Lett., 16: 191.

[9] Elizarova, G. L., Zhidomirov, G. M., Parmon, V. N. (2000) Catal. Today, 58: 71.

[10] Ikeda, S., Takata, T., Kondo, T., Hitoki, G., Hara, M., Kondo, J. N., Domen, K., Hosono, H., Kawazoe, H., Tanaka, A. (1998) Chem. Commun., 1998: 2185.

[11] de Jongh, P. E., Vanmaekelbergh, D., Kelly, J. J. (1999) Chem. Commun., 1999: 1069.

[12] Walker, A. V., Yates, J. T. (2000) J. Phys. Chem. B, 104: 9038.

[13] Chen, Z., Meyer, T. J. (2013) Angew. Chem. Int. Ed., 52: 700.

[14] Zhang, M.-T., Chen, Z., Kang, P., Meyer, T. J. (2013) J. Am. Chem. Soc., 135: 2048.

[15] Weinberg, D. R., Gagliardi, C. J., Hull, J. F., Murphy, C. F., Kent, C. A., Westlake, B. C., Paul, A., Ess, D. H., McCafferty, D. G., Meyer, T. J. (2012) Chem. Rev., 112: 4016.

[16] Hammes-Schiffer, S. (2010) Chem. Rev., 110 (special issue): December.

[17] Mayer, J. M. (2011) Acc. Chem. Res., 44: 36.

[18] Schrauben, J. N., Cattaneo, M., Day, T. C., Tenderholt, A. L., Mayer, J. M. (2012) J. Am. Chem. Soc., 134: 16 635.

[19] Kodera, M., Kano, K. (2007) Bull. Chem. Soc. Jpn., 80: 662.

[20] Hu, Z., Willimans, R. D., Tran, D., Spiro, T. G., Gorun, S. M. (2004) J. Am. Chem. Soc., 122: 3556.

[21] Lynch, W. E., Kurtz, D. M. J., Wang, S., Scott, R. A. (2005) Catal. Today, 110: 303.

[22] Kitajima, N., Fujisawa, K., Fujimoto, C., Moro-oka, Y., Hashimoto, S., Kitagawa, T., Toriumi, K., Tatsumi, K., Nakamura, A. (1992) J. Am. Chem. Soc., 114: 1277.

[23] Mahapatra, S., Halfen, J. A., Wilkinson, E. C., Que, L. J., Tolman, W. B. (1994) J. Am. Chem. Soc., 116, 9785.

[24] Tahsini, L., Kotani, H., Lee, Y.-M., Cho, J., Nam, W., Karlin, K. D., Fukuzumi, S. (2012) Chem. Eur. J., 18: 1084.

[25] Osako, T., Ohkubo, K., Taki, M., Tachi, Y., Fukuzumi, S., Itoh, S. (2003) J. Am. Chem. Soc., 125: 11 027.

[26] Rolff, M., Schottenheim, J., Decker, H., Tuczek, F. (2011) Chem. Soc. Rev., 40: 4077.

[27] Warren, J. J., Tronic, T. A., Mayer, J. M. (2010) Chem. Rev., 110: 6961.

[28] Manner, V. W., Markle, T. F., Freudenthal, J. H., Roth, J. P., Mayer, J. M. (2008) Chem. Commun.: 256.

[29] Waidmann, C. R. (2009) PhD thesis, University of Washington, Seattle, WA, USA.

[30] Chaudhuri, P., Ventur, D., Wieghardt, K., Peters, E., Peters, K., Simon, A. (1985) Angew. Chem. Int. Ed., 24: 57.

[31] Cole, A. P., Mahadevan, V., Mirica, L. M., Ottenwaelder, X., Stack, T. D. P. (2005) Inorg. Chem., 44: 7345.

[32] Lee, S. C., Holm, R. H. (1993) Inorg. Chem., 32: 4745.

[33] Halfen, J. A., Mahapatra, S., Wilkinson, E. C., Gengenbach, A. J., Young, V. G., Que, L., Jr., Tolman, W. B. (1996) J. Am. Chem. Soc., 118: 763.

[34] Mahapatra, S., Halfen, J. A., Wilkinson, E. C., Pan, G., Wang, X., Young, V. G., Cramer, C. J., Que, L. Jr., Tolman, W. B. (1996) J. Am. Chem. Soc., 118: 11 555.

[35] Brink, J. M., Rose, R. A., Holz, R. C. (1996) Inorg. Chem., 35: 2878.

[36] Henson, M. J., Mukherjee, P., Root, D. E., Stack, T. D. P., Solomon, E. I. (1999) J. Am. Chem. Soc., 121: 10 332.

[37] Cole, A. P., Root, D. E., Mukherjee, P., Solomon, E. I., Stack, T. D. P. (1996) Science, 273: 1848.

[38] Root, D. E., Henson, M. J., Machonkin, T., Mukherje, P., Stack, T. D. P., Solomon, E. I. (1998) J. Am. Chem. Soc., 120: 4982.

[39] DuBois, J. L., Mukherjee, P., Stack, T. D. P., Hedman, B., Solomon, E. I., Hodgson, K. O. (2000) J. Am. Chem. Soc., 122: 5775.

[40] Op't Holt, B. T., Vance, M. A., Mirica, L. M., Heppner, D. E., Stack, T. D. P., Solomon, E. I. (2009) J. Am. Chem. Soc., 131: 6421.

[41] Mirica, L. M., Vance, M., Rudd, D. J., Hedman, B., Hodgson, K. O., Solomon, E. I., Stack, T. D. P. (2002) J. Am. Chem. Soc., 124: 9332.

[42] Mirica, L. M., Vance, M., Rudd, D. J., Hedman, B., Hodgson, K. O., Solomon, E. I., Stack, T. D. P. (2005) Science, 308: 1890.

[43] Mirica, L. M., Rudd, D. J., Vance, M. A., Solomon, E. I., Hodgson, K. O., Hedman, B., Stack, T. D. P. (2006) J. Am. Chem. Soc., 128: 2654.

[44] Eberson, L., Larsson, B. (1986) Acta Chem. Scand., 1986: 210.

[45] Mahpatra, S., Halfen, J. A., Tolman, W. B. (1996) J. Am. Chem. Soc., 118: 11 575.

[46] Scheuermann, M. L., Osako, T., Mayer, J. M., unpublished results.

[47] Hong, S., Hill, L. M. R., Gupta, A. K., Naab, B. D., Gilroy, J. B., Hicks, R. G., Cramer, C. J., Tolman, W. B. (2009) Inorg. Chem., 48: 4514.

[48] Hill, L. M. R., Gherman, B. F., Aboelella, N. W., Cramer, C. J., Tolman, W. B. (2006) Dalton. Trans., 2006: 4933.

[49] Spencer, D. J. E., Reynolds, A. M., Holland, P. L., Jazdzewski, B. A., Duboc-Toia, C., Le Pape, L., Yokota, S., Tachi, Y., Itoh, S., Tolman, W. B. (2002) Inorg. Chem., 41: 6307.

[50] Spencer, D. J. E., Aboelella, N. W., Reynolds, A. M., Holland, P. L., Tolman, W. B. (2002) J. Am. Chem. Soc., 124: 2108.

[51] Scheuermann, M. L., Fekl, U., Kaminsky, W., Goldberg, K. I. (2010) Organometallics, 29: 4749.

[52] Dai, X., Warren, T. H. (2001) Chem. Commun., 2001: 1998.

[53] Shimokawa, C., Yokata, S., Tachi, Y., Nishiwaki, N., Ariga, M., Itoh, S. (2003) Inorg. Chem., 42: 8395.

[54] Shimokawa, C., Teroaka, J., Tachi, Y., Itoh, S. (2006) J. Inorg. Biochem., 100: 1118.

[55] Laitar, D. S., Mathison, C. J. N., Davis, W. M., Sadighi, J. P. (2003) Inorg. Chem., 42: 7354.

[56] Warren, T. H. (2008) Personal communication. Georgetown University, Washington, DC, USA.

[57] Hong, S., Hill, L. M. R., Gupta, A. K., Naab, B. D., Gilroy, J. B., Hicks, R. G., Cramer, C. J., Tolman, W. B. (2009) Inorg. Chem., 48: 4514.

[58] Warren, T. H. International patent WO2008073781.

[59] Pearson, D. E., Buehler, C. A. (1973) Chem. Rev., 74: 45.

[60] Rakowski DuBois, M. R., DuBois, D. L. (2009) Chem. Soc. Rev., 38: 62.

[61] Chen, Z., Concepcion, J. J., Luo, H., Hull, J. F., Paul, A., Meyer, T. J. (2010) J. Am. Chem. Soc., 132: 17670.

[62] Nord, G., Pedersen, B., Bjergbakke, E. (1983) J. Am. Chem. Soc., 105: 1913.

[63] Ghosh, P. K., Brunschwig, B. S., Chou, M., Creutz, C., Sutin, N. (1984) J. Am. Chem. Soc., 106: 4772.

[64] Lay, P. A., Sasse, W. H. F. (1985) Inorg. Chem., 24: 4707.

[65] Hurst, J. K. (2005) Coord. Chem. Rev., 240: 313.

[66] Newkome, G. R., Patri, A. K., Holder, E., Schubert, U. S. (2004) Eur. J. Org. Chem., 2004: 235.

[67] Harris, C., Sinn, E., Walker, W., Woolliams, P. (1968) Aust. J. Chem., 21: 631.

[68] Kolbe, H. (1849) Justus Liebigs Annalen der Chemie, 69: 257.

[69] Walter, M. G., Warren, E. L., McKone, J. R., Boettcher, S. W., Mi, Q., Santori, E. A., Lewis, N. S. (2010) Chem. Rev., 110: 6446.

[70] Izgorodin, A., Izgorodina, E., MacFarlane, D. R. (2012) Energy Environ. Sci., 5: 9496.

[71] Bard, A. J., Faulkner, L. R. (2001) Electrochemical Methods: Fundamentals and Applications. John Wiley & Sons, Ltd, New York, NY, USA.

[72] Zanello, P. (2003) Inorganic Electrochemistry: Theory, Practice, Application. Royal Society of Chemistry, Cambridge, UK.

[73] Savéant, J.-M. (2008) Chem. Rev., 108: 2348.

[74] Hull, J. F., Balcells, D., Blakemore, J. D., Incarvito, C. D., Einstein, O., Brudvig, G. W., Crabtree, R. H. (2009) J. Am. Chem. Soc., 131: 8730.

[75] Wasylenko, D. J., Ganesamoorthy, C., Borau-Garcia, J., Berlinguette, C. P. (2011) Chem. Commun., 47: 4249.

[76] Dsandthy, C., Borau-Garcia, J. Duan, L., Bozoglian, F., Mandal, S., Stewart, B., Privalov, T., Llobet, A., Sun, L. (2012) Nat. Chem., 4: 418.

[77] Ellis, W. C., McDaniel, N. D., Bernhard, S., Collins, T. J. (2010) J. Am. Chem. Soc., 132: 10990.

[78] Fábián, I. (1989) Inorg. Chem., 28: 3805.

[79] Prenesti, E. (2006) Polyhedron, 25: 2815.

[80] Garribba E., Micera G., Sanna D., Strinna-Erre L. (2000) Inorg. Chim. Acta, 299: 253.

[81] Drago, R. S. (1992) Physical Methods for Chemists, 2nd edition. Saunders College, Philadelphia, PA, USA.

[82] Crabtree, R. H. (2012) Chem. Rev., 112: 1536.

[83] Korpi, H., Figiel, P. J., Lankinen, E., Ryan, P., Leskelä, M., Repo, T. (2007) Eur. J. Inorg. Chem., 2007: 2465.

[84] Ciani, L., Branciamore, S., Romanelli, M., Martini, G. (2003) Appl. Magn. Reson., 24: 55.

[85] Winkler, J. R., Gray, H. B. (2012) Electronic Structures of Oxo-Metal Ions Struct Bond. In: Mingos, D. M. P., Day, P., Dahl, J. P. (eds) Molecular Electronic Structures of Transition Metal Complexes I. Springer, Heidelberg/New York.

11

Polyoxometalate Water Oxidation Catalytic Systems

Jordan M. Sumliner, James W. Vickers, Hongjin Lv, Yurii V. Geletii, and Craig L. Hill
Department of Chemistry, Emory University, Atlanta, GA, USA

11.1 Introduction

The development of strategies to produce fuel from renewable sources has become one of the most targeted and researched endeavors in recent years. This situation follows from several points. First, we will run out of economically accessible fossil fuel reserves in a few decades as the demand for energy globally grows dramatically from both population and standard-of-living increases. Hydraulic fracturing, known colloquially as "fracking," and related technologies buy us some time but there simply will not be sufficient fossil fuels relatively soon, even with such production innovations. The environmental cost of continued and escalating fossil fuel use is a separate but very serious issue that will not be further addressed here. Second, unlike solar electricity (photovoltaic technology), there is no viable large-scale solar fuel technology to date, and thus no industry at present. The rapidly decreasing price and increasing amount of commercial solar electricity, in part from competition between many firms, largely in East Asia is quite encouraging. However, we are not close yet to a commercial industry for solar fuel. Third, the need for fuel is more acute than that for electricity. Only fuel has the energy density required for nearly all large-scale transportation needs [1]. Fourth, the amount of energy we need to power our civilization given demographic projections is enormous and beyond what all the renewable sources of energy, except solar, can provide collectively [2]. The energy potentially derivable from optimized biomass harvesting, hydroelectric generation, wind electricity, tidal fluctuations, geothermal sources, and other renewables combined falls far short of the staggering requirements we will face in 3 or 4 decades and thereafter. Only solar energy has the capacity to meet our projected planetary energy requirements [2].

These and other realities define our acute need for the development of solar fuel production technology. Central to solar fuel science is the development of faster, more selective, and more robust catalysts for reduction of H_2O or CO_2 (based on water splitting or the reversing of combustion; Equations 11.1 and 11.2) and for the oxidation of water (Equation 11.3). Also needed are better photosensitizers (PSs) and efficient, integrated photochemical systems (PSs and both catalysts). Our group is working on all of these issues, but the primary focus to date has been the water oxidation problem and the development of water oxidation catalysts (WOCs) that are very fast, operate at low overpotential (η), and are stable to water (hydrolysis), heat, and air (oxidation).

$$2H_2O + h\nu \rightarrow 2H_2 + O_2 \tag{11.1}$$

$$2CO_2 + 4H_2O + h\nu \rightarrow 2CH_3OH + 3O_2 \tag{11.2}$$

$$2H_2O \rightarrow O_2 + 4H^+ + 4e^- \tag{11.3}$$

Before we entered the WOC development arena, there were two classes of WOC, each with their own strengths and weaknesses: heterogeneous WOCs, typically metal oxides because the anions of other salts (e.g. sulfides, selenides, phosphides, etc.) are susceptible to oxidation, and homogeneous WOCs. Heterogeneous WOCs have the advantage that they are usually robust, inexpensive, and readily available; they have the disadvantages that most are very slow and heterogeneous catalyst surfaces can easily get poisoned. The traditional homogeneous WOCs have been coordination complexes, beginning in 1982 with the well-studied "blue dimer" of T. J. Meyer and coworkers [3]. These WOCs have the huge advantage that they can be studied easily by spectroscopic, electrochemical, and computational methods, and thus their electronic structures, mechanisms of action, and so on can be elucidated at a high level of sophistication. However, they have the major drawback that their organic ligands are thermodynamically unstable with respect to oxidative degradation and formation of CO_2 and H_2O. All coordination-complex WOCs inactivate by oxidative decomposition of their organic ligands after a modest number of turnovers. While the record for lifetime in terms of turnover number (TON) has increased from 10 or less a few years ago to several thousands today, this level of stability is orders of magnitude lower than what will be needed for viable solar fuel production structures and devices. Examples of conventional coordination-compound WOCs are discussed in several other chapters in this book.

Our approach was to develop a class of WOCs that had the many advantages of soluble catalysts and simultaneously the stability advantages of metal oxides. To this end, we began to investigate and develop polyoxometalates (POMs) containing d-electron metal centers with suitable potentials as WOCs. POMs (Figure 11.1) comprise clusters of d^0 early transition metals, most commonly tungsten(VI), bridged by oxide ions, so they are not susceptible to oxidative degradation like organic structures. At the same time, POMs represent a huge and rapidly growing class of complexes whose catalysis-relevant properties (potentials, charges, acid–base chemistry, etc.) can be tuned through the preparation of families of synthetically accessible derivatives [4–6].

Our group and that of Bonchio and Sartorel published the first POM WOC, $[Ru_4(\mu\text{-}O)_4(\mu\text{-}OH)_2(H_2O)_4(\gamma\text{-}SiW_{10}O_{36})_2]^{10-}$ (**Ru₄SiPOM**; Figure 11.1b) at the same time in 2008

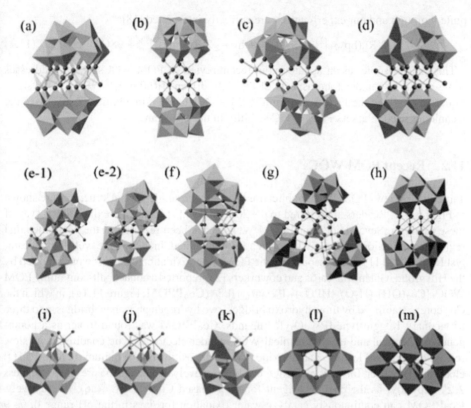

Figure 11.1 X-ray crystal structure of polyoxometalates (POMs) in combined ball-and-stick and polyhedral representations corresponding to Table 11.1. Red, O; magenta, Ru; blue, Co; green, Ni; yellow, Ir; dark teal, Cl; Purple, Bi; orange tetrahedra, $PO_4/SiO_4/GeO_4$; white tetrahedra or polyhedra, ZnO_4/ZnO_6; light-blue polyhedra, WO_6; grey polyhedra, MoO_6. See color plate

[7, 8], and since then many papers have been published by our team and others, examining the redox potentials, physicochemical properties, water oxidation features, and immobilization chemistry of these soluble but robust catalysts [9]. At the time **Ru₄SiPOM** was first reported, it was the fastest homogeneous WOC, exhibiting typical turnover frequencies (TOFs) of about 1 s^{-1}. Most coordination-compound WOCs (all based on Ru) turn over far slower (one to three orders of magnitude slower). Subsequently, we reported the first POM WOC composed of all earth-abundant elements, $[Co_4(H_2O)_2(\alpha-PW_9O_{34})_2]^{10-}$ (henceforth "**Co₄PPOM**"; Figure 11.1d). It catalyzes water oxidation in both sacrificial oxidant-driven [10] and visible light-driven [11] reactions at faster rates (~ 6 s^{-1} for $[Ru^{III}(bpy)_3]^{3+}$, where bpy = 2, 2'-bipyridine), which made it the fastest homogeneous catalyst at that time (2010). This POM WOC has been used by many other groups and its water oxidation chemistry and other properties have been discussed extensively [12–17]. There was some controversy about the stability and activity of **Co₄PPOM** when two other groups challenged some of the interpretations and conclusions of our initial studies [12, 15]. However, a recent, exhaustive study concludes that all our original publications on **Co₄PPOM** were

quite thorough and our experiments were correctly interpreted [18].

$$2[Ru(bpy)_3]^{2+} + S_2O_8^{2-} + h\nu \rightarrow 2[Ru(bpy)_3]^{3+} + 2SO_4^{2-} \qquad (11.4)$$

This chapter serves as an update to our recent review of POM WOCs [9] and addresses some of the less understood aspects of the light-driven ($[Ru(bpy)_3]^{2+}/S_2O_8^{2-}$) system, Equation 11.4 and the factors that limit POM WOC stability. In addition, it discusses the techniques used to assess POM WOC stability in these systems.

11.2   Recent POM WOCs

Table 11.1 reviews POM WOC publications to date; the single-crystal structures are shown in Figure 11.1. Readers are referred to the 2012 review [9] for a more exhaustive survey of these catalysts. Several additional POM WOCs have been reported in the year since that review's publication, and many have been further studied, including the well-known catalyst **Ru₄SiPOM** [19–22]. Inspired by the POM water oxidation catalysis work reported by the Hill group, Galán-Mascarós and coworkers [23] reported a nonacobalt-containing POM WOC, $[Co_9(OH)_3(H_2O)_6(HPO_4)_2(PW_9O_{34})_3]^{16-}$ (**Co₉PPOM**; Figure 11.1g), in which the Co_9 core is stabilized by three hydroxo bridges, two hydrogen phosphate bridges, and three tri-lacunary Keggin-type $[PW_9O_{34}]^{9-}$ ligands. **Co₉PPOM** was found to act as a homogeneous chemical and electrochemical WOC. Under electrocatalytic conditions, its slow release of Co^{2+}(aq) leads to cobalt oxide (CoO_x) deposition on the indium tin oxide (ITO) electrode. However, formation of CoO_x can be effectively avoided in the presence of excess 2, 2′ -bipyridyl as the chelating agent for the released Co^{2+} or Co^{3+}(aq). Impressively, **Co₉PPOM** can continuously catalyze water oxidation for days in the pH range of 7−9 using hypochlorite as the chemical oxidant, without obvious fatigue or decomposition. Under the experimental conditions (Table 1.1, entry 20), the authors achieved a maximum TON of 20, with an initial TOF of 42 h^{-1} and a chemical yield of 40%. Subsequently, the same group prepared an insoluble Cs$^+$ salt of **Co₉PPOM** (Table 11.1, entry 21) [24], which was incorporated into an amorphous carbon paste electrode and used to perform heterogeneous electrocatalytic water oxidation at $\eta \geq 540$ mV.

Recently, a di-cobalt-containing Keggin POM, $K_7[Co^{II}Co^{III}(H_2O)W_{11}O_{39}]$ (**Co^{II}Co^{III} POM**), was evaluated as an efficient WOC under both light-induced and thermal conditions [25]. As shown in Figure 11.1k, this new catalyst contains a central $Co^{III}O_4$ tetrahedral unit and a peripheral $Co^{II}O_6$ octahedral unit. The authors systematically studied different variables in order to understand the impact of each and confirmed, using isotope labeling, that the oxygen evolved during catalysis comes from water. Under optimal turnover conditions, a TON of 361 with TOF$_{initial}$ of 0.5 s^{-1} was obtained. The stability of **Co^{II}Co^{III}POM** was confirmed using a series of complementary techniques (see later). A series of Keggin-type cobalt-containing POMs were also evaluated as potential WOCs; only **Co^{II}Co^{III}POM** displayed catalytic activity, which might derive from the mixed-valence electronic features.

Guo and coworkers prepared a di-cobalt–di-bismuth-containing POM, $Na_{14}[Co_2Bi_2$ (B-CoW₉O₃₄)₂] (**Co₂Bi₂CoPOM**), characterized by single-crystal X-ray diffraction (SC-XRD; Figure 11.1h), infrared (IR), X-ray diffraction (XRD), and thermogravimetric analysis (TGA) [45]. The catalytic activity of **Co₂Bi₂CoPOM** was studied by cyclic voltammetry (CV) in the presence of $[Ru(bpy)_3]^{2+}$ and under photochemical conditions.

Table 11.1 Summary of published POM WOCs to date [9]

Entry No.	POM catalyst	Representative reaction conditions	Turnover frequency (TOF)	POM structures	[Ref.] (year)
1	$Na_{14}[Ru^{III}_2Zn_2(H_2O)_2(ZnW_9O_{34})_2]$	Electrochemical water oxidation using pulsed voltammetry in 0.1 M sodium phosphate buffer (pH 8.0) with 2 µM catalyst	No data	Figure 11.1a	[27] (2004)
2a	$Rb_8K_2[\{Ru_4O_4(OH)_2(H_2O)_4\}(\gamma\text{-}SiW_{10}O_{36})_2]$	$[Ru(bpy)_3](ClO_4)_3$ as oxidant, 20 mM sodium phosphate buffer (pH 7.2)	$0.45\text{–}0.60\ s^{-1}$	Figure 11.1b	[8] (2008)
2b	$Rb_8K_2[\{Ru_4O_4(OH)_2(H_2O)_4\}(\gamma\text{-}SiW_{10}O_{36})_2]$	Xe lamp, 420–520 nm, 1.0 mM [Ru(bpy)$_3$]$^{2+}$, 5.0 mM Na$_2$S$_2$O$_8$, 5.0 µM catalyst, 20 mM sodium phosphate buffer (pH 7.2)	$8 \times 10^{-2}\ s^{-1}$	Figure 11.1b	[28] (2009)
2c	$Rb_8K_2[\{Ru_4O_4(OH)_2(H_2O)_4\}(\gamma\text{-}SiW_{10}O_{36})_2]$	1.15 mM $[Ru(bpy)_3](ClO_4)_3$ as oxidant, 20 mM sodium phosphate buffer (pH 7.2)	No data	Figure 11.1b	[29] (2009)
2d	$Rb_8K_2[\{Ru_4O_4(OH)_2(H_2O)_4\}(\gamma\text{-}SiW_{10}O_{36})_2]$	All calculations performed using the Gaussian 03 program	N/A	Figure 11.1b	[30] (2009)
2e	$Rb_8K_2[\{Ru_4O_4(OH)_2(H_2O)_4\}(\gamma\text{-}SiW_{10}O_{36})_2]$	All reported calculations performed using the TURBOMOLE software, version 5.10	N/A	Figure 11.1b	[31] (2010)
3	$Cs_9[(\gamma\text{-}PW_{10}O_{36})_2Ru^{IV}_4O_5(OH)(OH_2)_4]$	Xe lamp (420–520 nm), 5.1 µM catalyst, 1.0 mM [Ru(bpy)$_3$]Cl$_2$, 5 mM Na$_2$S$_2$O$_8$, 20 mM Na$_2$SiF$_6$ buffer (pH 5.8)	$TOF_{initial} = 0.13\ s^{-1}$	Figure 11.1b	[32] (2010)
4	$K_{14}[(IrCl_4)KP_2W_{20}O_{72}]$	1.4 mM [Ru(bpy)$_3$]$^{3+}$, 20 µM catalyst, 20 mM sodium phosphate buffer (pH 7.2)	No data	Figure 11.1c	[33] (2009)
5a	$Na_{10}[Co_4(H_2O)_2(\alpha\text{-}PW_9O_{34})_2]$	1.5 mM $[Ru(bpy)_3](ClO_4)_3$ as oxidant, 30 mM NaPi buffer (pH 8.0)	$>5\ s^{-1}$	Figure 11.1d	[10] (2010)

(continued)

Table 11.1   (continued)

Entry No.	POM catalyst	Representative reaction conditions	Turnover frequency (TOF)	POM structures	[Ref.] (year)
5b	Na$_{10}$[Co$_4$(H$_2$O)$_2$(α-PW$_9$O$_{34}$)$_2$]	Xe lamp (420–470 nm), 1.0 mM [Ru(bpy)$_3$]Cl$_2$, 5.0 mM Na$_2$S$_2$O$_8$, 80 mM sodium borate buffer (pH 8.0)	No data	Figure 11.1d	[11] (2011)
6	K$_{10.2}$Na$_{0.8}$[{Co$_4$(μ-OH)(H$_2$O)$_3$}(Si$_2$W$_{19}$O$_{70}$)]	Xe lamp (420–520 nm), 10 μM catalyst, 1.0 mM [Ru(bpy)$_3$]Cl$_2$ 5 mM Na$_2$S$_2$O$_8$, 25 mM sodium borate buffer (pH 9.0)	TOF = 0.1 s^{-1}	Figure 11.1e-1 Figure 11.1e-2	[34] (2012)
7	K$_{10}$H$_2$[Ni$_5$(OH)$_6$(OH$_2$)$_3$](Si$_2$W$_{18}$O$_{66}$)]	LED lamp, 455 nm, 1.0 mM [Ru(bpy)$_3$]Cl$_2$, 5 mM Na$_2$S$_2$O$_8$, 80 mM sodium borate buffer (pH 8.0)	No data	Figure 11.1f	[35] (2012)
8	Cs$_{10}$[Ru$_4$(μ-O)$_4$(μ-OH)$_2$(H$_2$O)$_4$(γ-SiW$_{10}$O$_{36}$)$_2$]/Li$_{10}$[Ru$_4$(μ-O)$_4$(μ-OH)$_2$(H$_2$O)$_4$(γ-SiW$_{10}$O$_{36}$)$_2$]	4.3 mM catalyst, 0.172 M (NH$_4$)$_2$[Ce^{IV}(NO$_3$)$_6$], in H$_2$O (pH 0.6)	TOF$_{max}$ > 0.125 s^{-1}	Figure 11.1b	[7] (2008)
9	Li$_{10}$[Ru$_4$(μ-O)$_4$(μ-OH)$_2$(H$_2$O)$_4$(γ-SiW$_{10}$O$_{36}$)$_2$]	15.9 μmol catalyst at 1.15 V, pH 0.6; reticulated vitreous carbon as working electrode, reference electrode:SSCE; counter electrode: platinum wire, T = 298 K	N/A	Figure 11.1b	[36] (2009)
10a	Li$_{10}$[Ru$_4$(μ-O)$_4$(μ-OH)$_2$(H$_2$O)$_4$(SiW$_{10}$O$_{36}$)$_2$]@Dendron-MWCNT	ITO deposited with catalyst as working electrode, aqueous sodium phosphate buffer solution (pH 7.0); scan rate: 20 mV/s in the range 0–1.6 V; reference electrode: Ag/AgCl (3 M KCl); counter electrode: platinum wire, T = 298 K	0.010–0.085 s^{-1} at η = 0.35–0.6 V	Figure 11.1b	[26] (2010)

10b	$Li_{10}[Ru_4(\mu\text{-}O)_4(\mu\text{-}OH)_2(H_2O)_4(\gamma\text{-}SiW_{10}O_{36})_2]$@ Dendron-MWCNT	ITO deposited with catalyst as working electrode, aqueous sodium phosphate buffer solution (pH 7.0); scan rate: 20 mV/s in the range 0–1.6 V; reference electrode: Ag/AgCl (3 M KCl); counter electrode: platinum wire, T = 298 K	$0.010\text{–}0.085\,s^{-1}$ at η $= 0.35\text{–}0.6\,V$	Figure 11.1b	[37] (2011)
11	$Li_{10}[Ru_4(\mu\text{-}O)_4(\mu\text{-}OH)_2(H_2O)_4(\gamma\text{-}SiW_{10}O_{36})_2]$@MWCNT	A disk screen-printed carbon (DSC) electrode deposited with catalyst as working electrode, aqueous sodium phosphate buffer solution (pH 7.0); scan rate: 20 mV/s in the range 0–1.4 V; reference electrode: Ag/AgCl (3 M KCl); counter electrode: platinum wire, T = 298 K	No data	Figure 11.1b	[38] (2011)
12a	$Cs_{10}[Ru_4(\mu\text{-}O)_4(\mu\text{-}OH)_2(H_2O)_4(\gamma\text{-}SiW_{10}O_{36})_2]$	Surelite continuum surelite II Nd:YAG laser (excitation at 355 and 532 nm, half-width 8 ns), 47.6 μM $[Ru(bpy)_3]^{2+}$, 5.0 mM $Na_2S_2O_8$, varying [cat.], 10 mM sodium phosphate buffer (pH 7.0), or TiO_2 film sensitized with $[Ru(bpy)_2(dpb)]^{2+}$	N/A	Figure 11.1b	[39] (2010)
12b	$Cs_{10}[Ru_4(\mu\text{-}O)_4(\mu\text{-}OH)_2(H_2O)_4(\gamma\text{-}SiW_{10}O_{36})_2]$	50 W halogen lamp ($\lambda > 550$ nm), 60 μM catalyst, 0.1 mM $[Ru\{(\mu\text{-}dpp)Ru(bpy)_2\}_3](PF_6)_8$, 10 mM $Na_2S_2O_8$ and 50 mM Na_2SO_4, 10 mM potassium phosphate buffer (pH 7.2)	TOF $= 8 \times 10^{-3}\,s^{-1}$ at 60 μM catalyst	Figure 11.1b	[40] (2010)
12c	$Cs_{10}[Ru_4(\mu\text{-}O)_4(\mu\text{-}OH)_2(H_2O)_4(\gamma\text{-}SiW_{10}O_{36})_2]$	Surelite continuum surelite II Nd:YAG laser (excitation at 355 and 532 nm, half-width 6-8 ns), 0.1 mM $[Ru(bpy)_3]^{2+}$, 5.0 mM $Na_2S_2O_8$, varying [cat.], 80 mM sodium phosphate buffer (pH 7.0)	TOF $= 280\,s^{-1}$ at 0.5 μM catalyst	Figure 11.1b	[41] (2012)

(continued)

Table 11.1 (continued)

Entry No.	POM catalyst	Representative reaction conditions	Turnover frequency (TOF)	POM structures	[Ref.] (year)
13	$Cs_5[Ru^{III}(H_2O)SiW_{11}O_{39}]$	0.3 mM catalyst, 6 mM $(NH_4)_2[Ce^{IV}(NO_3)_6]$ in 0.1 M HNO_3	No data	Figure 11.1i	[42] (2011)
14	$Cs_5[Ru^{III}(H_2O)GeW_{11}O_{39}]$	0.3 mM catalyst, 6 mM $(NH_4)_2[Ce^{IV}(NO_3)_6]$ in 0.1 M HNO_3	No data	Figure 11.1i	[42] (2011)
15	$\alpha\text{-}K_6Na[\{Ru_3O_3(H_2O)Cl_2\}(SiW_9O_{34})]\cdot17H_2O$	LED lamp (465 nm), 50 μM catalyst, 1 mM $[Ru(bpy)_3]Cl_2$, 5 mM $Na_2S_2O_8$, 20 mM Na_2SiF_6 buffer (pH 5.8)	$TOF_{initial} = 0.7\,s^{-1}$	Figure 11.1j	[43] (2012)
16	$K_{11}Na_1[Co_4(H_2O)_2(SiW_9O_{34})_2]$	LED lamp (470 nm), 42 μM catalyst, 1 mM $[Ru(bpy)_3]Cl_2$, 5 mM $Na_2S_2O_8$, 20 mM Na_2SiF_6 buffer (pH 5.8)	$TOF_{initial} = 0.4\,s^{-1}$	Figure 11.1d	[43] (2012)
17	$(NH_4)_3[CoMo_6O_{24}H_6]\cdot7H_2O$	300 W Xe lamp (400–490/800 nm), 20 μM catalyst, 0.4 mM $[Ru(bpy)_3](NO_3)_2$, 3 mM $Na_2S_2O_8$, 0.1 M sodium borate buffer solution (pH 8.0)	$TOF_{initial} = 0.11\,s^{-1}$	Figure 11.1l	[44] (2012)
18	$(NH_4)_6[Co_2Mo_{10}O_{38}H_4]\cdot7H_2O$	300 W Xe lamp (400–490/800 nm), 10 μM catalyst, 0.4 mM $[Ru(bpy)_3](NO_3)_2$, 3 mM $Na_2S_2O_8$, 0.1 M sodium borate buffer solution (pH 8.0)	$TOF_{initial} = 0.16\,s^{-1}$	Figure 11.1m	[44] (2012)
19	$Na_{10}[Co_4(H_2O)_2(\alpha\text{-}PW_9O_{34})_2]/MCN$	ITO deposited with catalyst as working electrode, 0.1 M sodium phosphate buffer solution at pH 7, scan rate 20 mV/s in the range 0–1.5 V, all potentials referred to Ag/AgCl (3 M KCl) electrode	$\sim0.3\,s^{-1}$	Figure 11.1d	[17] (2012)
20	$Na_8K_8[Co_9(OH)_3(H_2O)_6(HPO_4)_2(PW_9O_{34})_3]\cdot43H_2O$	1 mM catalyst, 100 mM NaClO chemical oxidants, 0.9 M sodium phosphate buffer solutions (pH 8), reaction time 1.7 hours	$TOF_{initial} = 42\,h^{-1}$	Figure 11.1g	[23] (2012)

21	$Cs_{15}K[Co_9(OH)_3(H_2O)_6(HPO_4)_2(PW_9O_{34})_3] \cdot 41H_2O$	POM-modified amorphous carbon (POM-C) working electrode, controlled potential electrolysis at +1.3 V (versus NHE) in a two-compartment cell with a 50 mM pH 7 sodium phosphate buffer, 1 M NaNO$_3$ electrolyte	N/A	Figure 11.1g	[24] (2013)
22	$K_7[Co^{II}Co^{III}(H_2O)W_{11}O_{39}] \cdot 15H_2O$	LED lamp ($\lambda > 420$ nm, 16 mW), 1 μM catalyst, 1.0 mM [Ru(bpy)$_3$]Cl$_2$, 5 mM Na$_2$S$_2$O$_8$, 80 mM sodium borate buffer (pH 9.0)	TOF$_{initial}$ = 0.5 s^{-1}	Figure 11.1k	[25] (2013)
23	$Na_{14}[Co_2Bi_2(B-CoW_9O_{34})_2] \cdot 48H_2O$	300 W Xe lamp ($\lambda > 400$ nm), 5.6 μM catalyst, 100 mL solution of 1.0 mM [Ru(bpy)$_3$]Cl$_2$ and 5 mM Na$_2$S$_2$O$_8$, 50 mM sodium phosphate buffer (pH 7.4)	No data	Figure 11.1h	[45] (2013)
24	$Rb_8K_2[\{Ru_4O_4(OH)_2(H_2O)_4\}(\gamma\text{-}SiW_{10}O_{36})_2]$@Graphene	Rotating-ring disk (Pt collection ring for O$_2$ detection; glassy carbon generation disk modified with POM-graphene catalyst) working electrode (RRDE), 0.1 M sodium borate buffer (pH 7.50), 1 M Ca(NO$_3$)$_2$ electrolyte; reference electrode: Ag/AgCl (3 M NaCl); counter electrode: platinum wire	TOF = 0.82–8.01 s^{-1} at η = 0.35–0.81 V	Figure 11.1b	[19] (2013)
25	$[Ru_4(\mu\text{-}O)_4(\mu\text{-}OH)_2(H_2O)_4(\gamma\text{-}SiW_{10}O_{36})_2]^{10-}$@d-G; d-G: dendron functionalized graphene	ITO electrodes or carbon-based screen-printed microelectrodes (SPE) deposited with catalyst as working electrode, 0.2 M aqueous sodium phosphate buffer solution (pH 7.0); reference electrode: Ag/AgCl (3 M KCl); counter electrode: platinum wire	TOF = 1.73 h^{-1} at η = 0 V	Figure 11.1b	[21] (2013)

The results show that **Co$_2$Bi$_2$CoPOM** can oxidize water to oxygen both electrocatalytically and photocatalytically (Table 11.1, entry 23), but this work lacked some control experiments and did not fully study catalyst stability, so the identity of the true catalyst is not unequivocal.

By choosing a graphene nanoplatform, Bonchio and coworkers immobilized **Ru$_4$SiPOM** electrostatically on a positively charged PAMAM dendron-functionalized graphene (d-G) platform [21]. **Ru$_4$SiPOM**@d-G films were then cast on ITO electrodes and carbon-based screen-printed microelectrodes (SPE) for electrochemical experiments and were characterized by CV. **Ru$_4$SiPOM**@d-G-modified SPEs catalyze oxygen production at $\eta \geq 0.3$ V at neutral pH with stable performance for two 4 hour electrolyses (Table 11.1, entry 25). The calculated TOF is one order of magnitude higher than that of the isolated catalyst and two-fold higher than the **Ru$_4$SiPOM**@MWCNT system [26].

Continuing the efforts to develop stable, robust immobilized POM WOCs for electrocatalytic water oxidation, Bond, Hill, and coworkers constructed a stable modified electrode by confining the POM WOC, **Ru$_4$SiPOM**, in a highly porous wet graphene film electrochemically deposited on the glassy carbon electrode surface [19]. The authors demonstrated the need for high concentrations of supporting electrolyte in neutral pH conditions in order to diminish the electrical double-layer effect caused by the high negative charge of **Ru$_4$SiPOM** on the electrode surface. **Ru$_4$SiPOM**@graphene exhibits excellent activity and is stable under turnover conditions (neutral pH, 1.0 M Ca(NO$_3$)$_2$). The TOF of the **Ru$_4$SiPOM**@graphene electrocatalyst was measured using a rotating ring-disk electrode (modified glassy carbon disk for H$_2$O oxidation and Pt ring for O$_2$ detection). Under experimental conditions (Table 11.1, entry 24), the TOF values of water oxidation reached 0.82 and 8.01 s^{-1} at η of 0.35 and 0.81 V, respectively, which is over 80 times more active than the **Ru$_4$SiPOM**@MWCNT functionalized by PAMAM dendrimer (Table 11.1, entry 10a), even at a moderate $\eta = 0.35$ V [26].

11.3 Assessing POM WOC Reactivity

POM WOC reactivity has been evaluated in three distinct types of system: those that employ sacrificial chemical oxidants; those that use light, with a PS coupled with a sacrificial electron acceptor; and those that use an applied potential via an electrode. Continued development of WOCs and POM WOCs has set a high threshold for thorough studies of these catalysts and the systems in which they reside. A thoroughly examined case of this involves **Co$_4$PPOM** [18]; while the catalytic activity of this POM was shown with a sacrificial oxidant [10] and in a light-driven system [11], it has been shown to decompose in an electrochemical system under different experimental conditions [15, 16]. Clearly, WOC stability and reactivity are highly dependent on the specific system used, and meaningful evaluation and comparison between these three systems is frequently not trivial.

This dependence extends to all the components in a WOC system. For example, we found that in light-driven water oxidation with **Co$_4$PPOM**, the O$_2$ yield (water oxidation activity) strongly depends on the nature of the PS, [Ru(**L**)$_2$]$^{4+}$ where **L** = 4'-(4-pyridyl)-2, 2':6', 2''-terpyridine and its four N-alkylated derivatives, where N = benzyl, ethyl, allyl or 4-cyanobenzyl [27]. This ligand set changes the "selectivity" of the WOC from water oxidation to PS degradation. In WOC systems where Ru(bpy)$_3$$^{2+/3+}$ is used,

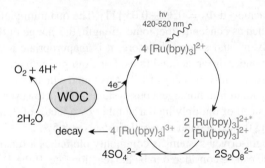

Scheme 11.1 *Principle processes in a homogeneous light-driven water oxidation/ O$_2$ evolution system*

the bpy ligand is a potential target for oxidation by the catalyst. This results in an O$_2$ yield per oxidant lower than theoretically predicted. We will soon discuss the factors that control the stability and reactivity of the widely used light-driven water oxidation system (Equation 11.4, Scheme 11.1). First, however, we have to understand what controls the rate and yields of O$_2$ formation in such systems [46].

11.4   The Ru(bpy)$_3$$^{2+}$ / S$_2$O$_8$$^{2-}$ System

This system of light-driven water oxidation requires the use of a sacrificial electron acceptor, as H$_2$ is not formed. Photosensitized water oxidation was first described more than two decades ago [47–51] and was revisited by our studies of POM WOCs with [Ru(bpy)$_3$]Cl$_2$ as a PS and Na$_2$S$_2$O$_8$ (persulfate) as a sacrificial electron acceptor [11, 28]. The photochemistry of this system has been thoroughly studied and described in several publications [52, 53], and will therefore not be discussed here. Despite this work, the data collected using this system are often misinterpreted, and for that reason we wish to briefly describe it here.

Photocatalytic O$_2$ evolution in this system proceeds through the well-established reaction mechanism presented in Scheme 11.1 [28, 32, 40]. The reaction is initiated upon the absorption of a photon by [Ru(bpy)$_3$]$^{2+}$. In most cases, due to the high extinction coefficient of [Ru(bpy)$_3$]$^{2+}$($\epsilon_{454} = 1.4 \times 10^4$ M^{-1} cm^{-1}), all the incident photons are captured in the solution by Ru(bpy)$_3$$^{2+}$. The excited [Ru(bpy)$_3$]$^{2+*}$ is quenched by S$_2$O$_8$$^{2-}$ through both bimolecular and unimolecular electron transfer pathways [53–55]. The products [Ru(bpy)$_3$]$^{3+}$ and SO$_4$$^{\cdot-}$, E°(SO$_4$$^{\cdot-}$/SO$_4$$^{2-}$) ~ 2.4 V [56], are both strong oxidants, and the latter oxidizes [Ru(bpy)$_3$]$^{2+}$ to form a second equivalent of [Ru(bpy)$_3$]$^{3+}$ (Equation 11.4) [57].

Thus, this light-driven water oxidation system utilizes [Ru(bpy)$_3$]$^{3+}$ as an intermediate oxidant generated *in situ*. In natural photosynthesis, four photons are consumed to form one O$_2$ molecule and the overall photon-to-O$_2$ generation quantum yield is defined as $\Phi_4 = [\Delta(\text{O}_2)/\Delta(h\nu)]$, where $\Delta(\text{O}_2)$ and $\Delta(h\nu)$ are the change in the total amount of O$_2$ produced and the number of photons absorbed, respectively. In the light-driven system with S$_2$O$_8$$^{2-}$, one photon generates two molecules of [Ru(bpy)$_3$]$^{3+}$; therefore, the quantum yield should be calculated using the expression $\Phi_2 = 2\Delta[\Delta(\text{O}_2)/\Delta(h\nu)]$. In the case of **Co$_4$PPOM**, we achieved $\Phi_2 = 0.30 \pm 0.05$ [11]. The quantum yield strongly depends on the conditions, such as concentration, ionic strength, the charge of the PS, the presence of an organic cosolvent, and so on. Therefore, it is inappropriate to compare the quantum yields obtained under different conditions. For example, Sun and coworkers recently claimed that "a high quantum efficiency of 17% was found which is a *new record* for visible light-driven water oxidation in homogeneous systems" [58]. However, in comparing their values with ours, instead of multiplying our number by a factor of 2, they divided it by 2. In other words, using the expression above, their Φ_2 is $\sim 8.5\%$.

The O$_2$ yield in light-driven systems is commonly plotted as a function of time. In this case, the initial slope can be considered to be the reaction rate. If the TON = (O$_2$)/(WOC) is plotted versus time then the slope is interpreted as the TOF = TON/(Δtime). However, the correct assignment of the slope is only possible if either the dependence of the slope on light intensity or a steady-state [Ru(bpy)$_3$]$^{2+}$/[Ru(bpy)$_3$]$^{3+}$ ratio is known. At high light intensity, all PSs should be in the [Ru(bpy)$_3$]$^{3+}$ form, the overall reaction rate is independent of light intensity, and the rate is controlled by the reaction of water oxidation by [Ru(bpy)$_3$]$^{3+}$. In this case, TOF = TON/(Δtime). At low light intensity, the slope increases with light and the reaction rate is controlled by the rate of [Ru(bpy)$_3$]$^{3+}$ generation. Under these conditions, the slope of TON versus time is proportional to the quantum yield. This is commonly used to quantify the initial quantum yield. In other words, either TOF or quantum yield can be determined, but not both simultaneously.

11.5 Ru(bpy)$_3$$^{3+}$ as an Oxidant for POM WOCs

Several sacrificial oxidants have been used in water oxidation catalysis with POMs, although the most commonly used and thus best studied are [Ru(bpy)$_3$]$^{3+}$ and cerium(IV) ammonium nitrate (CAN), both one-electron oxidants. Oxidants including periodate [58], Ir(Cl$_6$)$^{2-}$ [32], [Fe(phen)$_3$]$^{3+}$ [61, 62], and hypochlorite have also been used to a much lesser extent. [Ru(bpy)$_3$]$^{3+}$ is the most widely used in POM WOC systems as it relates to the light-driven system in which [Ru(bpy)$_3$]$^{2+}$ is the PS and, ultimately, [Ru(bpy)$_3$]$^{3+}$ acts as the oxidant itself. There are two options for monitoring reaction progress by UV/vis, making quantification of kinetics straightforward. Either formation of the reduced [Ru(bpy)$_3$]$^{2+}$ (intense absorption band at 454 nm) or the loss of [Ru(bpy)$_3$]$^{3+}$ (weak band with $\epsilon_{670} \approx 420\,M^{-1}\,cm^{-1}$) can be followed. While [Ru(bpy)$_3$]$^{3+}$ does self-reduce at pH values other than strong acid, its use is still feasible for relatively fast catalytic reactions at a pH of 8 or greater. Another possible limitation of [Ru(bpy)$_3$]$^{3+}$ is that its high positive charge and relatively large steric size make it a good precipitating counterion for POMs in aqueous solution at somewhat low catalyst concentrations [11], so precipitate formation must be assessed. While CAN has a strong absorption band in the visible, it is only stable below pH 1, which limits the catalysts that can be used and prevents studies at more relevant pH values for water oxidation (neutral to basic). A thorough review of these oxidants and several others for water oxidation catalysis (not necessarily with POM WOCs) was recently published by Crabtree and Brudvig [59].

When studying the kinetics of [Ru(bpy)$_3$]$^{3+}$ -based water oxidation catalyzed by a POM by UV/vis, it is absolutely crucial to be aware of the family of side reactions first discussed in the pioneering work of Sutin and coworkers [63]. Aside from the decomposition of $[Ru(bpy)_3]^{3+}$ at higher pH values, there is a set of self-reduction reactions that take place; however, the rates of these are typically much slower than catalytic water oxidation. It can be tempting to attribute all of the observed loss of oxidant to catalytic production of O_2, but this is dangerous without direct measurement of O_2. Thus water oxidation with $[Ru(bpy)_3]^{3+}$ can be studied by two different methods: either via direct measurement of the final O_2 yield, where little kinetic information is possible given the limitations in O_2 detection methods, or by fast mixing using UV/vis stopped-flow spectroscopy, with no measurement of O_2. Being able to conduct both these measurements simultaneously for the same reaction would be as insightful as it is difficult. No such system exists at present.

Incorporating an O_2 detector with the mixing portion of a stopped-flow apparatus facilitates accurate quantification of final O_2 yields for water oxidation reactions. As employed in a recent paper from our group [35], this method enables fast and uniform mixing in less than a second. The resulting data are not only useful in the evaluation of a WOC, but when matched with UV/vis data provide a good estimate of what percentage of the oxidant is utilized for catalytic water oxidation (assessed by O_2 yield based on oxidant consumption).[1]

When the kinetics of O_2 formation are directly measured, usually by gas chromatography (GC), the kinetic information obtained is limited due to the relative speed of the catalytic reaction and the delay of detection; therefore, the final oxygen yield of the reaction constitutes the majority of the useful information. This is an effective way of evaluating the TON of a catalytic reaction, but the limiting factor will likely be the amount of oxidant and not the performance of the catalyst, so this will be a low estimate. To avoid the problem of limited amounts of oxidant, a POM WOC can be reused in a catalytic system in generally one of two ways: the additional oxidant can be added to the same reaction solution as used in the previous experimental run, or the catalyst can be isolated by either extraction or precipitation and added to a fresh solution containing additional oxidant. Isolating the catalyst after reaction can be problematic because the amount of catalyst can be quite low (μM or even nM concentrations). Simply adding additional oxidant is more straightforward but care must be taken to keep the pH constant, either by starting with a much higher buffer concentration or by refreshing the buffer.

Despite the drawback of not detecting real-time oxygen formation for POM-catalyzed water oxidations with $[Ru(bpy)_3]^{3+}$, the reaction kinetics using UV/vis provides a wealth of information in addition to TON determination. Such kinetic data, especially when using a stopped-flow apparatus, can yield very early time points (ms timescale), which in turn can elucidate subtle kinetic attributes not accessible on the longer timescales mandated by slow O_2 detection methods. One important illustration of the value of early-timescale water oxidation is the marked difference in Co^{2+}(aq) -catalyzed versus POM WOC-catalyzed processes: the former always produce a marked and reproducible induction period (first 5 seconds), whereas the latter reproducibly do not. This was one of several key factors in differentiating water oxidation by **Co₄PPOM** from water oxidation by the likely decomposition product and known WOC, Co^{2+}(aq). The high sensitivity and reproducibility of kinetics data afforded by the use of a stopped-flow apparatus also make this technique ideal for examining the impact on the rate behavior of minor changes to the key reaction parameters: ionic strength, buffer, buffer concentration, and counterions. Such kinetic data are

---

[1] However, it is important not to put too much weight on these results, because the data are not collected truly simultaneously, even though the conditions may be quite similar. Consequently, the experimental error can be high.

of general value and can facilitate acquisition of rate constants useful in conjunction with theoretical rate calculations [8, 64].

11.6   Additional Aspects of WOC System Stability

Many of the factors that affect POM WOC stability, which were thoroughly addressed in our recent JACS paper, were previously largely ignored in the literature [18]. The identity of the buffer and its concentration, capacity, and pH are crucial for a functioning system, due to the water oxidation half-reaction stoichiometry (Equation 11.3): generation of four equivalents of protons per O_2 molecule. However, as was shown in the specific case of **Co₄PPOM**, this WOC has a specific interaction with phosphate buffer that appears to be absent when other buffers are used, even at the same concentration and pH. Thus, the selection of a specific buffer must be addressed on a case-by-case basis. The simple measure of stability provided by UV/vis kinetics data over extended times under the turnover reaction conditions can be sufficient for initial buffer selection if careful attention is paid to the analysis, particularly early time kinetics. Typically, the species in greatest concentration for a given WOC system will function as the buffer, so an effort must be made not to have overly high buffer concentrations [18–22]. Additionally, maintaining the buffer capacity may seem trivial but insufficient buffer can lead to destruction of the catalyst and/or PS. Monitoring of the reaction pH before, during, and after catalytic water oxidation will clearly indicate whether the buffer capacity is sufficient or not. The exact role of these factors in the water oxidation mechanism and POM stability is complicated; it is best understood currently for **Co₄PPOM** in phosphate and borate buffers [18]. As POM WOCs rely largely on electrostatics to facilitate interaction with the oxidant during turnover, ionic strength can greatly affect the performance of a system. A systematic analysis of each of these factors will provide a cursory or working understanding of the way in which each impacts overall performance.

11.7   Techniques for Assessing POM WOC Stability

As noted before, the issue of WOC stability, and specifically POM WOC stability, should be fully addressed for any newly reported catalyst. In cases where instability is observed, attempts to identify these decomposition products and to determine the species responsible for the observed catalysis should be made [65]. This has rarely been done in the homogeneous POM WOC studies to date. In this section, we describe some common techniques that have been used to assess POM WOC stability.

No single technique unequivocally addresses stability. Instead, researchers have relied on the use of multiple, complementary techniques, including: X-ray crystallography, Fourier transform infrared (FT-IR), Raman, electron paramagnetic resonance (EPR), energy-dispersive X-ray (EDX), nuclear magnetic resonance (NMR), and UV/vis spectroscopies, dynamic light scattering (DLS), small-angle X-ray scattering (SAXS), voltammetry, and even high-performance liquid chromatography (HPLC). In addition, destructive methods such as atomic absorption spectroscopy (AAS), atomic emission spectroscopy (AES), and inductively coupled plasma mass spectrometry (ICP-MS) help differentiate homogeneous and heterogeneous catalysts and their decomposition products. Semidestructive stripping voltammetry and colorimetry can also assess the hydrolytic stability of a POM WOC. Since the stability of **Co$_4$PPOM** has been studied using most of these techniques, it will serve as an exemplary POM WOC.

Decomposition of a POM WOC through hydrolysis of the redox-active metal centers has been observed in several cases, as shown generically and in a simplified manner in Equation 11.5. Since these hydrated metal cations can be WOCs themselves or act as precursors for metal-oxide WOCs, both possibilities should be assessed. Logically, such studies should start with identification of all the POM WOC decomposition products.

$$M_nPOM \rightarrow M_x + M_{n-x}POM \qquad (11.5)$$

The extent of metal-cation hydrolysis can be quantified by one of the following techniques: AAS, AES, ICP-MS, stripping voltammetry, and colorimetric methods. Since AAS, AES, and ICP-MS are destructive methods, it is necessary to remove the remaining intact POM from the solution prior to analysis. It is known that POMs can be quantitatively extracted into hydrophobic organic solvents such as toluene using a phase-transfer agent such as tetra- *n* -heptylammonium bromide [66]. The organic layer, which now contains all of the POM, can then be removed, and the remaining metal species in the aqueous layer can be quantified via AAS or AES. This approach was used to quantify the amount of hydrolyzed Co^{2+}(aq) in solution after aging **Co$_4$PPOM** under catalytic conditions [18].

Stripping voltammetry has been used in trace-metal analysis for decades [67], but was only recently applied to the study of POM WOC stability. This analysis relies on measuring the current caused by voltammetrically stripping off a complex formed between the hydrolyzed metal cation and an organic ligand that has been adsorbed on to an electrode surface. It was used to study hydrolysis of the cobalt metal centers in **Co$_4$PPOM** [15, 16, 18]. Since complexation of the hydrolyzed metal cation shifts the equilibrium in Equation 11.5, measurements performed in the presence of the POM must be conducted shortly after addition of the chelator. Alternatively, the POM can be extracted by the previously mentioned method, which would eliminate this issue.

An alternate to AAS, AES, ICP-MS, and stripping voltammetry for determining the amount of hydrolyzed metal cation in a POM WOC solution is colorimetry. Here, a strong chelator is added, which binds the metal cation, forming a highly colored complex. UV/vis is then used to quantify the concentration of the highly colored complex in solution. As in stripping voltammetry, formation of the colored metal complex may well shift the equilibrium in Equation 11.5; removal of intact POM will eliminate this issue. This method tends to be quite practical as UV/vis spectrometers are readily available, whereas ICP-MS and atomic spectroscopy instruments frequently are not.

If metal cations are detected, it is likely that their corresponding metal oxides or hydroxides will form. Classic work by Shafirovich and coworkers [51] and more recently studies from Nocera's group [68] have shown that Co^{2+}(aq) is able to form catalytically active CoO_x during catalytic water oxidation. DLS can be used to detect the formation of colloidal species and their presence after catalysis, but cannot identify their composition. The Mo-based POM WOCs, $[CoMo_6O_{24}H_6]^{6-}$ and $[Co_2Mo_{10}O_{38}H_4]^{6-}$, were both reported to be stable after catalysis via DLS, even though particles were detected as the quantity of the PS, $[Ru(bpy)_3]^{2+}$, was increased [44]. Only after using EDX was the identity of the precipitate confirmed to be a salt between the POM and PS. Multiple studies on **Co$_4$PPOM** have found that no particles of CoO_x form when $[Ru(bpy)_3]^{2+}$ or $[Ru(bpy)_3]^{3+}$ is used as either a PS or a stoichiometric oxidant.

Voltammetric techniques have been used to monitor the hydrolytic stability of several POM WOCs. **Ru_4SiPOM** has been shown to be oxidatively stable by multiple voltammetric studies by many research groups; by far the most prominent example involving a hydrated transition-metal cation dissociating from a POM substituted with the same metal is **Co_4PPOM**. In this case, the amount of cobalt that dissociates from the POM was quantitatively assessed using the techniques discussed before. However, the resulting POM species were not isolated or fully characterized *in situ*, so there remains some ambiguity over the final fate of this POM. Using CV, Goberna-Ferrón and coworkers found that CoO_x films did not form from **Co_9PPOM** in the presence of 2, 2′ -bipyridine during electrochemical water oxidation [23]. This ligand chelates the Co^{2+} or Co^{3+}(aq) released from the POM during catalysis. The resulting $Co(bpy)_3^{3+}$ precipitates **Co_9PPOM** and, presumably, **$Co_{9-x}PPOM$**; however, these species were not quantitated.

The techniques discussed so far have focused on the detection and quantification of hydrolyzed metal cations and their metal oxides, but the POM ligands themselves may rearrange if the POM is thermodynamically unstable towards hydrolysis under those conditions (pH, buffer, etc.) and/or precipitates out. Again, no single technique is sufficient to assess the stability of a POM WOC, so a combination must be used to address this key issue. Isolation of the initially dissolved POM catalyst after use in a catalytic run is a strong argument that the POM is stable. However, the age-old problem in catalysis pertains: even if an initial catalyst structure appears stable under turnover conditions, there might be a small quantity of catalyst decomposition product present that can explain the observed catalysis.[2]

X-ray crystallography, FT-IR, and Raman spectroscopies have been used to characterize POMs after catalysis. Usually, a catalyst is not quantitatively isolated, so these techniques may not clarify whether or not some POM decomposition has occurred during catalysis. In addition, these techniques do not fully address the identity of the actual operating catalyst in a WOC system. That is, what is isolated after the reaction may not represent the catalytically active species, even if this is what was initially added to the system.

Car and coworkers found that during photochemical water oxidation with α-$K_6Na[\{Ru_3 O_3(H_2O)Cl_2\}(SiW_9O_{34})]$ (Figure 11.1j) and $K_{11}Na[M_4(H_2O)_2(SiW_9O_{34})_2]$, where M = Co^{2+}, Ni^{2+} (Figure 11.1d), the POMs precipitate out as catalytically active salts with the PS, $[Ru(bpy)_3]^{2+}$ [43]. The composition of these salts was confirmed by attenuated total reflectance (ATR) FT-IR after isolation from the reaction solution, which matched authentically prepared samples of the PS-POM salts. Although this implies no decomposition of the POM during the reaction, the authors did not note the yield of the isolated PS-POM salt. *In situ* FT-IR and Raman studies enabling assessment of any changes in the POM itself during turnover have not been performed on any POM WOCs to date. These methods may be particularly useful for decomposition studies when the POM WOC is supported on an electrode.

As an initial probe, UV/vis can be used to verify the hydrolytic stability of the redox-active metal sites in a POM WOC. It is a fast, inexpensive method, but suffers from broad, often weak absorption bands, which tend to overlap with those of the corresponding

---

[2] However, the other techniques described in this chapter, when used in conjunction with this post-reaction catalyst isolation and characterization experiment, can effectively rule out this "trace material is the active catalyst" scenario.

decomposition products. Further, most POM WOCs tend to absorb weakly in the visible region, so UV/vis measurements are often performed at higher POM concentrations. For example, $Co^{II}Co^{III}POM$ ($\epsilon_{620} = 210$ $M^{-1}cm^{-1}$) was used at $16\,\mu M$ under catalytic conditions, but its stability was checked by UV/vis at $100\,\mu M$ [25]. The hydrolytically unstable POM $[\{Co_4(\mu\text{-}OH)(H_2O)_3\}(Si_2W_{19}O_{70})]^{11-}$ displayed catalytic activity for water oxidation, but UV/vis data found it underwent slow hydrolysis over a range of pH values and buffer conditions [55]. Equilibration in plain water takes 40 days; it is faster in pH 4.8 acetate buffer. Crystals isolated from the resulting solution were found to contain two kinetically stable POMs, $[\{Co(H_2O)_4\}(\mu\text{-}H_2O)_2K\{Co(H_2O)_4\}(Si_2W_{18}O_{66})]^{11-}$ and $[Co(H_2O)SiW_{11}O_{39}]^{6-}$. The decomposition reactions were discussed and decomposition models were supported by high-fidelity fitting of the UV/vis spectra. Since these studies were conducted under conditions different from those used in catalytic water oxidation, the authors could not probe the stability of the higher oxidation states of the catalyst. HPLC with UV absorbance detection was recently applied in a study of Co_4PPOM to quantify its stability at pH 5.8 and 8.0 [16]. The authors argued that in the time period examined, the POM was hydrolytically stable, since there was no significant change in the UV band at 240 nm. When a 1.4 V versus Ag/AgCl potential was applied, the POM WOC clearly became unstable. They also noted that this technique could be used to probe speciation of the POM, as it loses cobalt ions, but they did not perform such a study in depth.

NMR spectroscopy has long been used to investigate the stability of POMs and, more recently, Co_4PPOM. We checked the stability of Co_4PPOM by ^{31}P NMR over a wide pH range and found no changes in the NMR shift during aging in buffer [10]. While it is now accepted that the cobalt centers in this POM undergo slow hydrolytic release from the parent POM, ^{31}P NMR was insensitive to this. This defines a limit for the effective use of NMR in POM WOC stability studies.

11.8   Conclusion

The stability of catalytic systems based on POMs for water oxidation depends on multiple factors: the hydrolytic stability of POMs, the oxidative stability of associated subsystems containing organic molecules, and the reaction conditions (pH, buffer identity, and concentration and ionic strength). Since water oxidation is a multi-electron process, the hydrolytic stability of the POM in different oxidation states should be evaluated. In contrast to the oxidatively impervious POM WOCs, organic ligand-containing PSs can readily self-decompose via ligand oxidation. The selectivity of POM WOCs for water oxidation versus organic ligand oxidation, as for all WOCs, is important to elucidating their mechanism and critical for their ultimate practical use. A workable reaction mechanism can be established by analyzing the reaction kinetics of oxidant consumption and, simultaneously, O_2 formation.

Additional studies of the hydrolytic stability of all families of POM WOCs are needed. These studies entail measurement of the kinetics of O_2 formation from the suspected decomposition products and spectroscopic identification of these products. In the case of Co_4PPOM, one product of its hydrolysis is $Co^{2+}(aq)$. The kinetics of water oxidation catalyzed by $Co^{2+}(aq)$ has a very distinctive feature: an induction period and swift self-inhibition. In contrast, catalytic water oxidation by Co_4PPOM does not exhibit this feature. The concentration of metal cations can be determined by stripping voltammetry, colorimetry, and other methods. In addition, the negatively charged POM can be quantitatively distinguished from hydrated metal cations (Co^{2+}(aq) in the case of **Co₄PPOM**) and metal oxide (CoO_x) particles by hydrophobic cation extraction of the POM from the water into an organic solvent: the POM extracts completely, whereas the hydrated metal-cation and metal-oxide particles do not extract at all. Subsequent to such as extraction, the WOC activities of different phases are readily quantified; they are in the case of **Co₄PPOM** versus Co_{aq}^{2+} and CoO_x particles. The pH dependencies of different species with WOC activity should also be assessed, as they are frequently different.

Acknowledgments

CLH thanks the Department of Energy Solar Photochemistry program (grant DE-FG02-07ER-15906) for support.

References

[1] Walter, M. G., Warren, E. L., McKone, J. R., Boettcher, S. W., Mi, Q., Santori, E. A., Lewis, N. S. (2010) Chem. Rev., 110: 6446–6473.

[2] Lewis, N. S., Nocera, D. G. (2006) Proc. Natl. Acad. Sci., 103(43): 15 729–15 735.

[3] Gersten, S. W., Samuels, G. J., Meyer, T. J. (1982) J. Am. Chem. Soc., 104: 4029–4030.

[4] Pope, M. T. (2004) In Comprehensive Coordination Chemistry II: From Biology to Nanotechnology, Vol. 4. Elsevier, Oxford, UK.

[5] Hill, C. L. (2004) In Comprehensive Coordination Chemistry II: From Biology to Nanotechnology, Vol. 4. Elsevier, Oxford, UK.

[6] Long, D.-L., Tsunashima, R., Cronin, L. (2010) Angew. Chem. Int. Ed., 49: 1736–1758.

[7] Sartorel, A., Carraro, M., Scorrano, G., Zorzi, R. D., Geremia, S., McDaniel, N. D., Bernhard, S., Bonchio, M. (2008) J. Am. Chem. Soc., 130: 5006–5007.

[8] Geletii, Y. V., Botar, B., Kögerler, P., Hillesheim, D. A., Musaev, D. G., Hill, C. L. (2008) Angew. Chem. Int. Ed., 47: 3896–3899.

[9] Lv, H., Geletii, Y. V., Zhao, C., Vickers, J. W., Zhu, G., Luo, Z., Song, J., Lian, T., Musaev, D. G., Hill, C. L. (2012) Chem. Soc. Rev., 41: 7572–7589.

[10] Yin, Q., Tan, J. M., Besson, C., Geletii, Y. V., Musaev, D. G., Kuznetsov, A. E., Luo, Z., Hardcastle, K. I., Hill, C. L. (2010) Science, 328: 342–345.

[11] Huang, Z., Luo, Z., Geletii, Y. V., Vickers, J., Yin, Q., Wu, D., Hou, Y., Ding, Y., Song, J., Musaev, D. G., Hill, C. L., Lian, T. (2011) J. Am. Chem. Soc., 133: 2068–2071.

[12] Natali, M., Berardi, S., Sartorel, A., Bonchio, M., Campagna, S., Scandola, F. (2012) Chem. Commun., 48, 8808–8810.

[13] Lieb, D., Zahl, A., Wilson, E. F., Streb, C., Nye, L. C., Meyer, K., Ivanović-Burmazović, I. (2011) Inorg. Chem., 50: 9053–9058.

[14] Ohlin, C. A., Harley, S. J., McAlpin, J. G., Hocking, R. K., Mercado, B. Q., Johnson, R. L., Villa, E. M., Fidler, M. K., Olmstead, M. M., Spiccia, L., Britt, R. D., Casey, W. H. (2011) Chem. Eur. J., 17: 4408–4417.

[15] Stracke, J. J., Finke, R. G. (2011) J. Am. Chem. Soc., 133: 14 872–14 875.

[16] Stracke, J. J., Finke, R. G. (2013) ACS Catal., 3: 1209–1219.

[17] Wu, J., Liao, L., Yan, W., Xue, Y., Sun, Y., Yan, X., Chen, Y., Xie, Y. (2012) ChemSusChem, 5: 1207–1212.

[18] Vickers, J. W., Lv, H., Sumliner, J. M., Zhu, G., Luo, Z., Musaev, D. G., Geletii, Y. V., Hill, C. L. (2013) J. Am. Chem. Soc., 135: 14 110–14 118.

[19] Guo, S.-X., Liu, Y., Lee, C.-Y., Bond, A. M., Zhang, J., Geletii, Y. V., Hill, C. L. (2013) Energy Environ. Sci., 6: 2654–2663.

[20] Liu, Y., Guo, S.-X., Bond, A. M., Zhang, J., Geletii, Y. V., Hill, C. L. (2013) Inorg. Chem., in press.

[21] Quintana, M., López, A. M., Rapino, S., Toma, F. M., Iurlo, M., Carraro, M., Sartorel, A., Maccato, C., Ke, X., Bittencourt, C., Da Ros, T., Van Tendeloo, G., Marcaccio, M., Paolucci, F., Prato, M., Bonchio, M. (2013) ACS Nano, 7: 811–817.

[22] Xiang, X., Fielden, J., Rodríguez-Córdoba, W., Huang, Z., Zhang, N., Luo, Z., Musaev, D. G., Lian, T., Hill, C. L. (2013) J. Phys. Chem. C, 117: 918–926.

[23] Goberna-Ferrón, S., Vigara, L., Soriano-López, J., Galán-Mascarós, J. R. (2012) Inorg. Chem., 51: 11 707–11 715.

[24] Soriano-López, J., Goberna-Ferrón, S., Vigara, L., Carbó, J. J., Poblet, J. M., Galán-Mascarós, J. R. (2013) Inorg. Chem., 52: 4753–4755.

[25] Song, F., Ding, Y., Ma, B., Wang, C., Wang, Q., Du, X., Fu, S., Song, J. (2013) Energy Environ. Sci., 6: 1170–1184.

[26] Toma, F. M., Sartorel, A., Iurlo, M., Carraro, M., Parisse, P., Maccato, C., Rapino, S., Gonzalez, B. R., Amenitsch, H., Ros, T. D., Casalis, L., Goldoni, A., Marcaccio, M., Scorrano, G., Scoles, G., Paolucci, F., Prato, M., Bonchio, M. (2010) Nature Chem., 2: 826–831.

[27] Howells, A. R., Sankarraj, A., Shannon, C. J. (2004) Am. Chem. Soc. 126, 12258–12259.

[28] Geletii, Y. V., Huang, Z., Hou, Y., Musaev, D. G., Lian, T., Hill, C. L. (2009) J. Am. Chem. Soc., 131: 7522–7523.

[29] Geletii, Y. V., Besson, C., Hou, Y., Yin, Q., Musaev, D. G., Quinonero, D., Cao, R., Hardcastle, K. I., Proust, A., Kögerler, P., Hill, C. L. (2009) J. Am. Chem. Soc., 131, 17360–17370.

[30] Kuznetsov, A. E., Geletii, Y. V., Hill, C. L., Morokuma, K., Musaev, D. G. (2009) J. Am. Chem. Soc. 131, 6844–6854.

[31] Quiñonero, D., Kaledin, A. L., Kuznetsov, A. E., Geletii, Y. V., Besson, C., Hill, C. L., Musaev, D. G. (2010 J. Phys. Chem. A 114, 535–542.

[32] Besson, C., Huang, Z., Geletii, Y. V., Lense, S., Hardcastle, K. I., Musaev, D. G., Lian, T., Proust, A., Hill, C. L. (2010) Chem. Commun.: 2784–2786.

[33] Cao, R., Ma, H., Geletii, Y. V., Hardcastle, K. I., Hill, C. L. (2009) Inorg. Chem., 48, 5596–5598

[34] Zhu, G., Geletii, Y. V., Kögerler, P., Schilder, H., Song, J., Lense, S., Zhao, C., Hardcastle, K. I., Musaev, D. G., Hill, C. L. (2012) Dalton Trans., 41: 2084–2090.

[35] Zhu, G., Glass, E. N., Zhao, C., Lv, H., Vickers, J. W., Geletii, Y. V., Musaev, D. G., Song, J., Hill, C. L. (2012) Dalton Trans., 41: 13 043–13 049.

[36] Sartorel, A., Miro, P., Salvadori, E., Romain, S., Carraro, M., Scorrano, G., Valentin, M. D., Llobet, A., Bo, C., Bonchio, M., (2009) J. Am. Chem. Soc., 131, 16051–16053.

[37] Toma, F. M., Sartorel, A., Carraro, M., Bonchio, M., Prato, M., (2011) Pure Appl. Chem., 83, 1529–1542.

[38] Toma, F. M., Sartorel, A., Iurlo, M., Carraro, M., Rapino, S., Hoober Burkhardt, L., Ros, T. D., Marcaccio, M., Scorrano, G., Paolucci, F., Bonchio, M., Prato, M. (2011) ChemSusChem, 4, 1447–1451.

[39] Orlandi, M., Argazzi, R., Sartorel, A., Carraro, M., Scorrano, G., Bonchio, M., Scandola, F. (2010) Chem. Commun., 46, 3152–3154.

[40] Puntoriero, F., Ganga, G. L., Sartorel, A., Carraro, M., Scorrano, G., Bonchio, M., Campagna, S. (2010) Chem. Commun., 46: 4725–4227.

[41] Natali, M., Orlandi, M., Berardi, S., Campagna, S., Bonchio, M., Sartorel, A., Scandola, F. (2012) Inorg. Chem., 51, 7324–7331.(

[42] Murakami, M., Hong, D., Suenobu, T., Yamaguchi, S., Ogura, T., Fukuzumi, S. (2011) J. Am. Chem. Soc., 133, 11605–11613.(

[43] Car, P.-E., Guttentag, M., Baldridge, K. K., Albertoa, R., Patzke, G. R. (2012) Green Chem., 14: 1680–1688.

[44] Tanaka, S., Annaka, M., Sakai, K. (2012) Chem. Commun., 48: 1653–1655.

[45] Guo, D., Teng, S., Liu, Z., You, W., Zhang, L. (2013) J. Cluster Sci. 24, 549–558.

[46] Lv, H., Rudd, J. A., Zhuk, P. F., Lee, J. Y., Constable, E. C., Housecroft, C. E., Hill, C. L., Musaev, D. G., Geletii, Y. V. (2013) RSC Adv., 3: 20647–20654.

[47] Amouyal, E. (1995) Sol. Energy Mater. Sol. Cells, 38: 249–276.

[48] Lehn, J. M., Sauvage, J. P., Ziesel, R. (1979) Nouveau J. Chim., 3: 423–442.

[49] Alstrum-Acevedo, J. H., Brennaman, M. K., Meyer, T. (2005) J. Inorg. Chem., 44: 6802–6082.

[50] Bard, A. J., Fox, M. A. (1995) Acc. Chem. Res., 28: 141–145.

[51] Shafirovich, V. Y., Khannanov, N. K., Strelets, V. V. (1980) Nouveau J. Chim., 4: 81–84.

[52] Huang, Z., Geletii, Y. V., Musaev, D. G., Hill, C. L. (2012) Ind. Eng. Chem. Res., 51: 11 850–11 859.

[53] White, H. S., Becker, W. G., Bard, A. J. (1984) J. Phys. Chem., 88: 1840–1846.

[54] Kaledin, A. L., Huang, Z., Geletii, Y. V., Lian, T., Hill, C. L., Musaev, D. G. (2010) J. Phys. Chem. A, 114: 73–80.

[55] Kaledin, A. L., Huang, Z., Yin, Q., Dunphy, E. L., Constable, E. C., Housecroft, C. E., Geletii, Y. V., Lian, T., Hill, C. L., Musaev, D. G. (2010) J. Phys. Chem. A, 114: 6284–6297.

[56] Stanbury, D. M. (1989) In Advanced Inorganic Chemistry, Vol. 33. Academic Press, Waltham, MA, USA.

[57] Henbest, K., Douglas, P., Garley, M. S., Mills, A. (1994) J. Photochem. Photobiol A: Chem., 80: 299–305.

[58] Wang, L., Duan, L., Tong, L., Sun, L. (2013) J. Catal., 306: 129–132.

[59] Parent, A. R., Crabtree, R. H., Brudvig, G. W. (2013) Chem. Soc. Rev., 42: 2247–2252.

[60] Parent, A. R., Brewster, T. P., De Wolf, W., Crabtree, R. H., Brudvig, G. W. (2012) Inorg. Chem., 51, 6147–6152.

[61] Shafirovich, V. Y., Strelets, V. V. (1982) Nouveau J. de Chimie, 6, 183–186.

[62] Shafirovich, V. Y., Strelets, V. V. (1980) Bulletin of the Academy of Sciences of the USSR. Division of Chemical Sciences, pp. 7–12.

[63] Ghosh, P. K., Brunschwig, B. S., Chou, M., Creutz, C., Sutin, N. (1984) J. Am. Chem. Soc., 106: 4772–4783.

[64] Vickers, J., Lv, H., Zhuk, P. F., Geletii, Y. V., Hill, C. L. (2012) MRS Proceedings, 1387: mrsf11-1387-e1302-1301.

[65] Artero, V., Fontecave, M. (2012) Chem. Soc. Rev., 42: 2338–2356.

[66] Katsoulis, D. E., Pope, M. T. (1984) J. Am. Chem. Soc., 106: 2737–2738.

[67] Kissinger, P. T., Heineman, W. R. (1996) Laboratory Techniques in Electroanalytical Chemistry. CRC Press, Boca Raton, FL, USA.

[68] Kanan, M. W., Nocera, D. G. (2008) Science, 321: 1072–1075.

12

Quantum Chemical Characterization of Water Oxidation Catalysts

Pere Miró, Mehmed Z. Ertem, Laura Gagliardi, and Christopher J. Cramer
Department of Chemistry, Chemical Theory Center and Supercomputing Institute,
University of Minnesota, Minneapolis, MN, USA

12.1 Introduction

Higher plants, green algae, and cyanobacteria use light energy to couple the formation of molecular oxygen to the fixation of carbon dioxide. This process generates an aerobic atmosphere and a readily usable carbon pool, both of which are essential to sustaining almost all life on our planet. The formation of molecular oxygen occurs in a membrane-embedded protein, photosystem II (PSII), which catalyzes one of the most thermodynamically demanding reactions in biology: the photoinduced oxidation of water [1–11]. Initially, light energy is transformed into chemical energy, which is used to oxidize water into molecular oxygen, four protons, and four electrons (Equation 12.1). This process is thermodynamically uphill in free energy, with a redox potential of 1.23 V versus the standard hydrogen electrode (SHE) and a highly complex mechanism involving the serial removal of four protons and four electrons with concomitant O−O bond formation [12].

$$2H_2O_{(l)} \rightarrow O_{2(g)} + 4H^+_{(aq)} + 4e^-_{(g)} \tag{12.1}$$

In PSII, water oxidation takes place in the oxygen-evolving complex (OEC), which contains a μ-oxo-bridged tetramanganese cluster with a calcium ion [13–15]. The PSII-OEC is able to mediate the required four electron−four proton process while undergoing a stepwise oxidation in a series of different redox states [16, 17]. The design of synthetic analogs of the OEC capable of water oxidation in aqueous solution under ambient conditions is a highly relevant challenge. Synthetic analogs would provide an alternative way of studying

Molecular Water Oxidation Catalysis: A Key Topic for New Sustainable Energy Conversion Schemes,
First Edition. Edited by Antoni Llobet.
© 2014 John Wiley & Sons, Ltd. Published 2014 by John Wiley & Sons, Ltd.

the fundamental aspects of water oxidation without the complexity of the enzyme. Furthermore, artificial photosynthetic systems designed to convert solar energy into chemical energy in an environmentally benign way are likely to require efficient synthetic analogs of the OEC for water splitting.

The first attempts to achieve an artificial OEC were focused on manganese complexes like that found in the natural OEC [18]. Although several manganese-based complexes have been studied, only a few have been shown to be capable of catalyzing water oxidation [19]. The active manganese complexes usually have very small turnover numbers (TONs), and there remains some controversy over the fact that most of the oxidants employed have been hydrogen peroxide or peroxide derivatives.

The problems associated with the manganese complexes prompted the search for alternative transition-metal centers in synthetic OECs. Among these, ruthenium-based complexes require special attention, due to their robustness and well-characterized activity. In 1982, Gersten and coworkers presented the first ruthenium-based water oxidation catalyst (WOC), cis, cis-[(bpy)$_2$(H$_2$O)–Ru–O–Ru–(H$_2$O)(bpy)$_2$]$^{4+}$, also called Meyer's blue dimer [20]. Catalyst activation is achieved electrochemically or in the presence of a bulk oxidant like Ce^{IV}, but deactivation occurs after only a few turnovers. The O$_2$ formation mechanism in the blue dimer still remains controversial [7, 21–23].

The sequential oxidation of the catalyst by 4e$^-$ is suggested by the loss of four protons, which generates the highly reactive cis, cis-[(bpy)$_2$(O)–Ru–O–Ru–(O)(bpy)$_2$]$^{4+}$ species and leads to O–O bond formation via either a water nucleophilic attack (WNA) on one of the Ru^{IV}–O$^{•-}$ groups or the direct coupling of two Ru^{IV}–O$^{•-}$ (Figure 12.1). Isotope labeling studies support the former pathway as the major route but are not conclusive on the presence of other competitive pathways.

Since the initial work of Meyer and coworkers, other ruthenium-based catalysts with polypyridyl ligands and different nitrogen-based ligands have been explored. In particular, Llobet and coworkers reported the synthesis and characterization of several ruthenium-based WOCs, including cis-[Ru^{II}(bpy)$_2$(H$_2$O)$_2$]$^{2+}$, [Ru^{II}(damp)(bpy)(H$_2$O)]$^{2+}$, and in, in-[(Ru^{II}(trpy)(H$_2$O))$_2$(μ-bpp)]$^{3+}$ among others [8, 24–28].

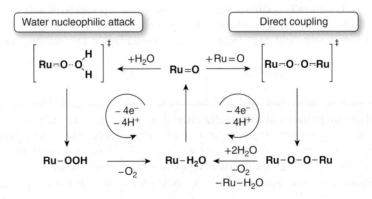

Figure 12.1 *Schematic representation of the two possible water oxidation mechanisms: water nucleophilic attack (left) and intra-or intermolecular direct-coupling O–O bond formation (right)*

The mononuclear cis-[Ru^{II}(bpy)$_2$(H$_2$O)$_2$]$^{2+}$ catalyst has two aquo-ligands coordinated to a Ru^{II} center, which, in the presence of Ce^{IV}, can be oxidized to a formal O $=$ Ru^{VI} $=$ O moiety [25]. Formation of the O–O bond was experimentally inferred through ^{18}O labeling to occur via WNA. In a different mononuclear example, [Ru^{II}(damp)(bpy)(H$_2$O)]$^{2+}$(damp $=$ diaminopropane) ultimately follows a WNA mechanism involving its unique Ru^{II}–H$_2$O moiety. Experimentally, three Ce^{IV} equivalents are required to generate the Ru^{IV}–O$^{\bullet-}$ moiety, but the O$_2$ bubbling occurs only when the fourth equivalent is added [26].

The dinuclear catalyst in,in-[(Ru^{II}(trpy)(H$_2$O))$_2$(μ-bpp)]$^{3+}$ features two Ru^{II} centers, bridged through a 2,6-bis(pyridyl)pyrazolate (hbpp) and coordinated by terpyridine (trpy) with aquo-ligands to complete the octahedral geometry [24]. This catalyst was the first to employ something other than an oxo bridge to serve as a bridging ligand between metal centers. The two metal centers are situated sufficiently close to one another to accommodate an acetate linker between them, as observed by X-ray crystallography. In the presence of Ce^{IV} as the sacrificial oxidant, the acetate is replaced by two water molecules, releasing dioxygen and protons into the media. This unique catalyst takes advantage of an anionic dinucleating ligand, which places two Ru^{II}–H$_2$O units in close proximity to one another and in proper orientation to support water oxidation catalysis through intramolecular direct coupling, leading to O–O bond formation.

While the ruthenium-based catalysts are a perfect playground for exploring fundamental features of the water oxidation reaction, ruthenium is a rare and consequently highly expensive metal, which renders impractical its large-scale implementation. As a result, significant recent research effort has been focused on the synthesis and characterization of WOCs based on earth-abundant metals such as cobalt and iron. In these categories, Dogutan and coworkers have recently reported the synthesis and electrocatalytic water oxidation of a cobalt-based catalyst containing a β-octafluoro hangman corrole, CoH$^{\beta F}$–CO$_2$H, while Ellis and coworkers have reported a series of iron(III)-centered tetraamido macrocyclic ligand (Fe^{III}-TAML) catalysts capable of oxidizing water to molecular oxygen in the presence of an excess of sacrificial oxidant [29–31].

In this chapter, we review computational studies on the water oxidation mechanisms of mononuclear (cis-[Ru^{II}(bpy)$_2$(H$_2$O)$_2$]$^{2+}$ and [Ru^{II}(bpy)(damp)(H$_2$O)]$^{3+}$) and dinuclear (in,in-[(Ru^{II}(trpy)(H$_2$O))$_2$(μ-bpp)]$^{3+}$) ruthenium catalysts, the cobalt hangman corrole (CoH$^{\beta F}$–CO$_2$H) catalyst, and the iron-TAML catalysts (Fe^{III}-TAML) (Figure 12.2).

12.2 Computational Details

12.2.1 Density Functional Theory Calculations

All species were fully optimized through first-principles calculations performed on the basis of density functional theory (DFT). The exchange and correlation terms were those in the meta-generalized gradient approximation (mGGA) functional M06-L [32–35]. The calculations were performed using the Gaussian09 package [36]. For geometry optimization, the Stuttgart [8s7p6d2f ∣ 6s5p3d2f] ECP10MDF, [8s7p6d2f ∣ 6s5p3d2f] ECP28MWB, and [8s7p6d2f ∣ 6s5p3d2f] ECP10MDF contracted pseudopotential basis sets were used for cobalt, ruthenium, and iron centers, respectively [37, 38]. The 6-31G(d) basis set was

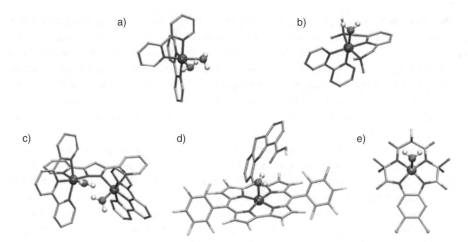

Figure 12.2 *WOCs studied: (a) cis-[$Ru^{II}(bpy)_2(H_2O)_2$]$^{2+}$, (b) [$Ru^{II}(damp)(bpy)(H_2O)$]$^{2+}$, (c) in, in-[($Ru^{II}(trpy)(H_2O)$)$_2(\mu$-bpp)]$^{3+}$, (d) CoH$^{\beta F}$-CO$_2$H, and (e) Fe^{III}-TAML. Only hydrogen atoms in water and carboxylic acid groups are shown; ^{t}Bu groups in the cobalt catalysts are not shown for clarity. Purple, ruthenium; ochre, cobalt; orange, iron; yellow, fluorine; green, chlorine; gray, carbon; white, hydrogen. See color plate*

used for all other atoms [39]. Integral evaluation made use of the grid defined as ultrafine in the Gaussian software package and an automatically generated density-fitting basis set was used within the resolution-of-the-identity approximation to speed the evaluation of Coulomb integrals.

The nature of all stationary points was verified by analytic computation of vibrational frequencies, which were also used for the computation of zero-point vibrational energies and molecular partition functions, and to determine the reactants and products associated with each transition-state structure (by following the normal modes associated with imaginary frequencies). Partition functions were used in the computation of 298 K thermal contributions to free energies employing the usual ideal-gas, rigid-rotator, quasiharmonic oscillator approximation [40]. In the quasiharmonic oscillator approximation, all frequencies below $50\,cm^{-1}$ were replaced by $50\,cm^{-1}$ when computing vibrational free energies, thereby avoiding complications associated with the breakdown of the harmonic oscillator approximation for very-low-frequency normal modes [41]. Free-energy contributions were added to single-point M06-L electronic energies computed with the Stuttgart/Dresden effective core potentials (SDD) basis set for the metal centers and the 6-311 + G(2df, p) basis set for all other atoms in order to compute our best free energies, which included free energies of aqueous solvation computed using the solvation model based on electron density (SMD) model for gas-phase optimized geometries [42, 43].

12.2.2 Multiconfigurational Calculations

The energies of selected species in the *cis*-[$Ru^{II}(bpy)_2(H_2O)_2$]$^{2+}$, [$Ru^{II}(damp)(bpy)(H_2O)$]$^{2+}$, *in, in*-[($Ru^{II}(trpy)(H_2O)$)$_2(\mu$-bpp)]$^{3+}$, and Fe^{III}-TAML catalytic cycles were also computed using complete active space self-consistent field (CASSCF) calculations with corrections from second-order perturbation theory (CASPT2) for the DFT-optimized geometry using the MOLCAS package [44–46]. Relativistic effects were included through the use of the scalar Douglas–Kroll–Hess (DKH) Hamiltonian [47]. All-electron ANO-RCC basis sets were used for all atoms [48, 49]. For all of the catalysts, the transition-metal center was treated with a basis set of triple-ξ quality. In the Ru catalysts, the C and H atoms were treated with a minimal basis set. For the cis-$[Ru^{II}(bpy)_2(H_2O)_2]^{2+}$ species, a basis set of triple-ξ quality was used for H_2O and a double-ξ basis for N. Likewise, in the $[Ru^{II}(damp)(bpy)(H_2O)]^{2+}$ species, a triple-ξ basis set for was used for both O and N. In the in,in-$[(Ru^{II}(trpy)(H_2O))_2(\mu\text{-bpp})]^{3+}$ species, O was treated at the triple-ξ level and N and H_2O with a double-ξ basis set. In Fe^{III}-TAML species, O and N were treated with a triple-ξ basis set and C and H with a double-ξ basis. The computational costs involved in generating the two-electron integrals were significantly reduced in all cases through the use of Cholesky decomposition combined with local exchange screening [50–52].

The CASSCF wave functions were constructed as linear combinations of the metal d (3d for iron and 4d for ruthenium) and oxygen 2p orbitals. The (14,11) and (16,12) active spaces were found to be well balanced for the cis-$[Ru^{II}(bpy)_2(H_2O)_2]^{2+}$ and in,in-$[(Ru^{II}(trpy)(H_2O))_2(\mu\text{-bpp})]^{3+}$ species, respectively. For the $[Ru^{II}(damp)(bpy)(H_2O)]^{2+}$ species, three different active spaces were considered, with either 10, 9, or 8 electrons in 8 orbitals corresponding to different ruthenium oxidation states (VI, V, and IV, respectively). Finally, for the Fe-TAML species, two active spaces, (12,12) and (14,14), were used.

12.3   Methodology

12.3.1   Solvation and Standard Reduction Potentials

As one might expect, almost all of the known molecular WOCs catalyze water evolution to dioxygen in aqueous solution or in mixtures of water with fully soluble solvents such as ethanol. As a consequence, water is both the solvent and the reactant in the catalytic cycle. The bulk solvation effects affecting the energetics of all of the species in the catalytic cycle are accounted for using the SMD aqueous-continuum solvation model [42, 43]. A 1 M standard state is used for all species in aqueous solution unless their experimental concentrations are otherwise fixed; 55.6 M is employed for water as solvent. Thus, for all solutes other than water, the free energy in aqueous solution is computed as the 1 atm gas-phase free energy, plus an adjustment for the 1 atm to 1 M standard-state concentration change of $RT \ln (24.5)$, which is 1.9 kcal/mol at 298 K, plus the solvation free energy computed from the SMD model. In the case of water, the 1 atm gas-phase free energy is adjusted by the sum of a 1 atm to 55.6 M standard-state concentration change, or 4.3 kcal/mol, and its experimental 1 M to 1 M solvation free energy, −6.3 kcal/mol. The experimental solvation free energy of −265.9 kcal/mol is used for the proton [53–56].

Computational prediction of redox potentials is a relatively straightforward procedure that allows estimation of potentials with deviations of about 200 mV with respect to experiment. For a redox reaction with the general form:

$$Ox_{(aq)} + ne^-_{(g)} \rightarrow Red_{(aq)} \tag{12.2}$$

where *Ox* and *Red* are the oxidized and reduced states of a general redox pair and *n* is the number of electrons in the redox reaction, the reduction potential relative to the SHE is expressed as:

$$E^0_{Ox/Red} = -\frac{\Delta G^0_{Ox/Red} - \Delta G^0_{SHE}}{nF} \tag{12.3}$$

where $\Delta G_{Ox/Red}^0$ and ΔG_{SHE}^0 are the free-energy change associated with Equation 12.2 and the SHE respectively, and F is the Faraday constant. The value of ΔG_{SHE}^0 has been experimentally measured with a generally accepted value of −4.28 V [55, 57, 58]. Our group and others have successfully used this protocol to compute the reduction potentials versus SHE for a wide number of organometallic species, including the ones presented in this chapter.

12.3.2 Multideterminantal State Energies

The generation of active species for the water oxidation process involves multiple proton-coupled electron transfers (PCETs), which often lead to the formation of intermediates that are not well described by a single determinant, such as antiferromagnetically coupled $Co^{IV}-O\cdot^-$ species or higher oxidation states in ruthenium dinuclear species. In such instances, standard Kohn–Sham DFT does not properly describe the low spin (LS) states as they are contaminated by higher ones [40, 59–61]. In the early 1980s, an approximate projection method known as broken-spin symmetry was devised. This procedure is valid for the computation of the energy of the spin-purified LS state in many circumstances [62, 63]. The energy of the LS state is expressed as:

$$^{LS}E = \frac{^{BS}E(^{HS}\langle S^2\rangle - ^{LS}\langle S^2\rangle) - ^{HS}E(^{BS}\langle S^2\rangle - ^{LS}\langle S^2\rangle)}{^{HS}\langle S^2\rangle - ^{BS}\langle S^2\rangle} \tag{12.4}$$

where HS is the single-determinantal high-spin coupled state that is related to the LS state and $<S^2>$ is the expectation value of the total spin operator applied to the appropriate determinant. For a broken-symmetry singlet contaminated by a triplet state, we would expect $^{HS}<S^2>$ to be 2 (the correct eigenvalue for a triplet state) and $^{BS}<S^2>$ to be about 1 (the average of a singlet and a triplet). In the same way, for a broken-symmetry doublet contaminated by a quartet state, we would expect $^{HS}<S^2>$ to be 3.75 (the correct eigenvalue for a quartet) and $^{BS}<S^2>$ to be 1.75 (the average of a doublet and a quartet). This approach has been proven effective for the prediction of energy-state splitting in many metal coodination compounds [59, 64–67].

12.4   Water Oxidation Catalysts

12.4.1   Ruthenium-Based Catalysts

12.4.1.1   Mononuclear Catalysts

The water oxidation mechanisms of two mononuclear ruthenium-based catalysts, *cis*-[Ru^{II}(bpy)$_2$(H$_2$O)$_2$]$^{2+}$ and [Ru^{II}(damp)(bpy)(H$_2$O)]$^{2+}$, have been investigated via

DFT and CASSCF/CASPT2 calculations [24, 26]. The cis-[Ru^{II}(bpy)$_2$(H$_2$O)$_2$]$^{2+}$ catalyst contains two Ru-H$_2$O bonds and can lose 4H$^+$/4e$^-$, leading to the formation of cis-[Ru^{VI}(bpy)$_2$(O)$_2$]$^{2+}$, in which the ruthenium oxidation state is formally VI when all oxidations are considered to be taking place at the metal center (Figure 12.2a). Experimentally, the addition of Ce^{IV} to cis-[Ru^{II}(bpy)$_2$(H$_2$O)$_2$]$^{2+}$ has confirmed its activity as a WOC capable of a few turnovers [25]. Even though the performance of this catalyst is limited, it is very interesting from a mechanistic point of view as the catalytically active intermediates can follow both known O–O bond-formation pathways: a WNA on one of the oxygen centers and a direct coupling between the two Ru = O moieties. Both pathways have been investigated by computing the relative energetics of all the relevant species in the catalytic cycle (Figure 12.3), but it has been unambiguously demonstrated through ^{18}O-labeling experiments that the only mechanism operating in this catalyst is WNA.

The sequential oxidation of cis-[Ru^{II}(bpy)$_2$(H$_2$O)$_2$]$^{2+}$ generates cis-[Ru^{IV}(bpy)$_2$(O)(H$_2$O)]$^{2+}$, cis-[Ru^V(bpy)$_2$(O)(OH)]2+, and cis-[Ru^{VI}(bpy)$_2$O$_2$]$^{2+}$ species with the [Ru = O] unit susceptible to WNA. We were able to locate WNA transition-state structures for cis-[Ru^V(bpy)$_2$(O)(OH)]$^{2+}$ and cis-[Ru^{VI}(bpy)$_2$O$_2$]$^{2+}$ with ΔG‡ of 28.8 and 38.2 kcal/mol, respectively (Figure 12.3). This indicates that although the formation of cis-[Ru^{VI}(bpy)$_2$O$_2$]$^{2+}$ is possible under the experimental conditions, O–O bond formation via the WNA route occurs with a lower activation free energy through intermediate cis-[Ru^V(bpy)$_2$(O)(OH)]$^{2+}$.

Another possible pathway for O–O bond formation might be the direct coupling between the two Ru = O moieties of cis-[Ru^{VI}(bpy)$_2$O$_2$]$^{2+}$ (Figure 12.3), but the associated free energy of activation has been found to be prohibitively high (ΔG‡ = 56.8 kcal/mol). The preference for the "bis-oxo" state over "peroxo" in cis-[Ru^{VI}(bpy)$_2$(O)$_2$]$^{2+}$ can be explained on the basis of the electronic structure of these structures. The molecular orbitals suggest that the Ru(O)$_2$ subsystem has substantial covalency between the ruthenium and oxygen centers and can be viewed as a three-center–six-electron fragment [25]. Ru^{II} formally has six 4d-electrons that, by symmetry, contribute to the Ru–O σ − bond (two) and to the π -system (four). On the other hand, each of the oxygen atoms contributes one electron to the σ -bond and another to the π -system. The molecular orbitals reveal that in the "bis-oxo" state, all of the unpaired spin density is in the π -system. Consequently, two oxygen σ lone-pair electrons must be driven together along the reaction coordinate during intramolecular direct-coupling O–O bond formation. Thus, the repulsive interaction between oxygen σ lone pairs disfavors intramolecular O–O bond formation.

The activation free energies for the WNA pathway (28.8 kcal/mol) and intramolecular direct-coupling O–O bond formation (56.8 kcal/mol) presented in Figure 12.3 differ by approximately 30 kcal/mol, firmly establishing the WNA as the water oxidation mechanism for cis-[Ru^{II}(bpy)$_2$(H$_2$O)$_2$]$^{2+}$. Gas-phase single-point calculations at the CASSCF/CASPT2 level agree with DFT results indicating a clear preference for the WNA mechanism with respect to the direct-coupling O–O bond formation [25]. The product, cis-[Ru^{III}(bpy)$_2$(OOH)(OH$_2$)]$^{2+}$, then undergoes a PCET at 0.65 V to generate cis-[Ru^{IV}(bpy)$_2$(OO)(OH$_2$)]$^{2+}$, which evolves O$_2$ with a ΔG‡ of 15.4 kcal/mol, regenerating the catalyst (Figure 12.3).

Another mononuclear ruthenium catalyst synthesized by Llobet and coworkers is the [Ru^{II}(damp)(bpy)(H$_2$O)]$^{2+}$ catalyst, which features a single Ru–H$_2$O moiety and thus cannot proceed through an intramolecular direct-coupling O–O bond formation mechanism

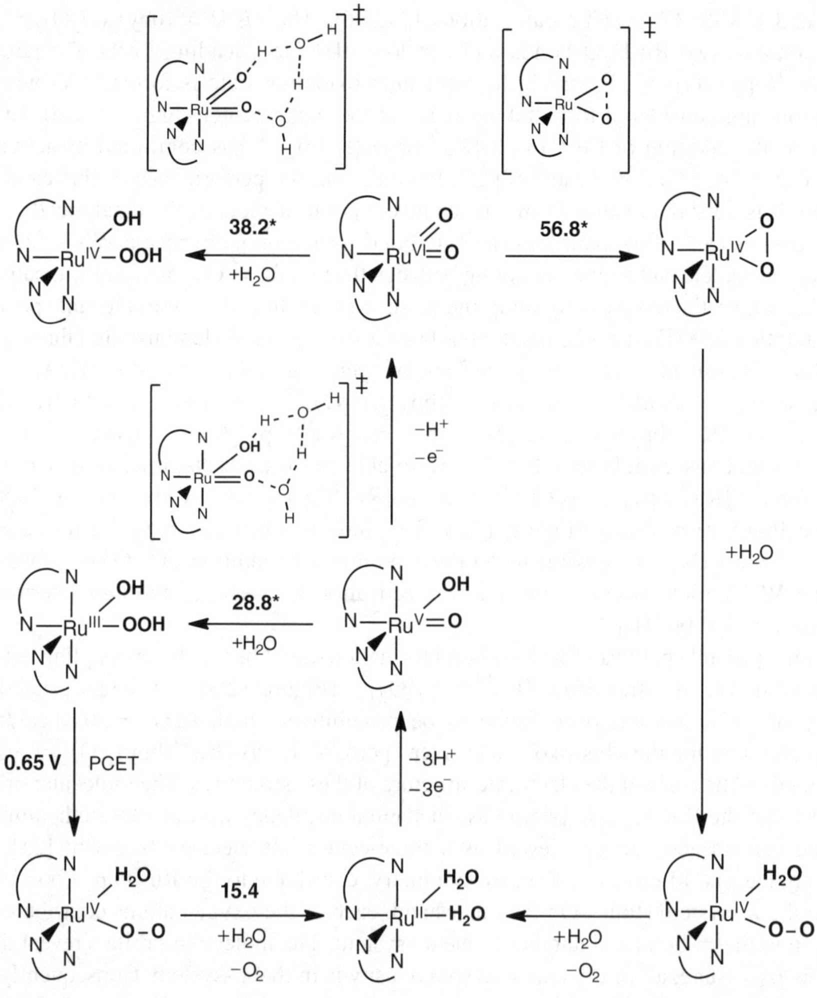

Figure 12.3 *Free energies of activation (kcal/mol) in aqueous solution for WNA (left) and intramolecular direct coupling O–O bond formation (right) mechanisms. Electrochemical potentials in volts. Free energies of activation are indicated with asterisks. Formal oxidation states are indicated for easy electron-counting purposes*

(Figure 12.2b) [26]. A possible intermolecular reaction between two catalyst molecules will be entropically unfavorable and is not compatible with experimental kinetics. We began the mechanistic investigation with the Ru^{II}–H_2O species and examined two consecutive steps that, combined, result in the formation of $Ru^{IV} = O$ (Figure 12.4, bottom). Three different possibilities were considered for each step: a deprotonation and subsequent oxidation; an oxidation followed by a deprotonation; and PCET, where both the oxidation and the deprotonation occur simultaneously. The most favorable process for Ru^{II} to Ru^{III} evolution at pH 1.0 is an oxidation of the metal center (0.46 V) and a deprotonation *a posteriori*

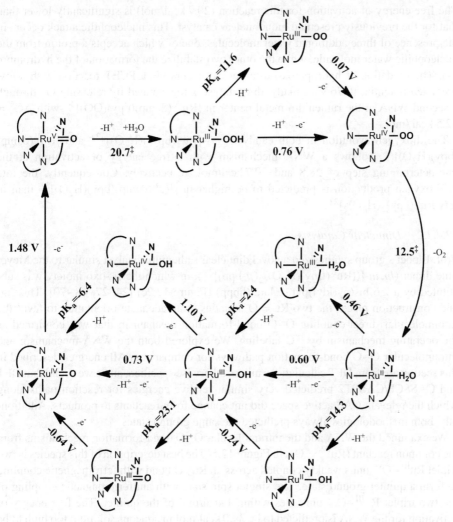

Figure 12.4 *Catalytic cycle for* $[Ru^{II}(damp)(bpy)(H_2O)]^{2+}$. *Free energies (kcal/mol) in aqueous solution for WNA mechanisms. Electrochemical potentials in volts, pK_a values in log units*

($pK_a = 2.4$). On the other hand, for an Ru^{III} to Ru^{IV} oxidation step at the same pH, PCET is the most favorable process, with a potential of 0.73 V.

Initially we considered $Ru^{IV} = O$ as a possible substrate for nucleophilic attack by a water molecule, but our attempts to locate a transition state for such a process were unsuccessful, as the approach of a water molecule to the oxo group was found to be repulsive even in the presence of additional water molecules to assist with proton removal. Our experience suggests that in general $Ru^{IV} = O$ fragments are not susceptible to WNA and instead higher oxidation states are required. In consequence, an additional oxidation (1.48 V) is required to reach the catalytically active $Ru^V = O$ species on which WNA occurs (Figure 12.4, top).

The free energy of activation for this reaction (20.7 kcal/mol) is significantly lower than that for the previously presented mononuclear catalyst. This nucleophilic attack occurs in the presence of three additional water molecules, one of which accepts a proton from the nucleophilic water molecule, while the other two stabilize the formation of the hydronium ion. The oxidation of hydroperoxo to a peroxo occurs via a PCET reaction with a low predicted potential (0.76 V). Finally, the catalyst is regenerated by releasing O_2 through a second WNA at the ruthenium metal center in $[Ru^{IV}(damp)(bpy)(OO)]^{3+}$ with $\Delta G^{\ddagger} = 12.5$ kcal/mol.

The dioxygen evolution in both catalysts, *cis*-$[Ru^{II}(bpy)_2(H_2O)_2]^{2+}$ and $[Ru^{II}(damp)(bpy)(H_2O)]^{2+}$, follows a WNA mechanism with a free energy of activation at the rate-determining step of 28.8 and 20.7 kcal/mol, respectively. Consequently, the rate of dioxygen production is predicted to be higher in $[Ru^{II}(damp)(bpy)(H_2O)]^{2+}$ than in $[cis$-$Ru^{II}(bpy)_2(H_2O)_2]^{2+}$.

12.4.1.2   Dinuclear Catalysts

Prof. Llobet's group synthesized a novel dinuclear ruthenium catalyst similar to the Meyer blue dimer (*in, in*-$[((Ru^{II}(trpy)(H_2O))_2(\mu$-bpp)$]^{3+}$), in which the μ -oxo moiety was substituted by a 2,6-bis(pyridyl)pyrazolate (hbpp) (Figure 12.2c) [24, 27, 68–70]. This ligand construction places the two Ru = O moieties in an adequate orientation to favor the intramolecular direct-coupling O–O bond-formation mechanism that was confirmed as the operating mechanism by ^{18}O labeling. We explored both the WNA-mechanism and intramolecular O–O bond-formation pathways for comparison. All energies described in this section are from DFT calculations unless otherwise specified, since we found that DFT and CASSCF/CASPT2 predicted very similar relative energies for reaction paths along which the relevant CAS active space did not change from reactants to products. Additionally, both methodologies always predicted the same ground states.

We examined the WNA and the intramolecular O–O bond-formation mechanisms from the common reactant $(Ru^{IV} = O)_2$ in Figure 12.5. The best description of this species is two triplet Ru^{III}–$O^{\cdot-}$ units weakly coupled across an Ru–O bond with ferromagnetic coupling to form a quintet ground state. The singlet spin state with antiferromagnetic coupling of the two triplet Ru^{III}–$O^{\cdot-}$ units lies within 1 kcal/mol of the quintet. The free energy of activation for the WNA is predicted to be 39.0 kcal/mol in aqueous solution; too high to be consistent with experiment. Additionally, this pathway is not consistent with the available ^{18}O isotope-labeling experiments, which show that the generated O_2 originates from the oxygen atoms of Ru–O units.

The intramolecular direct-coupling O–O bond formation between the two terminal oxo moieties has a $\Delta G^{\ddagger}$ of just 15.5 kcal/mol in aqueous solution (23.5 kcal/mol lower than in the WNA mechanism). The transition state involved has an O–O distance of 1.71 Å, which evolves to an η^2 peroxo intermediate with O–O distance of 1.39 Å (a singlet ground state). Next, we turn our attention to the steps leading to O_2 evolution from the peroxo intermediate. We found that the sequential displacement of both ends of the initial peroxo moiety from the two metal centers by individual water molecules to generate first an η^1 superoxo intermediate and then released dioxygen required high free energies of activation ($\Delta G^{\ddagger} = 28.6$ and 40.6 kcal/mol, respectively). As observed in the calculations of the mononuclear catalysts, however, explicit water molecules in the first solvation shell can

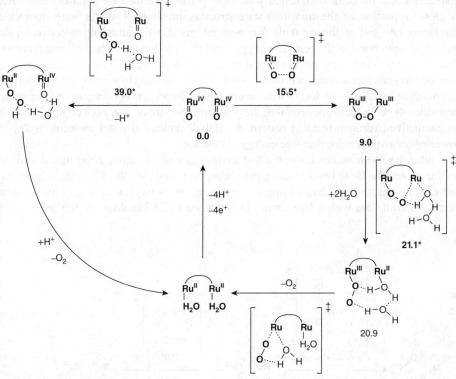

Figure 12.5 *Free energies (kcal/mol) in aqueous solution for WNA (left cycle) and intramolecular bond-formation (right cycle) mechanisms starting from a ($Ru^{IV} = O$)$_2$ species. Free energies of activation are indicated with asterisks. Formal oxidation states are indicated for easy electron-counting purposes*

play an important role in stabilizing intermediates or transition-state structures. When an additional water molecule was introduced in the transition-state structures, it was possible to optimize hydrogen-bonding patterns to the dioxo fragment that led to lower activation free energies (Figure 12.5). An additional water molecule effectively allowed the catalyst to enjoy special benefits associated with microsolvation not accounted for in the continuum solvation model [24, 71–73].

As a next step, we investigated the water oxidation pathways involving $O = Ru^{IV}\text{-}Ru^{V} = O$ obtained via one-electron oxidization of $O = Ru^{IV}\text{-}Ru^{IV} = O$. The predicted redox potential for this step was 1.28 V versus normal hydrogen electrode (NHE), so this oxidation should be spontaneous in the presence of ceric ammonium nitrate (CAN) as the sacrificial oxidant. Like in the less oxidized case, we found intramolecular O–O bond formation to be favored significantly over the WNA pathway ($\Delta G^{\ddagger} = 11.6$ and 27.4 kcal/mol, respectively; Figure 12.6). More importantly, the displacements of peroxo by a water molecule to yield the superoxo intermediate and subsequent O_2 evolution step were found to proceed with free energies of activation of 14.7 and 23.3 kcal/mol, respectively (Figure 12.6). Modeling of the transition-state structure for the water nucleophilic displacement at the ruthenium center with an η^2 peroxo intermediate requires some care. A close inspection of the transition-state structure involving a single water molecule displacing one end of the peroxide fragment reveals that the attacking water molecule has a hydrogen bond to the peroxide oxygen that remains bound to the ruthenium center. This hydrogen bond does stabilize this structure, but is not optimal as it is the departing peroxide oxygen that acquires a proton from the water, evolving to a hydroxo/hydroperoxo intermediate. Attempts to locate a structure with a hydrogen bonding to the departing peroxide oxygen were unsuccessful as the geometric constraints were not compatible with an alternative hydrogen-bonding pattern. A catalytic additional water molecule is key to lowering the overall activation free energy of this step.

In summary, we have demonstrated that the presence of a bridging hbbp ligand leads to an intramolecular O−O bond-forming path being preferred over the WNA path. Based on our calculations, the rate-determining step is the release of oxygen from the mixed-valent peroxo intermediate with a free energy of activation of 23.3 kcal/mol in the presence of

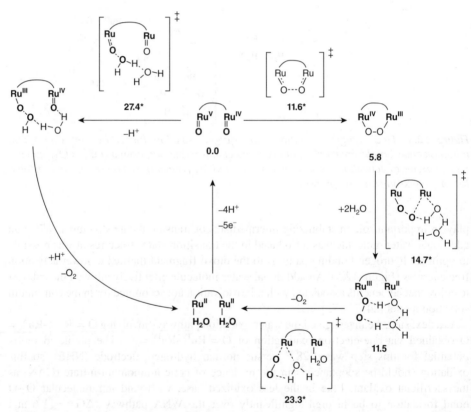

Figure 12.6 *Free energies (kcal/mol) in aqueous solution for WNA (left cycle) and intramolecular bond-formation (right cycle) mechanisms starting from a (Ru^{IV} = O)(Ru^{V} = O) species. Free energies of activation are indicated with asterisks. Formal oxidation states are indicated for easy electron-counting purposes*

abundant amounts of a sacrificial oxidant. This is in contrast to the mononuclear ruthenium catalysts, where WNA was the rate-determining step in each case.

12.4.2 Cobalt-Based Catalysts

Considerable effort has recently been focused on the design of next-generation WOCs based on inexpensive, earth-abundant metals such as cobalt. In this regard, Dogutan and coworkers have reported the synthesis of a β-octafluoro hangman corrole, $CoH^{\beta F}CX-CO_2H$, which shows electrocatalytic water oxidation activity (Figure 12.2d) [29, 30]. This species was designed to place a hanging carboxylic acid group over the face of the corrole macrocycle, positioned to accept or donate a proton during PCET steps for which oxidation or reduction takes place on the macrocycle and/or on the coordinated metal center. This intrinsic ability to act as both a proton and an electron sink/source helps this catalyst facilitate both the oxygen-reduction reaction (ORR) and the oxygen-evolving reaction (OER) given the four proton and electron transfers along their respective reaction coordinates.

We studied the generation of O_2 from the cobalt(III)-centered β-octafluoro hangman corrole catalyst. The speciation of the initial aquo-metal moiety in the $CoH^{\beta F}CX-CO_2H$ catalyst may be characterized through consideration of the coupled equilibria presented in Figure 12.7. The goal made manifest in this figure is the identification of an energetically feasible path from the initial complex, at upper left, to one or more oxidized intermediates that will react with a water molecule to form an O–O bond. In addition, the resulting

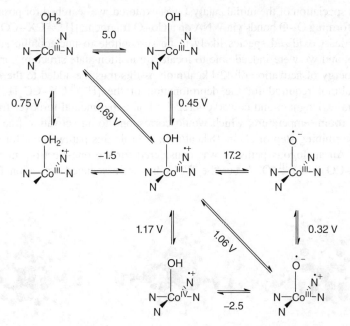

Figure 12.7 *Coupled equilibria connecting more highly protonated and reduced species (top left) to less highly protonated and oxidized species (bottom right), starting from $CoH^{\beta F}CX-CO_2H$ species. Free energies (kcal/mol) in aqueous solution for acid–base processes and electrochemical potentials in volts*

product of O–O bond formation must ultimately be able to release O_2 and regenerate the initial catalyst via an energetically feasible pathway. In Figure 12.7, horizontal steps represent acid–base equilibria and their free energies can be expressed as pK_a values or as free energies given a particular pH (here taken to be neutral, the value assessed experimentally). Vertical steps represent one-electron oxidation/reduction processes and their free energies can be expressed as standard reduction potentials. Diagonal steps are PCET reactions. The relevance of a first solvation-shell water molecule was considered through extensive conformational searches as a hydrogen-bonding network could stabilize the conjugate base and acid forms of the $–CO_2H$ moiety. In this chapter, we report only the free energy of the most stable conformer in each case; we found little evidence for significant conformational influence on our results.

The initial catalyst $[H^{\beta F}CX–CO_2H–Co^{III}\text{-}H_2O]$ is found in the top-left corner of Figure 12.7 and can evolve to $[H^{+\cdot\beta F}CX–CO_2H–Co^{III}\text{-}OH]$ and $[H^{+\cdot\beta F}CX–CO_2H–Co^{III}\text{-}O^{\cdot-}]$ through two PCET processes. We determined the ground state of these species to be triplet, doublet, and triplet, respectively. The–spin density plots of these three species are presented in Figure 12.8, revealing non-innocent behavior of the $H^{\beta F}CX–CO_2H$ ligand system. Thus, the PCET step starting from initial $[H^{\beta F}CX–CO_2H–Co^{III}–H_2O]$ yields $[H^{+\cdot\beta F}CX–CO_2H–Co^{III}–OH]$ species instead of $[H^{\beta F}CX–CO_2H–Co^{IV}–OH]$ ones, with oxidation taking place in the corrole ligand rather than at the metal center. Additionally, the spin distribution in the final intermediate at the bottom right of Figure 12.7 is not consistent with a $Co^{IV} = O^{2-}$ formulation, but rather $Co^{III}–O^{\cdot-}$ and a non-innocent corrole.

Once the speciation of the initial catalyst was explored, we searched for possible transition states forming O–O bonds via WNA on a Co–O fragment. $[H^{+\cdot\beta F}CX–CO_2H–Co^{III}–O^-]$ is the least oxidized species likely to be susceptible to nucleophilic attack at the oxyl group, and we were indeed able to locate a transition-state structure corresponding to a free energy of activation of 42.1 kcal/mol; as this must be added to the free energy of 17.2 kcal/mol required for the deprotonation of the $[H^{+\cdot\beta F}CX–CO_2H–Co^{III}–OH]$ species, and as Dogutan and coworkers reported an experimental turnover frequency of 0.8 s^{-1} at room temperature, which would correspond to an activation free energy for the rate-determining step of about 18 kcal/mol, we rule this pathway out for O–O bond formation. An analogous pathway was considered for the one-electron, more-oxidized $[H^{+\cdot\beta F}CX–CO_2H–Co^{III}–O^{\cdot-}]$, but the WNA has an associated activation free energy

Figure 12.8 *Spin-density plots of* $[H^{\beta F}CX–CO_2H–Co^{III}–H_2O]$ (³A) *(left),* $[H^{+\cdot\beta F}CX–CO_2H–Co^{III}–OH]$ (²A) *(center), and* $[H^{+\cdot\beta F}CX–CO_2H–Co^{III}–O^{\cdot-}]$ (³A) *(right). Orange, cobalt; gray, carbon; blue, nitrogen; red, oxygen; teal, fluoride; white, hydrogen. See color plate*

of 36.3 kcal/mol, which is still too high to be consistent with the experimental kinetics (Figure 12.9). We then turned our attention to a possible deprotonation of the pendant carboxylic acid group to yield $[H^{+\cdot\beta F}CX-CO_2-Co^{III}-O^{\cdot-}]^-$, since it has a predicted free energy change of only 7.3 kcal/mol. The subsequent WNA on this species has an associated free energy of activation of 15.2 kcal/mol, leading to a combined free energy of 22.5 kcal/mol, in reasonable agreement with the experimentally determined value. In consequence, the most plausible mechanism for O–O bond formation involves oxidation and deprotonation of the initial complex to reach $[H^{+\cdot\beta F}CX-CO_2-Co^{III}-O^{\cdot-}]^-$, followed by a WNA to generate the hydroperoxide species (Figure 12.9). One should note that the first solvent-shell water acts as a catalyst to shuttle a proton from the attacking water to the carboxylate group.

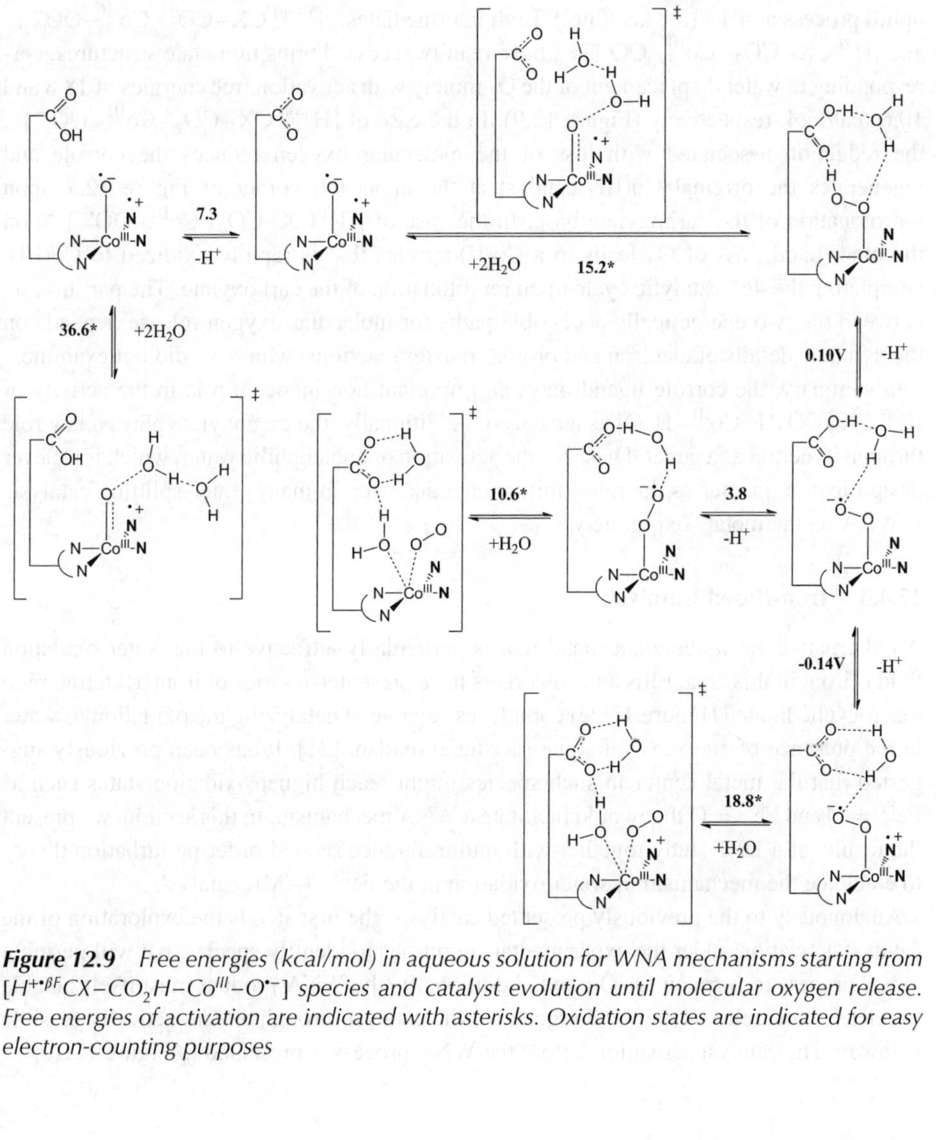

Figure 12.9 *Free energies (kcal/mol) in aqueous solution for WNA mechanisms starting from $[H^{+\cdot\beta F}CX-CO_2H-Co^{III}-O^{\cdot-}]$ species and catalyst evolution until molecular oxygen release. Free energies of activation are indicated with asterisks. Oxidation states are indicated for easy electron-counting purposes*

An interesting feature of the water oxidation mechanism in this catalyst is that it requires only two oxidation steps before the WNA step. On the other hand, many other transition metal-based homogeneous catalysts involve three to four oxidation steps before enough oxidizing power has been expended to render the WNA process feasible [7, 26, 71, 73–75]. The origin of this difference could be attributed to the intrinsic electrophilicity of the $[H^{+ \cdot \beta F}CX-CO_2-Co^{III}-O^{\cdot -}]^-$ species or to the proton transfer during the WNA, which occurs to the carboxylate group as a local general base rather than to a first-shell water molecule.

In order to regenerate the catalyst, the oxygen must subsequently be released. We determined that two consecutive PCET steps at very low potentials (0.10 and −0.14 V) generate $[H^{\beta F}CX-CO_2H-Co^{III}-OO^{\cdot -}]^-$ and $[H^{+ \cdot \beta F}CX-CO_2-Co^{III}-OO^{\cdot -}]^-$. The $[H^{\beta F}CX-CO_2H-Co^{III}-OO^{\cdot -}]^-$ species can also simply deprotonate through a slightly uphill process at pH 7 (3.8 kcal/mol). Both intermediates, $[H^{+ \cdot \beta F}CX-CO_2-Co^{III}-OO^{\cdot -}]^-$ and $[H^{\beta F}CX-CO_2-Co^{III}-OO^{\cdot -}]^{2-}$, have readily accessed transition-state structures, corresponding to water displacement of the O_2 moiety with activation free energies of 18.8 and 10.6 kcal/mol, respectively (Figure 12.9). In the case of $[H^{+ \cdot \beta F}CX-CO_2-Co^{III}-OO^{\cdot -}]^-$, the reduction associated with loss of the molecular oxygen reduces the corrole and regenerates the original Co(III) catalyst at the upper-left corner of Figure 12.7 upon reprotonation of the carboxylate base. In the case of $[H^{\beta F}CX-CO_2-Co^{III}-OO^{\cdot -}]^{2-}$, on the other hand, loss of O_2 leads to a Co(II) species that is rapidly oxidized to Co(III), completing the 4e$^-$ catalytic cycle upon reprotonation of the carboxylate. The partitioning between the two energetically accessible paths for molecular oxygen release depends on the dynamic details of electron and proton transfer reactions, which we did not examine.

In summary, the corrole ligand plays an important non-innocent role in the activity of $[H^{\beta F}CX-CO_2H-Co^{III}-H_2O]$ as a catalyst. Additionally, the carboxylate plays a key role through its action as a general base for the activation of nucleophilic water, which is a clever design feature insofar as the rate-limiting chemical step in many water-splitting catalysts is WNA on the metal–oxo moiety.

12.4.3 Iron-Based Catalysts

An alternative earth-abundant metal that is particularly attractive to the water oxidation field is iron. In this area, Ellis and coworkers have presented a series of iron(III) tetraamido macrocyclic ligated (Figure 12.2e) complexes capable of catalyzing the oxidation of water in the presence of an excess of some sacrificial oxidant [31]. It has been previously suggested that the metal center in such species might reach higher oxidation states such as $Fe^{IV} = O$ and $Fe^{V} = O$ that would facilitate a WNA mechanism. In this section, we present the results of a DFT study together with multireference second-order perturbation theory to elucidate the mechanism of water oxidation in the Fe^{III}-TAML catalyst.

Analogously to the previously presented catalysts, the first step is the exploration of the catalyst activation under the experimental conditions to identify species that will permit a WNA process to form an O–O bond. Additionally, the WNA product must itself be able to release molecular oxygen and regenerate the catalyst through an energetically feasible pathway. The catalyst activation before the WNA process is presented in Figure 12.10.

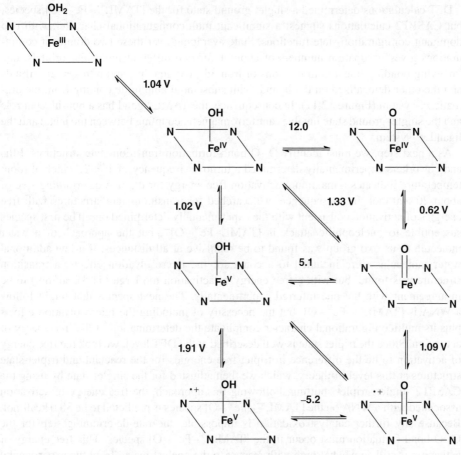

Figure 12.10 *Coupled equilibria connecting more highly protonated and reduced species (to left and top, respectively) to less highly protonated and oxidized species (to right and bottom, respectively), starting from the inital Fe^{III} – TAML catalyst. Free energies (kcal/mol) in aqueous solution for WNA mechanisms, electrochemical potentials in volts, pK$_a$ values in log units*

We first consider two PCETs that start from the initial [TAML–Fe^{III}–OH$_2$]$^-$ species and lead ultimately to [TAML–Fe^{V}–O]$^-$. These two processes occur at relatively low potentials (1.04 and 1.33 V, respectively) and are expected to be spontaneous under the experimental conditions, where CAN, with a standard reduction potential of 1.7 V, is present as the sacrificial oxidant. The electronic structures of the species formed during these steps, [TAML–Fe^{III}–OH$_2$]$^-$, [TAML–Fe^{IV}–OH]$^-$, and [TAML–Fe^{V}–O]$^-$, exhibit quartet, triplet, and doublet ground states, respectively. The oxidation of [TAML–Fe^{V}–O]$^-$ is predicted to be spontaneous in the presence of CAN, with a calculated potential of 1.09 V, generating a formal [TAML–Fe^{VI}–O] species; however, analysis of the electronic structure of the oxidized species makes clear that oxidation occurs at the TAML ligand instead of the metal center, leading to [TAML$^{+\bullet}$–Fe^{V}–O] being chosen as a more accurate formulation.

DFT calculations determined a singlet ground state for the $[\text{TAML}^{+\bullet}-\text{Fe}^{V}-\text{O}]$ species, but CASPT2 calculations suggest a significant multiconfigurational character, with two dominant configuration-state functions. State averaging over these two dominant configurations gives occupation numbers of about 1 for two different molecular orbitals, one involving bonding linear combinations of iron 3d-, oxygen 2p-, and nitrogen 2p-orbitals and the other delocalized on the ligand, with substantial amplitude coming from the aromatic π-system (Figure 12.11). In consequence, the TAML ligand has a non-innocent role and the singlet ground state involves antiferromagnetic coupling between the metal and the ligand π-system.

As a next step, we must identify O—O bond-formation transition-state structures. Ellis and coworkers experimentally determined a turnover frequency of 1 s^{-1}, which at room temperature indicates a maximum activation free energy for the rate-determining step of about 20 kcal/mol. In consequence, we searched for transition-state structures with free energies of activation consistent with the experimentally determined one. The first species susceptible to nucleophilic attack is $[\text{TAML}-\text{Fe}^{V}-\text{O}]^{-}$, but the approach of a water molecule to its oxo group was found to be repulsive at all distances. If a few additional water molecules were included to account for the microsolvation effects, a transition structure was found, but with a free energy of activation too large (41.5 kcal/mol) to be in agreement with the one inferred experimentally. The next species that might follow a WNA is $[\text{TAML}^{+\bullet}-\text{Fe}^{V}-\text{O}]$, but the necessity of including the microsolvation effects plus its multiconfigurational character complicate the determination of the free energy of activation. Since the triplet state is well described at the DFT level, we took the free energy of activation to be the difference in triplet free energies for the reactant and triplet-state structures at this level of theory, which we then adjusted for the singlet state by using the CASPT2 singlet–triplet splitting. Following this approach, the free energy of activation associated with a WNA on the $[\text{TAML}^{+\bullet}-\text{Fe}^{V}-\text{O}]$ species is predicted to be 30.0 kcal/mol. Because any further catalyst oxidation is impossible, the rate-determining step for the O—O bond formation must occur in the $[\text{TAML}^{+\bullet}-\text{Fe}^{V}-\text{O}]$ species. This free energy of activation is still somewhat high with respect to the one estimated from the experimental turnover frequency, but some additional discussion should be undertaken here.

In our experience, the M06-L DFT functional has performed very well in the modeling of water oxidation reactions when compared with experimental results or high-level

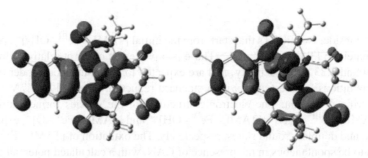

Figure 12.11 *SOMOs for the $[TAML^{+\bullet}-Fe^{V}-O]$ triplet state from the DFT triplet state that are analogous to the CASSCF orbitals. White, hydrogen; gray, carbon; blue, nitrogen; red, oxygen; teal, fluorine; green, chlorine; mauve, iron. See color plate*

multireference calculations [24, 25, 76–79]. However, for this particular reaction we observe enormous sensitivity to the choice of density functional. With triple-ξ plus two polarization functions basis sets, the electronic energy difference between reactants and the corresponding triplet-state structure on the triplet surface is 9.0 kcal/mol at the M06-L level, but it ranges from as low as 6.8 kcal/mol with the M06 functional to as high as 35.3 kcal/mol with the OPBE one. In all cases, the $< S^2 >$ operator values did not show any sign of spin contamination associated with a multideterminantal character and the quintet states always lied far above the triplet ones. Analysis of the spin distribution on this species revealed that the reactant is better described as a separated biradical, with one spin in the ligand π-system and one in the Fe $=$ O fragment, while in the triplet-state structure both spins are localized more on the Fe–OOH moiety. Consequently, it could be that the degree of localization is the origin of the functional sensitivity. Of course, another important consideration is that many of the functionals we surveyed did *not* include medium-range correlation energies very effectively (short-range dispersion, if you will). As such, the microsolvated triplet-state structure will appear anomalously high in energy compared to the continuum-only solvated reactant.

Several alternative pathways were considered, such as a possible ligand protonation, the addition of an aquo-, hydroxo-, or oxo-ligand at the iron center, counterion (NO_3^-) assistance in the proton translocation during the WNA process, and the influence of tunneling on the apparent rate constant [80], but none of these options led to lower-energy routes to O–O bond formation. We also studied the influence of the number of water molecules participating in the stabilization of the WNA transition-state structure and found no relevant change in the calculated free energy of activation. One possibility is that quantitative agreement with experiment might require a more complete sampling of all relevant microsolvated structures, but this task is obviously not trivial and it fell outside of the scope of our study. Hence, we must embrace our somewhat less than satisfactory agreement with the experimental results and assume that the WNA occurs in [TAML$^{+\bullet}$–Fe^V–O] species, leading to [TAML–Fe^{IV}–OOH]$^-$, which evolves towards [TAML–Fe^V–OO]$^-$ through a PCET process that occurs at only 0.38 V. The release of molecular oxygen occurs when a solvent molecule displaces the O_2 coordinated to the iron center. This process has an associated free energy of activation of 8.8 kcal/mol.

In summary, we have explored a wide range of mechanistic possibilities, since the calculated rate-determining step free energy (WNA in {TAML$^{+\bullet}$–Fe^V–O} species) is predicted to be several kcal/mol higher than the one determined experimentally. We cannot exclude the possibility of alternative mechanisms *not* explored in our study (such as the iron-based catalyst degrading to an unexpected nanoparticulate solid that acts as a heterogeneous catalyst for water splitting). However, we consider it more likely that the remaining disagreement here between theory and experiment is not associated with an incorrect mechanism, but rather is an electronic structure challenge posed by the most highly oxidized iron species.

12.5 Conclusion

We have studied and compared the water oxidation mechanisms of various transition metal-based WOCs. All of the mononuclear catalysts studied (*cis*-[Ru^{II}(bpy)$_2$(H$_2$O)$_2$]$^{2+}$,

[Ru^{II}(damp)(bpy)(H_2O)]$^{2+}$, $CoH^{\beta F}$ $-$ CO_2H, and Fe^{III}-TAML) follow a WNA mechanism, while the dinuclear catalyst (*in, in*-[(Ru^{II}(trpy)(H_2O))$_2$(μ-bpp)]$^{3+}$) follows an intramolecular O$-$O bond-formation mechanism. Furthermore, when non-pyridine-based ligands are uses, such as TAML or a hangman corrole, we have shown that the non-innocent role of the ligand plays a part in the water oxidation mechanism.

References

[1] Balzani, V., Credi, A., Venturi, M. (2008) ChemSusChem, 1: 26–58.

[2] Benniston, A. C., Harriman, A. (2008) Mat. Today, 11: 26–34.

[3] Betley, T. A., Surendranath, Y., Childress, M. V., Alliger, G. E., Fu, R., Cummins, C. C., Nocera, D. G. (2008) Phil. Trans. Royal Soc. B – Biol. Sci., 363: 1293–1303.

[4] Dau, H., Limberg, C., Reier, T., Risch, M., Roggan, S., Strasser, P. (2010) ChemCatChem, 2: 724–761.

[5] Dempsey, J. L., Esswein, A. J., Manke, D. R., Rosenthal, J., Soper, J. D., Nocera, D. G. (2005) Inorg. Chem., 44: 6879–6892.

[6] Eisenberg, R., Gray, H. B. (2008) Inorg. Chem., 47: 1697–1699.

[7] Liu, F., Concepcion, J. J., Jurss, J. W., Cardolaccia, T., Templeton, J. L., Meyer, T. J. (2008) Inorg. Chem., 47: 1727–1752.

[8] Romain, S., Vigara, L., Llobet, A. (2009) Acc. Chem. Res., 42: 1944–1953.

[9] Saveant, J.-M. (2008) Chem. Rev., 108: 2348–2378.

[10] Tinker, L. L., McDaniel, N. D., Bernhard, S. J. (2009) Mat. Chem., 19: 3328–3337.

[11] Yamazaki, H., Shouji, A., Kajita, M., Yagi, M. (2010) Coord. Chem. Rev., 254: 2483–2491.

[12] Huynh, M. H. V., Meyer, T. (2007) J. Chem. Rev., 107: 5004–5064.

[13] Guskov, A., Kern, J., Gabdulkhakov, A., Broser, M., Zouni, A., Saenger, W. (2009) Nat. Struct. Mol. Biol., 16: 334–342.

[14] Umena, Y., Kawakami, K., Shen, J. R., Kamiya, N. (2011) Nature, 473: 55–65.

[15] Ferreira, K. N., Iverson, T. M., Maghlaoui, K., Barber, J., Iwata, S. (2004) Science, 303: 1831–1838.

[16] Kok, B., Forbush, B., Mcgloin, M. (1970) Photochem. Photobiol., 11: 457–475.

[17] Forbush, B., Kok, B., Mcgloin, M. P. (1971) Photochem. Photobiol., 14: 307–321.

[18] Naruta, Y., Sasayama, M., Sasaki, T. (1994) Ang. Chem. Int. Ed., 33: 1839–1843.

[19] Mullins, C. S., Pecoraro, V. L. (2008) Coord. Chem. Rev., 252: 416–443.

[20] Gersten, S. W., Samuels, G. J., Meyer, T. J. (1982) J. Am. Chem. Soc., 104: 4029–4030.

[21] Bianco, R., Hay, P. J., Hynes, J. T. (2011) J. Phys. Chem. A, 115: 8003–8016.

[22] Concepcion, J. J., Jurss, J. W., Templeton, J. L., Meyer, T. J. (2008) Proc. Nat. Acad. Sci. U. S. A., 105: 17632–17635.

[23] Ozkanlar, A., Clark, A. E. (2012) J. Chem. Phys., 136: 204104.

[24] Bozoglian, F., Romain, S., Ertem, M. Z., Todorova, T. K., Sens, C., Mola, J., Rodriguez, M., Romero, I., Benet-Buchholz, J., Fontrodona, X., Cramer, C. J., Gagliardi, L., Llobet, A. (2009) J. Am. Chem. Soc., 131: 15176–15187.

[25] Sala, X., Ertem, M. Z., Vigara, L., Todorova, T. K., Chen, W., Rocha, R. C., Aquilante, F., Cramer, C. J., Gagliardi, L., Llobet, A. (2010) Ang. Chem. Int. Ed., 49: 7745–7747.

[26] Vigara, L., Ertem, M. Z., Planas, N., Bozoglian, F., Leidel, N., Dau, H., Haumann, M., Gagliardi, L., Cramer, C. J., Llobet, A. (2012) Chem. Sci., 3: 2576–2586.

[27] Romain, S., Bozoglian, F., Sala, X., Llobet, A. (2009) J. Am. Chem. Soc., 131: 2768 ff.

[28] Romain, S., Rich, J., Sens, C., Stoll, T., Benet-Buchholz, J., Llobet, A., Rodriguez, M., Romero, I., Clerac, R., Mathoniere, C., Duboc, C., Deronzier, A., Collomb, M.-N. (2011) Inorg. Chem., 50: 8427–8436.

[29] Dogutan, D. K., McGuire, R. Jr., Nocera, D. G. (2011) J. Am. Chem. Soc., 133: 9178–9180.

[30] Dogutan, D. K., Stoian, S. A., McGuire, R., Jr., Schwalbe, M., Teets, T. S., Nocera, D. G. (2011) J. Am. Chem. Soc., 133: 131–140.

[31] Ellis, W. C., McDaniel, N. D., Bernhard, S., Collins, T. J. (2010) J. Am. Chem. Soc., 132: 10990–10991.

[32] Zhao, Y., Truhlar, D. G. (2006) J. Chem. Phys., 125: 194101.

[33] Zhao, Y., Truhlar, D. G. (2008) Acc. Chem. Res., 41: 157–167.

[34] Zhao, Y., Truhlar, D. G. (2008) Theor. Chem. Acc., 120: 215–241.

[35] Zhao, Y., Truhlar, D. G. (2011) Chem. Phys. Lett., 502: 1–13.

[36] Frisch, M. J., Trucks, G. W., Schlegel, H. B., Scuseria, G. E., Robb, M. A., Cheeseman, J. R., Scalmani, G., Barone, V., Mennucci, B., Petersson, G. A., Nakatsuji, H., Caricato, M., Li, X., Hratchian, H. P., Izmaylov, A. F., Bloino, J., Zheng, G., Sonnenberg, J. L., Hada, M., Ehara, M., Toyota, K., Fukuda, R., Hasegawa, J., Ishida, M., Nakajima, T., Honda, Y., Kitao, O., Nakai, H., Vreven, T., Montgomery, J. A. Jr., Peralta, J. E., Ogliaro, F., Bearpark, M., Heyd, J. J., Brothers, E., Kudin, K. N., Staroverov, V. N., Kobayashi, R., Normand, J., Raghavachari, K., Rendell, A., Burant, J. C., Iyengar, S. S., Tomasi, J., Cossi, M., Rega, N., Millam, J. M., Klene, M., Knox, J. E., Cross, J. B., Bakken, V., Adamo, C., Jaramillo, J., Gomperts, R., Stratmann, R. E., Yazyev, O., Austin, A. J., Cammi, R., Pomelli, C., Ochterski, J. W., Martin, R. L., Morokuma, K., Zakrzewski, V. G., Voth, G. A., Salvador, P., Dannenberg, J. J., Dapprich, S., Daniels, A. D., Farkas, Ö., Foresman, J. B., Ortiz, J. V., Cioslowski, J., Fox, D. J. (2009) Gaussian 09, Revision A.1. Gaussian, Inc., Wallingford, CT, USA.

[37] Dolg, M., Wedig, U., Stoll, H., Preuss, H. (1987) J. Chem. Phys., 86: 866–872.

[38] Martin, J. M. L., Sundermann, A. (2001) J. Chem. Phys., 114: 3408–3420.

[39] Hehre, W. J., Radom, L., Schleyer, P. V. R., Pople, J. A. (1986) Ab Initio Molecular Orbital Theory. John Wiley & Sons, Ltd, New York, NY, USA.

[40] Cramer, C. J. (2004) Essentials of Computational Chemistry: Theories and Models, 2nd edition. John Wiley & Sons, Ltd, Chichester, UK.

[41] Ribeiro, R. F., Marenich, A. V., Cramer, C. J., Truhlar, D. G. (2011) J. Phys. Chem. B, 115: 14556–14562.

[42] Cramer, C. J., Truhlar, D. G. (2008) Acc. Chem. Res., 41: 760–768.

[43] Marenich, A. V., Cramer, C. J., Truhlar, D. G. (2009) J. Phys. Chem. B, 113: 6378–6396.

[44] Andersson, K., Malmqvist, P. A., Roos, B. O. (1992) J. Chem. Phys., 96: 1218–1226.

[45] Roos, B. O., Taylor, P. R., Siegbahn, P. E. M. (1980) Chem. Phys., 48: 157–173.

[46] Aquilante, F., De Vico, L., Ferre, N., Ghigo, G., Malmqvist, P.-A., Neogrady, P., Pedersen, T. B., Pitonak, M., Reiher, M., Roos, B. O., Serrano-Andres, L., Urban, M., Veryazov, V., Lindh, R. (2010) J. Comp. Chem., 31: 224–247.

[47] Hess, B. A. (1986)Phys. Rev. A, 33: 3742–3748.

[48] Roos, B. O., Lindh, R., Malmqvist, P. A., Veryazov, V., Widmark, P. O. (2004) J. Phys. Chem. A, 108: 2851–2858.

[49] Roos, B. O., Lindh, R., Malmqvist, P. A., Veryazov, V., Widmark, P. O. (2005) J. Phys. Chem. A, 109: 6575–6579.

[50] Eichkorn, K., Weigend, F., Treutler, O., Ahlrichs, R. (1997) Theor. Chem. Acc., 97: 119–124.

[51] Eichkorn, K., Treutler, O., Ohm, H., Haser, M., Ahlrichs, R. (1995) Chem. Phys. Lett., 240: 283–289.

[52] Von Arnim, M., Ahlrichs, R. (1998) J. Comp. Chem., 19: 1746–1757.

[53] Bryantsev, V. S., Diallo, M. S., Goddard, W. A. III, (2008) J. Phys. Chem. B, 112: 9709–9719.

[54] Camaioni, D. M., Schwerdtfeger, C. A. (2005) J. Phys. Chem. A, 109: 10 795–10 797.

[55] Kelly, C. P., Cramer, C. J., Truhlar, D. G. (2006) J. Phys. Chem. B, 110: 16 066–16 081.

[56] Tissandier, M. D., Cowen, K. A., Feng, W. Y., Gundlach, E., Cohen, M. H., Earhart, A. D., Coe, J. V., Tuttle, T. R. (1998) J. Phys. Chem. A, 102: 7787–7794.

[57] Lewis, A., Bumpus, J. A., Truhlar, D. G., Cramer, C. J. (2004) J. Chem. Ed., 81: 596–604 (erratum 2007, 84: 934).

[58] Winget, P., Cramer, C. J., Truhlar, D. G. (2004) Theor. Chem. Acc., 112: 217–227.

[59] Cramer, C. J., Truhlar, D. G. (2009) Phys. Chem. Chem. Phys., 11: 10 757–10 816.

[60] Noodleman, L. (1981) J. Chem. Phys., 74: 5737–5743.

[61] Ziegler, T., Rauk, A., Baerends, E. (1977) J. Theor. Chim. Acta, 43: 261–271.

[62] Soda, T., Kitagawa, Y., Onishi, T., Takano, Y., Shigeta, Y., Nagao, H., Yoshioka, Y., Yamaguchi, K. (2000) Chem. Phys. Lett., 319: 223–230.

[63] Yamaguchi, K., Jensen, F., Dorigo, A., Houk, K. N. (1988) Chem. Phys. Lett., 149: 537–542.

[64] Ciofini, I., Daul, C. A. (2003) Coord. Chem. Rev., 238: 187–209.

[65] Harvey, J. N. (2004) In Principles and Applications of Density Functional Theory in Inorganic Chemistry I, Vol. 112. Springer, Berlin, Germany, p. 151.

[66] Neese, F. (2009) Coord. Chem. Rev., 253: 526–563.

[67] Noodleman, L., Peng, C. Y., Case, D. A., Mouesca, J. M. (1995) Coord Chem. Rev., 144: 199–244.

[68] Garcia-Anton, J., Bofill, R., Escriche, L., Llobet, A., Sala, X. (2012) Eur. J. Inorg. Chem., 4775–4789.

[69] Francas, L., Sala, X., Escudero-Adan, E., Benet-Buchholz, J., Escriche, L., Llobet, A. (2011) Inorg. Chem., 50: 2771–2781.

[70] Mola, J., Dinoi, C., Sala, X., Rodriguez, M., Romero, I., Parella, T., Fontrodona, X., Llobet, A. (2011) Dalton Trans., 40: 3640–3646.

[71] Ertem, M. Z., Gagliardi, L., Cramer, C. (2012) J. Chem. Sci., 3: 1293–1299.

[72] Li, X., Chen, G., Schinzel, S., Siegbahn, P. E. M. (2011) Dalton Trans., 40: 11 296–11 307.

[73] Wang, L.-P., Wu, Q., Van Voorhis, T. (2010) Inorg. Chem., 49: 4543–4553.

[74] Concepcion, J. J., Jurss, J. W., Norris, M. R., Chen, Z., Templeton, J. L., Meyer, T. J. (2010) Inorg. Chem., 49: 1277–1279.

[75] Yang, X., Baik, M. H. (2006) J. Am. Chem. Soc., 128: 7476–7485.

[76] Cramer, C. J., Gour, J. R., Kinal, A., Wtoch, M., Piecuch, P., Shahi, A. R. M., Gagliardi, L. (2008) J. Phys. Chem. A, 112: 3754–3767.

[77] Hong, S., Huber, S. M., Gagliardi, L., Cramer, C. J., Tolman, W. B. (2007) J. Am. Chem. Soc., 129: 14 190–14 192.

[78] Huber, S. M., Ertem, M. Z., Aquilante, F., Gagliardi, L., Tolman, W. B., Cramer, C. (2009) J. Chem. Eur. J., 15: 4886–4895.

[79] Planas, N., Vigara, L., Cady, C., Miro, P., Huang, P., Hammarstrom, L., Styring, S., Leidel, N., Dau, H., Haumann, M., Gagliardi, L., Cramer, C. J., Llobet, A. (2011) Inorg. Chem., 50: 11 134–11 142.

[80] Skodje, R. T., Truhlar, D. G. (1981) J. Phys. Chem., 85: 624–628.

Index

Molecular Water Oxidation Catalysis: A Key Topic for New Sustainable Energy Conversion Schemes,
First Edition. Edited by Antoni Llobet.
© 2014 John Wiley & Sons, Ltd. Published 2014 by John Wiley & Sons, Ltd.